Oberwolfach Seminars
Volume 36

Joachim Cuntz
Ralf Meyer
Jonathan M. Rosenberg

Topological and Bivariant K-Theory

Birkhäuser
Basel · Boston · Berlin

Joachim Cuntz
Mathematisches Institut
Westfälische Wilhelms-Universität Münster
Einsteinstraße 2   6
48149 Münster
Germany
e-mail: cuntz@math.uni-muenster.de

Ralf Meyer
Mathematisches Institut
Georg-August-Universität Göttingen
Bunsenstraße 3–5
37073 Göttingen
Germany
e-mail: rameyer@uni-math.gwdg.de

Jonathan M. Rosenberg
Department of Mathematics
University of Maryland
College Park, MD 20742
USA
e-mail: jmr@math.umd.edu

2000 Mathematical Subject Classification: primary 19-XX, secondary 46L80, 46L85, 58J20, 81T75

Library of Congress Control Number: 2007929010

Bibliographic information published by Die Deutsche Bibliothek
Die Deutsche Bibliothek lists this publication in the Deutsche Nationalbibliografie;
detailed bibliographic data is available in the Internet at <http://dnb.ddb.de>.

ISBN 978-3-7643-8398-5 Birkhäuser Verlag, Basel – Boston – Berlin

© 2007 Birkhäuser Verlag AG
Basel · Boston · Berlin
P.O. Box 133, CH-4010 Basel, Switzerland
Part of Springer Science+Business Media
Printed on acid-free paper produced from chlorine-free pulp. TCF ∞

ISBN 78-3-7643-8398-5   e-ISBN 78-3-7643-8399-2

9 8 7 6 5 4 3 2 1   www.birkhauser.ch

Contents

Preface

The new field of noncommutative geometry (see [29, 63]) applies ideas from geometry to mathematical structures determined by noncommuting variables, and vice versa. Typically, a crucial part of the information is encoded in a noncommutative algebra whose elements represent these noncommuting variables. Such algebras are naturally associated — for instance as algebras of differential or pseudo-differential operators, algebras of intertwining operators for representations, Hecke algebras, algebras of observables in quantum mechanics — with many different geometric structures arising from subjects ranging from mathematical physics and differential geometry to number theory. The fundamental tools for the study of topological invariants attached to noncommutative structures are given by K-theory and cyclic homology. These generalised homology theories are naturally given as bivariant theories, that is, as functors of two variables. For instance, bivariant K-theory specialises both to ordinary topological K-theory and to its dual, K-homology.

This book grew out of an Oberwolfach Seminar organised by the three authors in May 2005. Our aim in this seminar was to introduce young mathematicians to the various forms of topological K-theory for (noncommutative) algebras without assuming too much background on the part of our audience. A second aim was to sketch some typical applications of these techniques, including bivariant versions of the Atiyah–Singer Index Theorem, twisted K-theory, some applications to mathematical physics, and the Baum–Connes conjecture.

An important part of our book is devoted to a complete and unified description of a formalism that has been developed over the past 10 years in [36, 37, 39], and which allows us to construct topological K-theory and associated bivariant theories with good properties for many different categories of algebras over $\mathbb{R}$ or $\mathbb{C}$ such as C^*-algebras, Banach algebras, locally convex algebras, Ind- or Pro-Banach algebras. Since the construction has to be adapted to the different possible categories, one first problem that we have to address is to fix the setting in which to present the construction. Here we have settled for the category of bornological algebras. This setting has been advocated in various contexts in [82, 84, 85]. It is particularly flexible and elegant and covers many interesting examples (for instance it is especially well suited for smooth group algebras). Another argument for this choice is the fact that the construction of bivariant K-theory for locally convex algebras is already available in published form in [36, 37, 39]. So we can use this opportunity to spell out the (minor) changes that have to be made in the bornological setting.

We start from scratch with a discussion of elementary topological K-theory. We do this in a bornological setting, which is more general than the one of Banach algebras. A good choice for this turns out to be the class of local Banach algebras. These algebras are essentially inductive limits of Banach algebras and provide a class of bornological algebras which allow functional calculus. For these algebras basic topological K-theory can be developed in complete analogy with the case of Banach algebras. We present proofs of Bott periodicity and of the Pimsner–Voiculescu exact sequence, and we show how K-theory can be computed in examples. We also briefly discuss the computation of K-theory for group C^*-algebras and the Baum–Connes conjecture. This topic is taken up again in Chapters 10 and 13.

The next chapters treat bivariant K-theory for bornological algebras following [36, 37, 39]. The original arguments have been improved and streamlined in various places. We have made an effort to present everything with complete technical detail. As a consequence, this book contains the most comprehensive and technically complete account to date of this approach to bivariant K-theory. We also introduce the framework of triangulated categories. It fits perfectly to describe the kind of bivariant theories we are discussing and helps to understand their nature. In fact, different bivariant K-theories can be described as different localisations of a version of stable homotopy.

An account of topological K-theory and its bivariant forms cannot be complete without a discussion of the situation for C^*-algebras, and notably of Kasparov's KK-theory — the origin of many of the ideas and concepts in the field. We survey some of the different theories and techniques that have been developed for C^*-algebras and some other theories — algebraic dual K-theory, homotopy-theoretic KK — that can be defined whenever one-variable topological K-theory is available.

We also discuss twisted K-theory in the setting of C^*-algebras as the K-theory of a bundle with fibres that are elementary C^*-algebras. This involves continuous-trace algebras, the Dixmier–Douady class, and related topics such as the Brauer group. In the setting of C^*-algebras we further discuss the K-theory of crossed products by $\mathbb{R}$ (Connes' Thom Isomorphism Theorem) and its relation to the Pimsner–Voiculescu sequence.

The last five chapters of the book are devoted to applications. These chapters are largely independent of one another, except that Chapters 9 and 10 are needed in Chapter 11. Readers interested in index theory may want to concentrate on Chapter 12, while Chapter 11 deals with mathematical physics (in particular with T-duality) and Chapter 13 treats the Universal Coefficient Theorem for KK and the Baum–Connes conjecture via localisation of triangulated categories. Some easier cases of Baum–Connes conjecture are already treated from a more down-to-earth point of view in Chapters 5 and 10.

We would like to thank the Director of the Mathematische Forschungsinstitut Oberwolfach, Professor Dr. Gert-Martin Greuel, his excellent and very professional staff, and all the participants in the 2005 Oberwolfach Seminar for their contributions to making this book possible.

The first two authors were supported by the EU-Network *Quantum Spaces and Noncommutative Geometry* (Contract HPRN-CT-2002-00280) and the *Deutsche Forschungsgemeinschaft* (SFB 478). The third author was supported by the U.S. National Science Foundation, grants DMS-0103647 and DMS-0504212. Any opinions, findings, and conclusions or recommendations expressed in this material are those of the authors and do not necessarily reflect the views of the National Science Foundation.

March, 2007

Joachim Cuntz (Münster)
Ralf Meyer (Göttingen)
Jonathan Rosenberg (College Park)

Chapter 1

The elementary algebra of K-theory

Originally, K-theory was the study of vector bundles on topological spaces. But it was soon realised that the notion of vector bundle can be formulated more algebraically: Swan's Theorem identifies the monoid of vector bundles over a compact space X with the monoid of finitely generated projective modules over the algebra $C(X)$ of continuous functions on X; we take real- or complex-valued functions here to get real or complex vector bundles, respectively.

The finitely generated projective modules over a unital ring R form a commutative monoid, which we denote by $\mathbf{V}(R)$. We describe $\mathbf{V}(R)$ more concretely using equivalence classes of idempotent elements in matrix rings over R; this is used in many proofs. The K-theory $K_0(R)$ of R is defined as the Grothendieck group of $\mathbf{V}(R)$. This group is usually much easier to handle than $\mathbf{V}(R)$ itself because we may use tools from homological algebra that do not work for monoids.

We extend K_0 to non-unital rings by a standard trick. We need this in order to formulate the exactness properties of K_0, which are crucial for many computations. We will see that K_0 is half-exact and split-exact.

We also study the failure of left-exactness of K_0. This leads us to the index map and hints at a definition of K_1: this group should classify invertible elements in matrix rings up to an appropriate equivalence relation. Now K-theory splits into two branches: the algebraic approach uses commutators or, equivalently, elementary matrices to generate the equivalence relation, whereas the topological approach uses homotopy. We will only study the topological version of K_1. This theory is *considerably* easier to compute than its algebraic counterpart, using various tools from algebraic topology. Since the definition requires a certain amount of functional analysis, we only treat K_1 in Chapter 2. Whereas higher algebraic K-theory is even more complicated than algebraic K_1, higher topological K-theory gives nothing essentially new by Bott periodicity.

1.1   Projective modules, idempotents, and vector bundles

Let R be a ring with unit. Let $\mathrm{Mod}(R)$ be the category of left R-modules with module homomorphisms as morphisms. As usual, we require the unit element of R to act identically on modules.

Definition 1.1. A left R-module M is called *finitely generated* if there exist finitely many elements $x_1, \ldots, x_n \in M$ such that the map

$$R^n \to M, \qquad (a_1, \ldots, a_n) \mapsto a_1 x_1 + \cdots + a_n x_n$$

is surjective.

Definition 1.2. A left R-module M is called *projective* if any surjective module homomorphism $p \colon N \to M$ for any left R-module N splits, that is, there is a module homomorphism $s \colon M \to N$ such that $p \circ s = \mathrm{id}_M$.

Definition 1.3. Let $\mathbf{V}(R)$ be the set of isomorphism classes of finitely generated projective left R-modules. (We will see below that this is a set, not just a class.)

The set $\mathbf{V}(R)$ contains the zero module and is closed under direct sums. Thus the direct sum operation turns $\mathbf{V}(R)$ into a commutative monoid.

Example 1.4. The category $\mathrm{Mod}(\mathbb{Z})$ is nothing but the category of Abelian groups. The classification of finitely generated Abelian groups implies that any finitely generated projective $\mathbb{Z}$-module is free. Thus we get a monoid isomorphism

$$\mathbf{V}(\mathbb{Z}) \cong \{[\mathbb{Z}^n] \mid n \in \mathbb{N}\} \cong (\mathbb{N}, +).$$

A similar argument yields $\mathbf{V}(R) \cong (\mathbb{N}, +)$ if R is a field and, more generally, if R is a principal ideal domain.

Now we describe $\mathbf{V}(R)$ using idempotents in matrix rings over R.

Definition 1.5. Let R be a (possibly non-unital) ring. We let $\mathbb{M}_n(R)$ for $n \in \mathbb{N}$ be the ring of $n \times n$-matrices with entries in R. For $m \leq n$, we view $\mathbb{M}_m(R)$ as a subring of $\mathbb{M}_n(R)$ via $x \mapsto \left(\begin{smallmatrix} x & 0 \\ 0 & 0 \end{smallmatrix}\right)$. Let $\mathbb{M}_\infty(R) := \bigcup_{n \in \mathbb{N}} \mathbb{M}_n(R)$.

Exercise 1.6. Let R^∞ be the direct sum of countably many copies of R. Let $\mathbb{M}_n(R)$ act on R^n by matrix-vector multiplication on the *right* for $n \in \mathbb{N} \cup \{\infty\}$.

Check that this identifies $\mathbb{M}_n(R)$ for $n \in \mathbb{N}$ with the ring $\mathrm{Hom}_R(R^n, R^n)$ of *left* R-module endomorphisms of R^n. Check that an R-module endomorphism of R^∞ belongs to $\mathbb{M}_\infty(R)$ if and only if it factors through $R^n \subseteq R^\infty$ for some $n \in \mathbb{N}$, if and only if it factors through some finitely generated R-module.

Definition 1.7. Let R be a possibly non-unital ring. An element $e \in R$ is called *idempotent* if $e^2 = e$. We let $\mathrm{Idem}\, R$ be the set of idempotent elements in R.

We call $e_1, e_2 \in \mathrm{Idem}\, R$ *equivalent* and write $e_1 \sim e_2$ if there are $v, w \in R$ with $vw = e_1$ and $wv = e_2$.

Proposition 1.8. *Any finitely generated projective left R-module is of the form $R^\infty e$ for some $e \in \mathrm{Idem}\, \mathbb{M}_\infty(R)$ and, conversely, all such modules are finitely generated and projective. Let $e_1, e_2 \in \mathrm{Idem}\, \mathbb{M}_\infty(R)$. There is an R-module isomorphism $R^\infty e_1 \cong R^\infty e_2$ if and only if $e_1 \sim e_2$.*

Thus $[e] \mapsto [R^\infty e]$ defines a bijection $\mathrm{Idem}\, \mathbb{M}_\infty(R)/{\sim} \cong \mathbf{V}(R)$.

If S is any ring, then $\sim$ defines an equivalence relation on $\mathrm{Idem}\, S$.

If $e_1, e_2 \in \mathrm{Idem}(S)$ are equivalent, then there are $v, w \in S$ with

$$vw = e_1, \quad wv = e_2, \quad e_1 v = v = v e_2, \quad e_2 w = w = w e_1. \tag{1.9}$$

Proof. Let M be a finitely generated projective module. Since M is finitely generated, we get a surjective module homomorphism $\pi\colon R^m \to M$ for some $m \in \mathbb{N}$. This map splits by a module homomorphism $\iota\colon M \to R^m$ because M is projective. Thus M is isomorphic to the range of the idempotent map $\iota \circ \pi\colon R^m \to R^m$. This map is of the form $x \mapsto x \cdot e$ for some idempotent element $e \in \mathbb{M}_m(R)$ by Exercise 1.6. Thus any finitely generated projective module is of the form $M \cong R^m e \cong R^\infty e$. We leave the proof of the converse as an exercise.

Let $e_1 \in \mathbb{M}_m(R)$, $e_2 \in \mathbb{M}_n(R)$ be idempotent. If $R^m e_1 \cong R^n e_2$, then we use the decomposition $R^m \cong R^m \cdot e_1 \oplus R^m \cdot (1 - e_1)$ to extend this isomorphism to a map $R^m \to R^n$: send elements of $R^m \cdot (1 - e_1)$ to 0. This map $R^m \to R^n$ is of the form $x \mapsto x \cdot v$ for some $v \in \mathbb{M}_{n \times m}(R) \subseteq \mathbb{M}_\infty(R)$ by Exercise 1.6. Similarly, the inverse isomorphism $R^n e_2 \to R^m e_1$ yields $w \in \mathbb{M}_{m \times n}(R)$. By construction, these matrices satisfy the relations in (1.9). Conversely, if v and w merely satisfy $vw = e_1$ and $wv = e_2$, then right multiplication by v and w defines maps $R^m e_1 \to R^n e_2 \to R^m e_1$ that are inverse to one another. Hence $R^\infty e_1 \cong R^\infty e_2 \iff e_1 \sim e_2$.

Along the way, this argument shows that the relations (1.9) can be achieved for equivalent idempotents in $\mathbb{M}_\infty(R)$ with unital R. A direct proof goes as follows. Suppose $vw = e_1$ and $wv = e_2$. Put $v' := e_1 v e_2$ and $w' := e_2 w e_1$, so that $e_1 v' = v' = v' e_2$ and $e_2 w' = w' = w' e_1$. Then

$$v'w' = e_1 v e_2^2 w e_1 = e_1 v (wv)^2 w e_1 = e_1 (vw)^3 e_1 = e_1^5 = e_1$$

and, similarly, $w'v' = e_2$.

Using this, we show that equivalence of idempotents is an equivalence relation. Reflexivity and symmetry are obvious. Let $e_1 \sim e_2 \sim e_3$ in $\mathrm{Idem}\, S$. Suppose that (v_1, w_1) and (v_2, w_2) implement the equivalences $e_1 \sim e_2$ and $e_2 \sim e_3$ and satisfy (1.9). The computations

$$v_1 v_2 w_2 w_1 = v_1 e_2 w_1 = v_1 w_1 = e_1, \qquad w_2 w_1 v_1 v_2 = w_2 e_2 v_2 = w_2 v_2 = e_3$$

show that $(v_1 v_2, w_2 w_2)$ provides an equivalence $e_1 \sim e_3$. $\qquad \square$

The monoid structure on $\mathbf{V}(R)$ translates to idempotents as follows. Write $M, N \in \mathbf{V}(R)$ as $M \cong R^m e_1$, $N \cong R^n e_2$ with $m, n \in \mathbb{N}$ and $e_1 \in \mathrm{Idem}\, \mathbb{M}_m(R)$, $e_2 \in \mathrm{Idem}\, \mathbb{M}_n(R)$. Then we have $M \oplus N \cong R^{m+n} \cdot (e_1 \oplus e_2)$, where

$$e_1 \oplus e_2 := \begin{pmatrix} e_1 & 0 \\ 0 & e_2 \end{pmatrix} \in \mathbb{M}_{m+n}(R).$$

Conversely, take $e_1, e_2 \in \operatorname{Idem} \mathbb{M}_\infty(R)$. We call e_1 and e_2 *orthogonal* if $e_1 e_2 = 0 = e_2 e_1$. This implies that $e_1 + e_2$ is an idempotent element as well and that $R^\infty e_1 \oplus R^\infty e_2 \cong R^\infty \cdot (e_1 + e_2)$. Thus $e_1 \oplus e_2 \sim e_1 + e_2$.

Now we have two equivalent definitions of the monoid $\mathbf{V}(R)$, each having its own virtues. Since finitely generated modules and projective modules play an important role in algebra, $\mathbf{V}(R)$ occurs rather naturally in many situations. The definition with idempotents in matrix algebras looks artificial: why should we bother to pass to matrices, and why use this particular equivalence relation? Since idempotents tend to be more concrete and more tractable than finitely generated projective modules, we often use this description in proofs. For instance, it shows immediately that $\mathbf{V}(R)$ is a set.

1.1.1   General properties

In order to get used to our two descriptions of $\mathbf{V}$, we consider some of its basic properties. Since these are not hard to prove, we leave all arguments as exercises.

First we discuss *functoriality*. Let $f \colon R \to S$ be a unital ring homomorphism. This induces a ring homomorphism $\mathbb{M}_\infty(f) \colon \mathbb{M}_\infty(R) \to \mathbb{M}_\infty(S)$ by applying f entry-wise to matrices. Using the definition of $\mathbf{V}$ via idempotents, we define a map

$$f_* \colon \mathbf{V}(R) \to \mathbf{V}(S), \qquad f_*(e) := \mathbb{M}_\infty(f)(e).$$

You should check that this turns $\mathbf{V}$ into a functor from the category of unital rings to the category of commutative monoids.

We view S as an S, R-bimodule via $s_1 \cdot s_2 \cdot r_3 = s_1 s_2 f(r_3)$ for $s_1, s_2 \in S, r_3 \in R$ and define $f_*(M) := S \otimes_R M$ for a finitely generated projective R-module M.

Exercise 1.10. Check that $f_*(R^\infty \cdot e) \cong S^\infty \cdot f_*(e)$. Thus $f_*(M)$ is again finitely generated and projective if M is, and both constructions above yield the same map $f_* \colon \mathbf{V}(R) \to \mathbf{V}(S)$.

Definition 1.11. A functor F on the category of (unital) rings is called *additive* if $F(R_1 \oplus R_2) \cong F(R_1) \times F(R_2)$ for any two (unital) rings R_1, R_2.

Exercise 1.12. Check that the functor $\mathbf{V}$ is additive in this sense.

Definition 1.13. The *opposite ring* R^{op} of a ring R is the ring that we get from R by reversing the order of the product: $a \bullet b := b \cdot a$.

Left R^{op}-modules are the same thing as right R-modules. Hence the following exercise shows that $\mathbf{V}$ does not care whether we use left or right modules.

Exercise 1.14. Check that the transposition of matrices defines a ring isomorphism $\mathbb{M}_\infty(R^{\mathrm{op}}) \cong (\mathbb{M}_\infty R)^{\mathrm{op}}$, and use this to construct a natural isomorphism $\mathbf{V}(R) \cong \mathbf{V}(R^{\mathrm{op}})$.

Check that, in terms of finitely generated projective modules, this isomorphism agrees with the *duality map* $M \mapsto \operatorname{Hom}_R(M, R)$, where we use the bimodule structure of R to view $\operatorname{Hom}_R(M, R)$ as a right R-module if M is a left R-module, and vice versa. You must also show that $\operatorname{Hom}_R(M, R)$ is again finitely generated and projective if M is.

Exercise 1.15. Show that the functor $\mathbf{V}$ commutes with inductive limits, that is, $\mathbf{V}(\varinjlim R_i) \cong \varinjlim \mathbf{V}(R_i)$ for any inductive system of unital rings $(R_i)_{i \in I}$.

Exercise 1.16. Recall that two unital rings are called *Morita equivalent* if their categories of (left) modules are equivalent. Show that $\mathbf{V}(R) \cong \mathbf{V}(S)$ if R and S are Morita equivalent. If you know about the characterisation of Morita equivalence using bimodules, try to describe this isomorphism explicitly.

1.1.2 Similarity of idempotents

Definition 1.17. Two idempotents e_1, e_2 in a unital ring R are called *similar* if there is an invertible element $u \in R$ with $u e_1 u^{-1} = e_2$.

It is clear that similarity is an equivalence relation; it is a more natural substitute for the one in Proposition 1.8.

Lemma 1.18. *Two idempotents $e_1, e_2 \in \mathbb{M}_m(R)$ are similar in $\mathbb{M}_m(R)$ if and only if both $R^m e_1 \cong R^m e_2$ and $R^m \cdot (1 - e_1) \cong R^m \cdot (1 - e_2)$. Thus similar idempotents are equivalent.*

Conversely, equivalent idempotents in $\mathbb{M}_m(R)$ become similar in $\mathbb{M}_{2m}(R)$.

Proof. If $u e_1 u^{-1} = e_2$, then right multiplication by u defines a map $R^m \to R^m$ that restricts to isomorphisms $R^m e_2 \cong R^m e_1$ and $R^m \cdot (1 - e_2) \cong R^m \cdot (1 - e_1)$. Conversely, such isomorphisms yield a module automorphism of R^m, which is of the form $x \mapsto x \cdot u$ for some invertible element $u \in \mathbb{M}_m(R)$ with $u e_1 u^{-1} = e_2$.

Now let $e_1, e_2 \in \operatorname{Idem} \mathbb{M}_m(R)$ be equivalent, that is, $R^m e_1 \cong R^m e_2$. Then we construct an invertible operator on R^{2m} as follows:

$$R^{2m} \cong R^m e_1 \ \oplus \ R^m \cdot (1_m - e_1) \ \oplus \ R^m \cdot (1_m - e_2) \ \oplus \ R^m e_2$$

$$R^{2m} \cong R^m e_2 \ \oplus \ R^m \cdot (1_m - e_2) \ \oplus \ R^m \cdot (1_m - e_1) \ \oplus \ R^m e_1$$

Thus $e_1 \oplus 0_m$ and $e_2 \oplus 0_m$ are similar in $\mathbb{M}_{2m}(R)$. Explicitly, if $e_1 \sim e_2$ is implemented by v, w satisfying (1.9), then $e_2 \oplus 0_m = u(e_1 \oplus 0_m)u^{-1}$ with

$$u := \begin{pmatrix} w & 1 - e_2 \\ 1 - e_1 & v \end{pmatrix}, \qquad u^{-1} = \begin{pmatrix} v & 1 - e_1 \\ 1 - e_2 & w \end{pmatrix}. \qquad \square$$

1.1.3 Relationship to vector bundles

Let X be a compact Hausdorff space and let $\mathbb{K}$ be $\mathbb{R}$ or $\mathbb{C}$ (we may also allow the algebra of quaternions). Consider the ring $C(X, \mathbb{K})$ of continuous functions $X \to \mathbb{K}$ with pointwise addition and multiplication. We want to identify $\mathbf{V}\big(C(X, \mathbb{K})\big)$ with the monoid of $\mathbb{K}$-vector bundles over X. First we recall what a $\mathbb{K}$-vector bundle is.

Definition 1.19. A $\mathbb{K}$-*vector bundle* over X is a topological space V, called the *total space* of the vector bundle, equipped with a continuous map $p\colon V \to X$ and $\mathbb{K}$-vector space structures on the *fibres* $V_x := p^{-1}(x)$ for all $x \in X$ such that the following *local triviality* condition holds: for any $x \in X$ there is $n \in \mathbb{N}$, a neighbourhood $U \subseteq X$ of x, and vector space isomorphisms $\varphi_y\colon V_y \xrightarrow{\cong} \mathbb{K}^n$ for all $y \in U$ that piece together to a homeomorphism $\varphi\colon p^{-1}(U) \to U \times \mathbb{K}^n$. A vector bundle is called *trivial* if such an isomorphism φ exists with $U = X$.

A *morphism of vector bundles* is a family of linear maps $f_x\colon V_x \to V'_x$ that piece together to a continuous map $f\colon V \to V'$.

Usually we denote a vector bundle just by its total space V and omit the remaining data from our notation.

Definition 1.20. We denote the set of isomorphism classes of $\mathbb{K}$-vector bundles over X by $\mathbf{V}_{\mathbb{K}}(X)$.

Definition 1.21. A *(continuous) section* of a $\mathbb{K}$-vector bundle is a continuous map $s\colon X \to V$ such that $p \circ s(x) = x$ for all $x \in X$. We may add sections and multiply them by continuous functions $X \to \mathbb{K}$. Thus the set of sections of V becomes a $C(X, \mathbb{K})$-module, which we denote by $\Gamma(V)$.

Theorem 1.22 (Swan's Theorem). *Let X be a compact Hausdorff space and let $\mathbb{K}$ be $\mathbb{R}$ or $\mathbb{C}$. Then any finitely generated projective module over $C(X, \mathbb{K})$ is isomorphic to $\Gamma(V)$ for a $\mathbb{K}$-vector bundle V that is unique up to isomorphism. Thus Γ defines a bijection $\mathbf{V}_{\mathbb{K}}(X) \cong \mathbf{V}\big(C(X, \mathbb{K})\big).$*

Even more, our proof shows that Γ is an equivalence of categories between the categories of vector bundles and of finitely generated modules. This contains the fact that $\Gamma(V_1 \oplus V_2) \cong \Gamma(V_1) \oplus \Gamma(V_2)$, where $V_1 \oplus V_2$ denotes the direct sum of vector bundles. Thus Γ is a monoid isomorphism.

Proof. Let $V \to X$ be a $\mathbb{K}$-vector bundle. Since X is compact, there is a finite open covering $X = U_1 \cup \cdots \cup U_k$ such that $V|_{U_j}$ is trivial for all $j = 1, \ldots, k$. The trivialisations $V|_{U_j} \cong U_j \times \mathbb{K}^{n_j}$ yield module isomorphisms

$$\Psi_j\colon \Gamma_0(V|_{U_j}) \xrightarrow{\cong} C_0(U_j, \mathbb{K})^{n_j};$$

here Γ_0 and C_0 denote spaces of sections and functions that vanish at infinity. Extending sections and functions by 0 outside U_j, we embed $\Gamma_0(V|_{U_j}) \subseteq \Gamma(V)$ and $C_0(U_j, \mathbb{K})^{n_j} \subseteq C(X, \mathbb{K})^{n_j}$. Let $n := n_1 + \cdots + n_k$.

There exist functions $\varphi_j\colon U_j \to [0, 1]$ such that the support of φ_j is contained in U_j and $\sum_{j=1}^{k} \varphi_j^2(x) = 1$ for all $x \in X$; these functions are a variant of a partition of unity subordinate to the covering (U_j). Now we define $C(X, \mathbb{K})$-module homomorphisms $\Gamma(V) \to C(X, \mathbb{K})^n \to \Gamma(V)$, sending $s \in \Gamma(V)$ to

$$\bigoplus_{j=1}^{k} \Psi_j(\varphi_j \cdot s) \in \bigoplus_{j=1}^{k} C(X, \mathbb{K})^{n_j} \cong C(X, \mathbb{K})^n$$

and sending $(f_j) \in \bigoplus_{j=1}^k C(X, \mathbb{K})^{n_j} \cong C(X, \mathbb{K})^n$ to $\sum_{j=1}^k \Psi_j^{-1}(\varphi_j \cdot f_j)$. Since Ψ_j is $C(X, \mathbb{K})$-linear, the composite map $\Gamma(V) \to C(X, \mathbb{K})^n \to \Gamma(V)$ is the identity map. Thus $\Gamma(V)$ is a direct summand of $C(X, \mathbb{K})^n$, which means that $\Gamma(V)$ is a finitely generated projective module over $C(X, \mathbb{K})$.

It is clear that $\Gamma(V_1) \cong \Gamma(V_2)$ if $V_1 \cong V_2$. Conversely, any module isomorphism $\Gamma(V_1) \cong \Gamma(V_2)$ is implemented by a *unique* vector bundle isomorphism $V_1 \cong V_2$. This is clear for trivial vector bundles and therefore holds locally; this implies the assertion globally, using a partition of unity to patch together the unique locally defined isomorphisms.

Conversely, take a finitely generated projective module over $C(X, \mathbb{K})$ and write it as $C(X, \mathbb{K})^m e$ for some $e \in \operatorname{Idem} \mathbb{M}_m(C(X, \mathbb{K}))$ by Proposition 1.8. There is a natural isomorphism $\mathbb{M}_m(C(X, \mathbb{K})) \cong C(X, \mathbb{M}_m(\mathbb{K}))$. Thus we may view e as a continuous function $\tilde{e}$ from X to the topological space $\operatorname{Idem} \mathbb{M}_m(\mathbb{K})$. For each $x \in X$, let $V_x \subseteq \mathbb{K}^m$ be the range of $\tilde{e}(x) \in \mathbb{M}_m(\mathbb{K})$. We topologise $\coprod_{x \in X} V_x$ as a subset of $X \times \mathbb{K}^m$. We claim that this defines a vector bundle over X; the only issue is local triviality.

Fix $x \in X$. The function that sends a matrix to its rank is locally constant on idempotent matrices. Therefore, the rank of $\tilde{e}(y)$ is locally constant. Since $\tilde{e}(x)\tilde{e}(y)\tilde{e}(x) \colon V_x \to V_x$ is invertible for $y = x$, it remains invertible for y in a neighbourhood of x. Thus we may find a neighbourhood U of x where the rank of $\tilde{e}(y)$ is constant and $\tilde{e}(x)\tilde{e}(y)(\mathbb{K}^m) = V_x$. Therefore, $\tilde{e}(x)$ restricts to an isomorphism $V_y \to V_x$ for all $y \in U$. This is the desired local trivialisation.

Thus any finitely generated projective module over $C(X, \mathbb{K})$ is isomorphic to $\Gamma(V)$ for a vector bundle V. $\qquad\square$

The following exercises may help you understand the proof of Swan's Theorem.

Exercise 1.23. For any vector bundle V over a compact space X, there is a vector bundle W for which $V \oplus W$ is a trivial vector bundle.

Exercise 1.24. Let $X = \mathbb{CP}^1$ be the complex projective space $\mathbb{C} \cup \{\infty\}$, which is diffeomorphic to the 2-sphere. Since points of X are 1-dimensional vector subspaces $x \subseteq \mathbb{C}^2$, there is a *canonical vector bundle* on X whose fibre at x is x. Find a corresponding idempotent in $C(X, \mathbb{M}_\infty(\mathbb{C}))$.

Exercise 1.25. We still have a canonical map $\mathbf{V}(C(X, \mathbb{K})) \to \mathbf{V}_\mathbb{K}(X)$ for any topological space X. This map is not surjective if $\mathbb{K} = \mathbb{C}$ and X is the projective space of a separable Hilbert space (use the canonical complex line bundle).

Exercise 1.26. Swan's Theorem still holds if X is a paracompact Hausdorff space with finite covering dimension.

Tangent and normal bundles

Vector bundles occur frequently in geometry. We briefly mention tangent and normal bundles here, which we will use when we study the topological index map.

Let X be a smooth manifold. Then its *tangent bundle* TX is an $\mathbb{R}$-vector bundle over X, which contains important information about the smooth structure on X. If X is a complex manifold, then TX is a $\mathbb{C}$-vector bundle.

Let X and Y be smooth manifolds and let $f\colon X \to Y$ be a smooth map. Then we may pull TY back to a vector bundle $f^*(TY)$ over X. We may view the differential of f as a morphism of vector bundles $Df\colon TX \to f^*(TY)$. If f is an immersion, then Df is injective and the fibrewise cokernels $f^*(TY)/TX$ form a vector bundle on X, which is called the *normal bundle* N_f of f. The bundle N_f is complex if X and Y are complex manifolds and f is holomorphic.

Let f be a closed embedding. The *tubular neighbourhood theorem* asserts that there are an open neighbourhood $U \subseteq Y$ of $f(X)$ and a diffeomorphism $U \cong N_f$ whose composition with f is the *zero section* $X \to N_f$. Roughly speaking, Y looks like a vector bundle over X near X. This is one of the reasons why vector bundles play an important role in differential topology.

1.2   Passage to K-theory

The monoids $\mathbf{V}_{\mathbb{C}}(X)$ are usually quite complicated, even if X is as simple as the n-dimensional torus $\mathbb{T}^n$ for $n \gg 0$. In order to get a tractable invariant, we complete $\mathbf{V}(R)$ to an Abelian group by the following general construction:

Definition 1.27. The *Grothendieck group* of a commutative semigroup M is an Abelian group $\mathrm{Gr}(M)$ together with a semigroup homomorphism $i\colon M \to \mathrm{Gr}(M)$ that has the universal property that any semigroup homomorphism from M into an Abelian group factors uniquely through $\mathrm{Gr}(M)$.

The Grothendieck group always exists and is unique up to natural isomorphism. It can be constructed as follows. Its elements are equivalence classes of pairs $(x_+, x_-) \in M^2$, where we consider $(x_+, x_-) \sim (y_+, y_-)$ if there are $z, z' \in M$ with $(x_+ + z, x_- + z) = (y_+ + z', y_- + z')$ (think of (x_+, x_-) as the formal difference $i(x_+) - i(x_-)$). The addition is defined by $(x_+, x_-) + (y_+, y_-) := (x_+ + y_+, x_- + y_-)$. One checks easily that this defines an Abelian group G; the inverse of (x_+, x_-) is (x_-, x_+). The map $M \to G$, $x \mapsto (x + x, x)$ is a semigroup homomorphism and has the required universal property for the Grothendieck group of M.

Exercise 1.28. The map $i\colon M \to G$ need not be injective: check that $i(x_1) = i(x_2)$ if and only if there is $y \in M$ with $x_1 + y = x_2 + y$ and find a monoid and $x_1 \neq x_2$ for which this happens.

Definition 1.29. Let R be a ring with unit. Its *K-theory* $\mathrm{K}_0(R)$ is defined as the Grothendieck group of the monoid $\mathbf{V}(R)$.

By definition, we have a monoid homomorphism $\mathbf{V}(R) \to \mathrm{K}_0(R)$ whose range generates $\mathrm{K}_0(R)$ as an Abelian group. We write elements of $\mathrm{K}_0(R)$ as formal differences (M_+, M_-) of finitely generated projective modules $M_\pm$ as above. Since there is always some complementary finitely generated projective module N with

$M_- \oplus N \cong R^k$, we have $(M_+, M_-) = (M_+ \oplus N, M_- \oplus N) = (M_+ \oplus N, R^k)$; that is, any element of $K_0(R)$ has the form (M_+, R^k) for some finitely generated projective module M_+ and some $k \in \mathbb{N}$.

We have $(M_+, M_-) = (M'_+, M'_-)$ if and only if $(M_+ \oplus M'_-, M_- \oplus M'_+) = 0$, if and only if $M_+ \oplus M'_- \oplus N \cong M_- \oplus M'_+ \oplus N$ for some $N \in \mathbf{V}(R)$. As above, we may restrict attention to N of the form R^k for some $k \in \mathbb{N}$.

Thus $M_1, M_2 \in \mathbf{V}(R)$ have the same image in $K_0(R)$ if and only if $M_1 \oplus R^k \cong M_2 \oplus R^k$ for some $k \in \mathbb{N}$. We call M_1 and M_2 *stably isomorphic* in this case; the associated idempotents are called *stably equivalent*.

We say that $\mathbf{V}(M)$ has the *cancellation property* if stable isomorphism implies isomorphism or, equivalently, if the map $\mathbf{V}(M) \to K_0(R)$ is injective. The following geometric example shows that this may fail.

Example 1.30. Let $\mathbb{S}^n$ be the n-dimensional sphere. It is easy to see that the normal bundle of the standard embedding $\mathbb{S}^n \subseteq \mathbb{R}^{n+1}$ is the trivial bundle $\mathbb{R}$ of dimension 1. Since $N_f \oplus TX \cong f^*(TY)$ for any smooth map $f \colon X \to Y$, we get $T\mathbb{S}^n \oplus \mathbb{R} \cong \mathbb{R}^{n+1} = \mathbb{R}^n \oplus \mathbb{R}$. Thus $T\mathbb{S}^n$ and $\mathbb{R}^n$ are stably isomorphic.

It sometimes happens that $T\mathbb{S}^n \cong \mathbb{R}^n$ is trivial. This is impossible, however, if n is even because then the Euler characteristic of the vector bundle $T\mathbb{S}^n$—which agrees with the Euler characteristic of the space $\mathbb{S}^n$—is $1 + (-1)^n = 2$. Hence $T\mathbb{S}^{2k}$ is stably trivial but not trivial.

The functor K_0 evidently inherits the properties of $\mathbf{V}$ considered in §1.1.1: it is functorial for unital ring homomorphisms, additive, commutes with inductive limits, and is invariant under Morita equivalence and passage to opposite rings.

1.2.1 Euler characteristics of finite projective complexes

Finitely generated projective modules yield elements of $K_0(R)$ by construction. We are going to attach elements of $K_0(R)$ to certain non-projective modules as well. We only sketch this construction rather briefly, assuming some familiarity with notions from homological algebra. You can find a more detailed account in [109, §1.7].

Definition 1.31. Let R be a unital ring. A *perfect chain complex* over R is an R-module chain complex of finite length $P_\bullet := (P_n, \delta_n)$ whose entries P_n are finitely generated and projective. Its *Euler characteristic* is defined by

$$\chi(P_\bullet) := \sum_{n \in \mathbb{Z}} (-1)^n [P_n] \in K_0(R).$$

Proposition 1.32. *Let $P_\bullet$ and $P'_\bullet$ be perfect chain complexes.*

If $P_\bullet$ is exact, then $\chi(P_\bullet) = 0$. The mapping cone of a chain map $f \colon P_\bullet \to P'_\bullet$ satisfies $\chi\big(C(f)\big) = \chi(P'_\bullet) - \chi(P_\bullet)$.

If $P_\bullet$ and $P'_\bullet$ are quasi-isomorphic, then $\chi(P_\bullet) = \chi(P'_\bullet)$.

Proof. If $P_\bullet$ is exact, then it is contractible because it is projective and bounded below. Using the contracting homotopy, we get direct sum decompositions $P_n \cong \ker \partial_n \oplus \ker \partial_{n-1}$ for all $n \in \mathbb{Z}$; the direct summands $\ker \partial_n$ are again finitely generated and projective. This implies

$$\chi(P_\bullet) := \sum_{n \in \mathbb{Z}} (-1)^n [P_n] = \sum_{n \in \mathbb{Z}} (-1)^n [\ker \partial_n] + (-1)^n [\ker \partial_{n-1}] = 0.$$

We have $\chi(C(f)) = \chi(P'_\bullet) - \chi(P_\bullet)$ because the mapping cone is defined by $C(f)_n := P'_n \oplus P_{n-1}$. If $f \colon P_\bullet \to P'_\bullet$ is a quasi-isomorphism, then $C(f)$ is exact, so that $\chi(C(f)) = 0$. $\qquad\square$

Definition 1.33. An R-module $M \in \mathrm{Mod}(R)$ has *type (FP)* if it has a perfect resolution $P_\bullet$; in this case, we define its *rank* by $\mathrm{rank}\, M := \chi(P_\bullet)$.

Since any two projective resolutions of a module are quasi-isomorphic, Proposition 1.32 shows that $\mathrm{rank}\, M$ does not depend on the resolution.

Let $K \rightarrowtail E \twoheadrightarrow Q$ be an extension of R-modules. The rank is additive for extensions of R-modules in the following sense: if two of K, E, Q have type (FP), then so does the third, and we have

$$\mathrm{rank}\, K - \mathrm{rank}\, E + \mathrm{rank}\, Q = 0.$$

Modules of type (FP) are finitely generated, even finitely presented. The converse need not hold in general. It does hold if the ring R is Noetherian and has finite cohomological dimension. Being *Noetherian* means that all left ideals in R are finitely generated R-modules. Having *finite cohomological dimension* means that any module has a projective resolution of finite length.

Example 1.34. Recall that $\mathbf{V}(\mathbb{Z}) \cong \mathbb{N}$ and that $\mathrm{Mod}(\mathbb{Z})$ is isomorphic to the category of Abelian groups. Thus $\mathrm{K}_0(\mathbb{Z}) \cong \mathbb{Z}$. Any finitely generated Abelian group is a finite direct sum of cyclic groups. Since cyclic groups admit a perfect resolution, any finitely generated Abelian group has type (FP). For a finite cyclic group, the resolution is of the form $\mathbb{Z} \to \mathbb{Z}$ and hence has vanishing Euler characteristic. Thus the rank of a finitely generated Abelian group M is the dimension of the $\mathbb{Q}$-vector space $M \otimes_{\mathbb{Z}} \mathbb{Q}$.

1.2.2   Definition of K_0 for non-unital rings

In order to discuss exactness properties of K_0, we have to extend it to rings without unit. If R is such a ring, we formally adjoin a unit and let $R^+ := R \oplus \mathbb{Z}$ with multiplication

$$(x_1, n_1) \cdot (x_2, n_2) := (x_1 \cdot x_2 + n_1 \cdot x_2 + n_2 \cdot x_1, n_1 \cdot n_2) \qquad \forall x_1, x_2 \in X, n_1, n_2 \in \mathbb{Z}.$$

(When we deal with algebras over $\mathbb{C}$, we often use $R_{\mathbb{C}}^+ := R \oplus \mathbb{C}$ with a similar multiplication.) By construction, we get a ring extension

$$R \rightarrowtail R^+ \twoheadrightarrow \mathbb{Z}, \tag{1.35}$$

which splits by the unique unital homomorphism $\mathbb{Z} \to R^+$.

Exercise 1.36. If R is already unital, then there is a ring isomorphism $R^+ \cong R \oplus \mathbb{Z}$ such that the maps in (1.35) are the coordinate embedding $R \to R \oplus \mathbb{Z}$ and projection $R \oplus \mathbb{Z} \to \mathbb{Z}$.

The map $K_0(R^+) \to K_0(\mathbb{Z})$ induced by the quotient map $R^+ \to \mathbb{Z}$ is split-surjective because the unit map $\mathbb{Z} \to R^+$ induces a section. We define

$$K_0(R) := \ker\big(K_0(R^+) \to K_0(\mathbb{Z})\big). \tag{1.37}$$

We have a short exact sequence $K_0(R) \rightarrowtail K_0(R^+) \twoheadrightarrow K_0(\mathbb{Z})$; it follows from Example 1.4 that $K_0(\mathbb{Z}) \cong \mathbb{Z}$.

A ring homomorphism $f \colon R \to S$ extends uniquely to a unital ring homomorphism $f^+ \colon R^+ \to S^+$, which induces a map $f_*^+ \colon K_0(R^+) \to K_0(S^+)$. Since the right square in the diagram

$$
\begin{array}{ccccc}
K_0(R) & \rightarrowtail & K_0(R^+) & \twoheadrightarrow & K_0(\mathbb{Z}) \\
\downarrow{\scriptstyle f_*} & & \downarrow{\scriptstyle f_*^+} & & \| \\
K_0(S) & \rightarrowtail & K_0(S^+) & \twoheadrightarrow & K_0(\mathbb{Z})
\end{array}
$$

commutes, we get an induced map $f_* \colon K_0(R) \to K_0(S)$. Thus K_0 is a functor from the category of non-unital rings to the category of Abelian groups.

We claim that our new definition of $K_0(R)$ agrees with the old one if R is already unital. For the proof, we temporarily write $K_0^{\mathrm{new}}(R)$ for the group defined in (1.37). Since R is unital, Exercise 1.36 yields an isomorphism $R^+ \cong R \oplus \mathbb{Z}$ that intertwines the quotient map $R^+ \to \mathbb{Z}$ and the projection onto the second coordinate. The additivity of K_0 for unital rings yields $K_0(R^+) \cong K_0(R) \oplus K_0(\mathbb{Z})$ and hence $K_0^{\mathrm{new}}(R) \cong K_0(R)$ as asserted. This isomorphism is natural, that is, if $f \colon R \to S$ is a unital ring homomorphism, then the following diagram commutes:

$$
\begin{array}{ccc}
K_0^{\mathrm{new}}(R) & \xrightarrow{\;\cong\;} & K_0(R) \\
\downarrow{\scriptstyle f_*^{\mathrm{new}}} & & \downarrow{\scriptstyle f_*} \\
K_0^{\mathrm{new}}(S) & \xrightarrow{\;\cong\;} & K_0(S).
\end{array}
$$

Thus we may identify the functors K_0 and K_0^{new} from now on.

Exercise 1.38. Let R be a ring. Show that any element in $K_0(R)$ is equal to $[e] - [1_n]$, where $e \in \mathrm{Idem}\, \mathbb{M}_\infty(R^+)$ maps to 1_n in $M_\infty(\mathbb{Z})$ and 1_n denotes the projection onto $(R^+)^n \subseteq (R^+)^\infty$ for some $n \in \mathbb{N}$.

Show also that $[e] - [1_n] = [e'] - [1_{n'}]$ in $K_0(R)$ if and only if there are $k, k' \in \mathbb{N}$ such that $e \oplus 1_k, e' \oplus 1_{k'} \in \mathrm{Idem}\, \mathbb{M}_\infty(R^+)$ are similar.

1.3   Exactness properties of K-theory

Definition 1.39. A ring *extension* is a diagram $I \overset{i}{\rightarrowtail} E \overset{p}{\twoheadrightarrow} Q$ with injective i, surjective p, and $\ker p = i(I)$. It is called *unital* if E and Q are unital and the map $E \to Q$ preserves the unit elements.

A *section* for an extension is a ring homomorphism $s\colon Q \to E$ with $p \circ s = \mathrm{id}_Q$. An extension with such a section is called *split*.

If $I \rightarrowtail E \twoheadrightarrow Q$ is a ring extension, then $i(I)$ is an *ideal* in R.

There are no interesting ring extensions with unital I:

Exercise 1.40. Any ring extension with unital I is isomorphic to a trivial extension of the form $I \rightarrowtail I \oplus Q \twoheadrightarrow Q$.

Definition 1.41. Let F be a covariant functor from the category of rings to an Abelian category. We call F *half-exact* if the sequence $F(I) \overset{i_*}{\rightarrow} F(E) \overset{p_*}{\rightarrow} F(Q)$ is exact at $F(E)$ for any ring extension. We call F *split-exact* if

$$0 \to F(I) \overset{i_*}{\longrightarrow} F(E) \overset{p_*}{\longrightarrow} F(Q) \to 0$$

is exact (at $F(I)$, $F(E)$, and $F(Q)$) for any *split* ring extension. Half-exactness and split-exactness of contravariant functors are defined similarly.

We will see that the functor K_0 is both half-exact and split-exact. In homological algebra, we usually consider functors that are even left- or right-exact. But there are no interesting functors on categories of rings that are more than half-exact.

1.3.1   Half-exactness of K_0

We need a preparatory lemma.

Lemma 1.42. *Let* $I \overset{i}{\rightarrowtail} E \overset{p}{\twoheadrightarrow} Q$ *be a unital ring extension. Let* M_+ *and* M_- *be finitely generated projective E-modules. Let* $u\colon p_*(M_+) \to p_*(M_-)$ *be a Q-module isomorphism. Then there is an E-module isomorphism* $\hat{u}\colon M_+ \oplus M_- \to M_- \oplus M_+$ *with* $p_*(\hat{u}) = u \oplus u^{-1}$.

In general, M_+ and M_- need not be isomorphic as E-modules; even if they are, we cannot expect u itself to lift to an invertible morphism.

Proof. The canonical maps

$$p_*\colon \mathrm{Hom}_E(M_\pm, M_\mp) \to \mathrm{Hom}_Q\big(p_*(M_\pm), p_*(M_\mp)\big)$$

are surjective because this is the case if $M_\pm$ are free modules. Hence there exist $v \in \mathrm{Hom}_E(M_+, M_-)$ and $w \in \mathrm{Hom}_E(M_-, M_+)$ with $p_*(v) = u$ and $p_*(w) = u^{-1}$. Define $\hat{u}\colon M_+ \oplus M_- \to M_- \oplus M_+$ and $\hat{u}^{-1}\colon M_- \oplus M_+ \to M_+ \oplus M_-$ by

$$\hat{u} := \begin{pmatrix} v & vw - 1 \\ 1 - wv & 2w - wvw \end{pmatrix}, \qquad \hat{u}^{-1} := \begin{pmatrix} 2w - wvw & 1 - wv \\ vw - 1 & v \end{pmatrix}. \qquad (1.43)$$

Check that $\hat{u}\hat{u}^{-1} = 1$ and $\hat{u}^{-1}\hat{u} = 1$ and that $p_*(\hat{u}) = u \oplus u^{-1}$. $\qquad\square$

Theorem 1.44. *The functor* K_0 *is half-exact.*

Proof. Half-exactness of K_0 for a ring extension $I \overset{i}{\rightarrowtail} E \overset{p}{\twoheadrightarrow} Q$ is equivalent to half-exactness for $I \rightarrowtail E^+ \twoheadrightarrow Q^+$. Hence we may assume that our extension is unital. We have $p_* \circ i_* = 0$ because the map $I^+ \to E \to Q$ factors through the projection $I^+ \to \mathbb{Z}$. It remains to show that $\ker p_* \subseteq \mathrm{K}_0(E)$ is contained in the range of i_*.

Elements of $\mathrm{K}_0(E)$ are equivalence classes of pairs (M_+, M_-) with $M_+, M_- \in \mathbf{V}(E)$. We have $p_*(M_+, M_-) = 0$ if and only if $p_*(M_+)$ and $p_*(M_-)$ are stably isomorphic. Equivalently, $p_*(M_+ \oplus E^n)$ and $p_*(M_- \oplus E^n)$ are isomorphic for sufficiently large n. Since $[(M_+, M_-)] = [(M_+ \oplus E^n, M_- \oplus E^n)]$, we may assume without loss of generality that $p_*(M_+) \cong p_*(M_-)$. Similarly, since $M_- \oplus N \cong E^m$ for some $N \in \mathbf{V}(E)$, $m \in \mathbb{N}$, we may also assume that $M_- = E^m$ (see also Exercise 1.38). Finally, we have $M_+ \cong E^k e$ for some $e \in \mathrm{Idem}\,\mathbb{M}_k(E)$, $k \in \mathbb{N}$.

Now we apply Lemma 1.42 to the isomorphism $p_*(M_+) \cong p_*(M_-)$ to get an invertible map $\hat{u}\colon M_+ \oplus M_- \to M_- \oplus M_+$. We have $M_+ \cong \hat{u}(M_+)$. The latter is a direct summand of $M_- \oplus M_+ \cong E^m \oplus E^k e$ and hence of the form $E^{m+k} e'$ for some $e' \in \mathrm{Idem}\,\mathbb{M}_{m+k}(E)$ with $e \sim e'$. We have $p_*(e') = 1_m$ because $p_*(\hat{u}) = u \oplus u^{-1}$ commutes with 1_m. Equivalently, $e' - 1_m \in \mathbb{M}_k(I)$, so that $(e', 1_m)$ defines a class in $\mathrm{K}_0(I)$. We have $i_*(e', 1_m) = (e, 1_m) = (M_+, M_-)$ because $e \sim e'$ in $\mathbb{M}_{m+k}(E)$. Thus $i_*\big(\mathrm{K}_0(I)\big) = \ker p_*$ as asserted. $\qquad\square$

1.3.2 Invertible elements and the index map

In order to prove the split-exactness of K_0, we must analyse the kernel of the map $i_*\colon \mathrm{K}_0(I) \to \mathrm{K}_0(E)$. This leads us to the important construction of the index map.

If R is a unital ring and $m \in \mathbb{N}$, we let $\mathrm{Gl}_m(R)$ be the set of invertible elements in $\mathbb{M}_m(R)$. We embed $\mathrm{Gl}_m(R) \to \mathrm{Gl}_{m+1}(R)$ by $u \mapsto u \oplus 1$ and form

$$\mathrm{Gl}_\infty(R) := \bigcup_{m=1}^{\infty} \mathrm{Gl}_m(R).$$

If R is a possibly non-unital ring, we let $\mathrm{Gl}_m(R^+, R) \subseteq \mathrm{Gl}_m(R^+)$ for $m \in \mathbb{N} \cup \{\infty\}$ be the kernel of the natural group homomorphism $\mathrm{Gl}_m(R^+) \to \mathrm{Gl}_m(\mathbb{Z})$; notice that this agrees with the kernel of the natural homomorphism $\mathrm{Gl}_m(E) \to \mathrm{Gl}_m(E/R)$ for any unital ring E containing R as an ideal. If R has a unit, then $\mathrm{Gl}_m(R^+) \cong \mathrm{Gl}_m(R) \times \mathrm{Gl}_m(\mathbb{Z})$, so that $\mathrm{Gl}_m(R^+, R) \cong \mathrm{Gl}_m(R)$. Hence we may abbreviate $\mathrm{Gl}_m(R) := \mathrm{Gl}_m(R^+, R)$ for non-unital rings.

Exercise 1.45. Identify $\mathrm{Gl}_\infty(R)$ with the group of all R^+-module automorphisms x of $(R^+)^\infty$ with $x - 1 \in \mathbb{M}_\infty(R)$.

Definition 1.46. The *index map* of a ring extension $I \overset{i}{\rightarrowtail} E \overset{p}{\twoheadrightarrow} Q$ is the map

$$\mathrm{ind} \colon \mathrm{Gl}_\infty(Q) \to \mathrm{K}_0(I)$$

that is defined as follows. Let $u \in \mathrm{Gl}_\infty(Q)$. Then $u \in \mathrm{Gl}_m(Q)$ for some $m \in \mathbb{N}$. Lift $u \oplus u^{-1}$ to $\hat{u} \in \mathrm{Gl}_{2m}(E)$, say, using (1.43). Then $\hat{u}$ commutes modulo $\mathbb{M}_{2m}(I)$ with the idempotent $1_m = 1_m \oplus 0_m \in \mathbb{M}_{2m}(\mathbb{Z})$, so that $\hat{u}1_m\hat{u}^{-1} - 1_m \in \mathbb{M}_{2m}(I)$. We let

$$\mathrm{ind}(u) := (\hat{u}1_m\hat{u}^{-1}, 1_m) \in \mathrm{K}_0(I).$$

Theorem 1.47. *The index map is well-defined, that is,* $\mathrm{ind}(u)$ *does not depend on auxiliary choices. We have* $\mathrm{ind}(u) = 0$ *if and only if* u *belongs to the range of* $p_* \colon \mathrm{Gl}_\infty(E) \to \mathrm{Gl}_\infty(Q)$, *and the range of* ind *is the kernel of* $i_* \colon \mathrm{K}_0(I) \to \mathrm{K}_0(E)$. *Thus we have an exact sequence of Abelian groups*

$$0 \to \frac{\mathrm{Gl}_\infty(Q)}{p_*(\mathrm{Gl}_\infty(E))} \overset{\mathrm{ind}}{\longrightarrow} \mathrm{K}_0(I) \overset{i_*}{\longrightarrow} \mathrm{K}_0(E) \overset{p_*}{\longrightarrow} \mathrm{K}_0(Q).$$

Proof. First we prove the independence of $\mathrm{ind}(u)$ from n, then the independence from $\hat{u}$. If we view $u \in \mathrm{Gl}_m(Q)$ as an element of $\mathrm{Gl}_n(Q)$ for some $n \geq m$, then our new lifting for $u \oplus u^{-1}$ can be taken to be $\hat{u} \oplus 1_{2(m-n)}$; thus we only add $(1_{n-m}, 1_{n-m})$ to $\mathrm{ind}(u)$, which has no effect on the class in $\mathrm{K}_0(I)$. If $\hat{u}, \hat{u}_2 \in \mathrm{Gl}_{2m}(E)$ are different liftings of $u \oplus u^{-1}$, then $\hat{u}_2^{-1}\hat{u} \in \ker p_* \cong \mathrm{Gl}_{2m}(I)$. Hence the resulting index idempotents are similar in $\mathbb{M}_{2m}(I^+)$ and hence yield the same class in $\mathrm{K}_0(I)$. This shows that ind is well-defined.

    We have $\mathrm{ind} \circ p_* = 0$ because if $u = p_*(\bar{u})$, then we may choose the lifting $\hat{u} = \bar{u} \oplus \bar{u}^{-1}$, which commutes with 1_m. Conversely, suppose that $\mathrm{ind}(u) = 0$. Thus the idempotents 1_m and $\hat{u}1_m\hat{u}^{-1}$ in $\mathbb{M}_{2m}(I^+)$ are stably equivalent. We may enlarge m so that they become similar in $\mathbb{M}_{2m}(I^+)$. Hence there is $y \in \mathrm{Gl}_{2m}(I)$ such that $y\hat{u}$ commutes with 1_m. Equivalently, $y\hat{u} = v_2 \oplus w_2$ for some $v_2, w_2 \in \mathrm{Gl}_m(E)$. Since $p_*(y) = 1_{2m}$, we must have $p_*(v_2) = u$, that is, u belongs to the range of $p_*(\mathrm{Gl}_\infty(E))$. Thus the kernel of ind is equal to $p_*(\mathrm{Gl}_\infty(E))$.

    We have $i_* \circ \mathrm{ind} = 0$ because 1_m and $\hat{u}1_m\hat{u}^{-1}$ are similar in $\mathbb{M}_{2m}(E^+)$. For the converse direction, we use Exercise 1.38 and represent a class in the kernel of i_* by $e \in \mathrm{Idem}\,\mathbb{M}_k(I^+)$ with $e - 1_m \in \mathbb{M}_k(I)$. The condition $i_*(e, 1_m) = 0$ means that 1_m and e become similar in $\mathbb{M}_{2m}(E^+)$ after stabilising; assuming this done, we get $\hat{u} \in \mathrm{Gl}_{2m}(E)$ with $\hat{u}1_m\hat{u}^{-1} = e$. Then $p_*(\hat{u})$ commutes with 1_m, so that, $p_*(\hat{u}) = u \oplus u_2$ for some $u, u_2 \in \mathrm{Gl}_m(Q)$. It does not matter whether u_2 is inverse to u because $u^{-1}u_2$ lifts to an invertible element in $\mathrm{Gl}_\infty(E)$ and because the lower right corner becomes irrelevant when we multiply with 1_m. Therefore, $\mathrm{ind}(u) = [e] - [1_m]$. $\square$

Corollary 1.48. *The functor* K_0 *is split-exact.*

Proof. Let $I \overset{i}{\rightarrowtail} E \overset{p}{\twoheadrightarrow} Q$ be a ring extension that splits by some homomorphism $s \colon Q \to E$. Then the maps $p_* \colon \mathrm{K}_0(E) \to \mathrm{K}_0(Q)$ and $p_* \colon \mathrm{Gl}_\infty(E) \to \mathrm{Gl}_\infty(Q)$ are split-surjective with sections induced by s. Now apply Theorem 1.47. $\square$

Exercise 1.49. Let R be an algebra over a unital ring $\mathbb{K}$. Let $R_{\mathbb{K}}^+$ be the unital $\mathbb{K}$-algebra generated by R; its underlying $\mathbb{K}$-module is $R \oplus \mathbb{K}$. Show that

$$K_0(R) \cong \ker\big(K_0(R_{\mathbb{K}}^+) \to K_0(\mathbb{K})\big).$$

We will often use this fact for $\mathbb{K} = \mathbb{C}$.

The group $\mathrm{Gl}_\infty(Q)/p_*(\mathrm{Gl}_\infty(E))$ combines the two algebras E and Q. We could rewrite the long exact sequence in Theorem 1.47 in the form

$$\mathrm{Gl}_\infty(E) \xrightarrow{p_*} \mathrm{Gl}_\infty(Q) \xrightarrow{\mathrm{ind}} K_0(I) \xrightarrow{i_*} K_0(E) \xrightarrow{p_*} K_0(Q).$$

But $R \mapsto \mathrm{Gl}_\infty(R)$ is a rather poor invariant. To get a reasonable theory, we need some equivalence relation on $\mathrm{Gl}_\infty(R)$. We will return to this issue in Chapter 2.

Exercise 1.50. Let R be a field. Let E be the ring of endomorphisms of R^∞. Recall that $\mathbb{M}_\infty(R) \subseteq E$ is the ideal of finite-rank operators.

An operator $F\colon R^\infty \to R^\infty$ is called a *Fredholm operator* if $\ker F$ and $\mathrm{coker}\, F$ are finite-dimensional. The *index* of such an operator is defined by

$$\mathrm{ind}(F) := \dim \ker F - \dim \mathrm{coker}\, F.$$

Show that F is Fredholm if and only if the image $\pi(F)$ of F in the quotient $E/\mathbb{M}_\infty(R)$ is invertible. We may lift $\pi(F)^{-1}$ to an operator $G \in E$ such that $1 - GF$ and $1 - FG$ are idempotents in $\mathbb{M}_\infty(R)$ whose ranges are isomorphic to $\ker F$ and $\mathrm{coker}\, F$, respectively. Conclude that the index map

$$\mathrm{Gl}_\infty\big(E/\mathbb{M}_\infty(R)\big) \xrightarrow{\mathrm{ind}} K_0\big(\mathbb{M}_\infty(R)\big) \cong K_0(R) \cong \mathbb{Z}$$

of Definition 1.46 maps $\pi(F)$ to $\mathrm{ind}(F) \in \mathbb{Z}$.

The situation in Exercise 1.50 is oversimplified. The operators that usually occur in index theory live on a Hilbert space like $L^2(M)$ or a nuclear Fréchet space like $C^\infty(M)$ for a smooth Riemannian manifold M; such spaces do not have a countable vector space basis. Moreover, we usually want to replace the ring E of *all* endomorphisms by a subring such as the ring of pseudo-differential operators on M. Nevertheless, the arguments that work in Exercise 1.50 remain valid in other situations.

1.3.3 Nilpotent extensions and local rings

We consider some special ring extensions; the basic assumption can be formulated in several ways:

Lemma 1.51. *Let* $I \xrightarrow{i} E \xrightarrow{p} Q$ *be a ring extension. The following assertions are equivalent:*

(1) $x \in E^+$ *is invertible if and only if* $p^+(x) \in Q^+$ *is invertible.*

(2) *All elements of the form $1 + x$ with $x \in I$ are invertible in I^+.*

(3) *The ideal I is contained in the Jacobson radical of E^+.*

(4) *For all $n \in \mathbb{N}_{\geq 1}$, $x \in \mathbb{M}_n(E^+)$ is invertible if and only if its image in $\mathbb{M}_n(Q^+)$ is invertible.*

Proof. Let $x \in I^+$ be invertible in E^+. Then $p^+(x) = \pm 1$ in Q^+, so that $p^+(x^{-1}) = \pm 1$ and hence $x^{-1} \in I^+$. This shows that (1)$\Longrightarrow$(2).

To prove the converse, take $x \in E^+$ with invertible $p^+(x) \in Q^+$. Choose $y \in E^+$ with $p^+(y) = p^+(x)^{-1}$. Then $xy - 1 \in I$ and $yx - 1 \in I$. By (2), xy and yx are invertible, so that x has both a left and a right inverse. Hence x is invertible.

The *Jacobson radical* $\mathrm{rad}(E^+)$ of E^+ is characterised in [109, Proposition 1.3.8] as the set of all $x \in E^+$ for which $1 - ax$ has a left inverse for all $a \in E^+$. Since I is an ideal, we get (2)$\Longrightarrow$(3). Conversely, Rosenberg shows in [109] that all elements of $1 + \mathrm{rad}(E^+)$ are invertible, so that (3)$\Longrightarrow$(2).

The Jacobson radical is stable in the sense that

$$\mathbb{M}_n\big(\mathrm{rad}(E^+)\big) = \mathrm{rad}\,\mathbb{M}_n(E^+)$$

(see [109, Remark after Proposition 1.3.7]). Therefore, if (3) holds for $I \subseteq E^+$, then it also holds for $\mathbb{M}_n(I) \subseteq \mathbb{M}_n(E^+)$. Therefore, (1) $\Longleftrightarrow$ (4). $\square$

Proposition 1.52. *Let $I \rightarrowtail E \twoheadrightarrow Q$ be a ring extension that satisfies the equivalent conditions of Lemma 1.51. Then $p_*\colon \mathrm{Gl}_\infty(E) \to \mathrm{Gl}_\infty(Q)$ is surjective, and the maps $p_*\colon \mathbf{V}(E^+) \to \mathbf{V}(Q^+)$ and $p_*\colon \mathrm{K}_0(E) \to \mathrm{K}_0(Q)$ are injective. The canonical map $\mathbf{V}(I^+) \to \mathbf{V}(\mathbb{Z})$ is an isomorphism and $\mathrm{K}_0(I) = 0$.*

Proof. Lemma 1.51.(4) shows that the map $\mathrm{Gl}_\infty(E) \to \mathrm{Gl}_\infty(Q)$ is surjective. Let $e_0, e_1 \in \mathbb{M}_\infty(E^+)$ be idempotents whose images in $\mathbb{M}_\infty(Q^+)$ are similar. We claim that $e_0 \sim e_1$. We have $e_0, e_1 \in \mathrm{Idem}\,\mathbb{M}_n(E^+)$ and there is $u \in \mathrm{Gl}_n(Q)$ with $up(e_0)u^{-1} = p(e_1)$ for sufficiently large $n \in \mathbb{N}_{\geq 1}$. Choose $x \in \mathbb{M}_n(Q^+)$ with $\mathbb{M}_n(p)(x) = u$ and let $\hat{u} := e_1 x e_0 + (1 - e_1)x(1 - e_0)$. Then

$$p(\hat{u}) = p(e_1)up(e_0) + \big(1 - p(e_1)\big)u\big(1 - p(e_0)\big) = up(e_0)^2 + u\big(1 - p(e_0)\big)^2 = u.$$

Hence $\hat{u} \in \mathrm{Gl}_n(E)$ by 1.51.(4). By construction, $\hat{u}e_0 = e_1 x e_0 = e_1\hat{u}$, so that e_0 and e_1 are similar via $\hat{u}$. Hence the map $\mathbf{V}(E^+) \to \mathbf{V}(Q^+)$ is injective.

Since condition 1.51.(2) only depends on the ideal I, the same argument for the ring extension $I \rightarrowtail I^+ \twoheadrightarrow \mathbb{Z}$ shows that the map $\mathbf{V}(I^+) \to \mathbf{V}(\mathbb{Z}) \cong (\mathbb{N}, +)$ is injective. Since it is obviously surjective, we get $\mathbf{V}(I^+) \cong \mathbf{V}(\mathbb{Z})$ and hence $\mathrm{K}_0(I^+) \cong \mathrm{K}_0(\mathbb{Z})$ and $\mathrm{K}_0(I) = 0$. Now the half-exactness of K_0 (Theorem 1.44) yields that the map $\mathrm{K}_0(E) \to \mathrm{K}_0(Q)$ is injective. $\square$

To conclude this section, we exhibit some cases where the map $\mathbf{V}(E) \to \mathbf{V}(Q)$ is surjective; this does not seem to follow from the conditions in Lemma 1.51.

A unital ring R is a *local ring* if and only if $R/\mathrm{rad}\,R$ is a skew-field by [109, Definition 1.3.3]. Typical examples are the rings $\mathbb{Z}_p$ of p-adic integers and $\mathbb{K}[\![t]\!]$ of formal power-series for a field $\mathbb{K}$.

Theorem 1.53. *Let R be a local ring. Then the unit map $\mathbb{Z} \to R$ induces an isomorphism $\mathbf{V}(\mathbb{Z}) \cong \mathbf{V}(R)$. That is, any finitely generated projective R-module is free, and two free modules are isomorphic if and only if they have the same rank.*

Proof. This is also proved in [109, §1.3]. Since $R/\operatorname{rad} R$ is a skew-field, we have $\mathbf{V}(R/\operatorname{rad} R) \cong (\mathbb{N}, +)$. The map $\mathbf{V}(R) \to \mathbf{V}(R/\operatorname{rad} R)$ is evidently surjective, and it is injective by Proposition 1.52. Hence $\mathbf{V}(R) \cong \mathbf{V}(R/\operatorname{rad} R) \cong (\mathbb{N}, +)$. $\qquad\square$

A (non-unital) ring I is called *nilpotent* if there is some $k \in \mathbb{N}_{\geq 1}$ with $I^k = \{0\}$, that is, $x_1 \cdots x_k = 0$ for all $x_1, \ldots, x_k \in I$.

Theorem 1.54. *Let $I \rightarrowtail E \twoheadrightarrow Q$ be a ring extension with nilpotent I. Then the equivalent assertions of Lemma 1.51 hold, and any idempotent $e \in \mathbb{M}_n(Q^+)$ lifts to an idempotent in $\mathbb{M}_n(E^+)$, which is unique up to similarity. The map $\mathbf{V}(E^+) \to \mathbf{V}(Q^+)$ is bijective.*

Proof. Formal computations with power series show that $(1 - x)^{-1}$ for $x \in I$ is given by the geometric series $\sum_{n=0}^{\infty} x^n$, which has only finitely many non-zero summands because $x^n = 0$ for all $n \geq k$. Hence 1.51.(2) holds, and Proposition 1.52 shows that the map $\mathbf{V}(E^+) \to \mathbf{V}(Q^+)$ is injective. It remains to lift idempotents in $\mathbb{M}_n(Q^+)$.

Since $\mathbb{M}_n(I)$ is nilpotent as well, we may replace our original extension by the extension $\mathbb{M}_n(I) \rightarrowtail \mathbb{M}_n(E^+) \twoheadrightarrow \mathbb{M}_n(Q^+)$. Therefore, we may assume that our extension is unital, and it suffices to lift idempotents in Q.

Let $e \in E$ be any lifting of an idempotent in Q. Then $x := e - e^2$ belongs to I. We want to find an idempotent $\hat{e} \in E$ with $e - \hat{e} \in I$. Our Ansatz is

$$\hat{e} = e + (2e - 1) \cdot \varphi(x)$$

for some power series $\varphi \in t\mathbb{Z}[\![t]\!]$. The right-hand side always defines an element of E because $x^k = 0$ for some $k \in \mathbb{N}_{\geq 1}$. A routine computation shows that

$$\hat{e}^2 - \hat{e} = \left(\varphi(x)^2 + \varphi(x) \right) \cdot (1 - 4x) - x.$$

Since $1 - 4x \in 1 + I$ is invertible, we can rewrite the condition $\hat{e}^2 = \hat{e}$ as

$$\varphi(x)^2 + \varphi(x) - \frac{x}{1 - 4x} = 0.$$

It suffices to solve this as an equation of formal power series. We get

$$\varphi = -\frac{1}{2} + \sqrt{\frac{1}{4} + \frac{t}{1 - 4t}} = -\frac{1}{2} + \frac{1}{2}(1 - 4t)^{-1/2}$$

$$= -\frac{1}{2} + \frac{1}{2} \sum_{n=0}^{\infty} \binom{-1/2}{n} (-4t)^n = \sum_{n=1}^{\infty} \binom{2n-1}{n} t^n.$$

Notice that the resulting power series has integral coefficients, although we use fractions along the way. As a result,

$$\hat{e} := e + (2e - 1) \sum_{n=1}^{\infty} \binom{2n-1}{n} (e - e^2)^n$$

is the desired idempotent lifting.   □

Chapter 2

Functional calculus and topological K-theory

The difference between algebraic and topological K-theory has its roots in analysis: K_0 behaves like topological K-theory as long as certain tools of functional analysis apply. A central issue is the holomorphic functional calculus, which underlies several important approximation lemmas in K-theory. We shall do functional analysis using *bornologies* instead of *topologies*. We prefer bornologies because bornological algebras are very close to Banach algebras and hence provide a very natural setting for the functional calculus.

We first recall some basic facts about analysis in bornological vector spaces. Then we introduce *local Banach algebras*; these are the bornological algebras in which we have a good functional calculus. We prove that K_0 is homotopy invariant for local Banach algebras. We also define higher K-theory groups $K_n(A)$ by taking, roughly speaking, the homotopy groups of $\mathrm{Gl}_\infty(A)$. These groups are related to K_0 by several long exact sequences.

We introduce isoradial homomorphisms and show that they induce isomorphisms in K-theory. For example, the embedding $C^\infty(M) \to C(M)$ for a smooth compact manifold M is isoradial. This implies that any vector bundle on M has a smooth structure. Other examples of isoradial embeddings that we discuss are completed inductive limits and stabilisations.

Throughout this chapter, we work with vector spaces and algebras over the real or complex numbers. We let $\mathbb{K} \in \{\mathbb{R}, \mathbb{C}\}$ be the field we are working over.

2.1 Bornological analysis

A semi-norm $\nu\colon V \to \mathbb{R}_+$ on an $\mathbb{R}$-vector space V is determined uniquely by its *closed unit ball* $B_\nu = \nu^{-1}([0,1]) \subseteq V$: if $x \in V$, then $\nu(x)$ is the smallest $r > 0$

with $x \in rB_\nu$. Conversely, a subset $B \subseteq V$ is the closed unit ball with respect to some semi-norm if and only if it satisfies the following conditions:

convex: $tx + (1-t)y \in B$ for all $x, y \in B$, $t \in [0,1]$;

circled: $\lambda \cdot B \subseteq B$ for all $\lambda \in \mathbb{K}$ with $|\lambda| \leq 1$;

contains its boundary: $B = \bigcap_{\varepsilon > 0}(1+\varepsilon) \cdot B$;

absorbing: $\mathbb{R}_+ \cdot B = V$.

A subset with the first three properties is called a *disk* in V. Thus the closed unit balls of semi-norms on V are precisely the absorbing disks in V. If $B \subseteq V$ is a disk, then $V_B := \mathbb{R}_+ \cdot B \subseteq V$ is a vector subspace of V, and B is an absorbing disk in V_B; thus V_B becomes a semi-normed space in a canonical way; we always equip V_B with this semi-norm. Recall that a normed space is called a *Banach space* if it is *complete* in the sense that any Cauchy sequence in V is convergent. A disk $B \subseteq V$ is called *complete* if V_B is a Banach space.

Before we define bornological vector spaces, we briefly explain the main idea by contrasting them with topological vector spaces. The topological and bornological point of view are indistinguishable for Banach spaces. Many important vector spaces such as $C^\infty(M)$ are not Banach spaces because we cannot capture the analysis in these spaces using a single norm or disk. There are two alternative ways of dealing with such spaces: the topological and the bornological approach.

On the one hand, we may focus on semi-norms and consider vector spaces that carry lots of them; this leads to *locally convex topological vector spaces*. Any separated locally convex topological vector space is a projective limit of normed spaces. Analytical constructions in separated locally convex topological vector spaces are reduced to the case of normed spaces in this fashion.

On the other hand, starting from closed unit balls we may consider vector spaces with a directed set of disks. These are the *convex bornological vector spaces*. We can equivalently describe their additional structure by a family of embeddings of semi-normed spaces: a disk B in V is equivalent to an embedding of a semi-normed space $V_B \to V$. Thus separated convex bornological vector spaces are in a canonical way inductive limits of normed spaces. Analytical constructions in separated convex bornological vector spaces are reduced to normed spaces using the embeddings $V_B \to V$. For instance, a sequence in V is *convergent* if and only if it is convergent in V_B for some B, and a function $X \to V$ from a compact space X to V is *continuous* if and only if it is continuous as a map to V_B for some B.

Bornological vector spaces went out of fashion some time ago, although they provide the most adequate setting for many problems in noncommutative geometry and representation theory. They tend to be easier to handle because inductive limits are easier than projective limits. A recent account of bornological vector spaces is contained in [86].

Definition 2.1. A *(complete convex) bornological* $\mathbb{K}$*-vector space* is a vector space V with a family $\mathfrak{S}$ of *bounded* subsets satisfying the following axioms:

(1) if $S_1 \subseteq S_2$ and $S_2 \in \mathfrak{S}$, then $S_1 \in \mathfrak{S}$;

(2) if $S_1, S_2 \in \mathfrak{S}$, then $S_1 \cup S_2 \in \mathfrak{S}$;

(3) $\{x\} \in \mathfrak{S}$ for all $x \in V$;

(4) if $S \in \mathfrak{S}$ and $c \in \mathbb{K}$, then $c \cdot S \in \mathfrak{S}$;

(5) any bounded subset of V is contained in a bounded complete disk.

We call $\mathfrak{S}$ a *(complete convex) bornology* on V if it satisfies these axioms.

The first three axioms define the notion of a bornological *set*; the last one comprises several properties, namely, compatibility of the bornology with the addition, convexity, and completeness. It is possible to weaken these requirements, but we shall not need this here. Hence we drop some adjectives from our notation and tacitly *require all bornological vector spaces to be complete and convex*. The more general notions can be found in [66, 84].

If V is a bornological vector space, then the bounded complete disks in V form a directed set with respect to inclusion, and these subsets already "generate" the bornology. Recall that the complete disks are exactly the unit balls of Banach subspaces of V. Thus a bornological vector space is nothing but an increasing union of Banach spaces, and we might also call these spaces *local Banach spaces*.

There are always lots of equivalent ways of writing a bornological vector space V as a union of Banach spaces. A similar situation occurs in the definition of a manifold structure: we may try to specify the structure by as small an atlas as possible; but then we have to explain when two of them generate the same manifold structure; we can avoid this by using a *maximal* atlas.

We now turn to some classes of examples of bornological vector spaces; we will discuss more specific examples later when we meet them in practice.

Example 2.2. If V is any vector space, then we can regard V as the union of its finite-dimensional vector subspaces; thus a subset of V is bounded if and only if it is contained in and bounded in the usual sense in some finite-dimensional vector subspace of V. This bornology is called the *fine bornology on V*. For example, this is a good choice of bornology on $\mathbb{M}_\infty = \bigcup_{n=1}^{\infty} \mathbb{M}_n$.

Example 2.3. If V is a complete locally convex topological vector space, we call a subset $S \subseteq V$ *von Neumann bounded* if $\nu(S) \subseteq \mathbb{R}_+$ is bounded for each continuous semi-norm $\nu\colon V \to \mathbb{R}_+$. This defines a complete convex bornology on V. If V is a Banach space, this bornology is generated by the closed unit ball of V.

If V is a Fréchet space, then its topology can be defined by an increasing sequence of continuous semi-norms; in contrast, its von Neumann bornology is enormous. Therefore, topological analysis in Fréchet spaces may seem more convenient than bornological analysis. Nevertheless, both approaches yield equivalent answers to many questions in this special case (see [84]). In particular, a linear map between two Fréchet spaces is bounded if and only if it is continuous, and a function $f\colon X \to V$ from a compact space into a Fréchet space is continuous if and only if it is continuous as a map $X \to V_B$ for some von Neumann bounded disk $B \subseteq V$.

The von Neumann bounded subsets are simply called bounded by most authors. We avoid this because there are other interesting bornologies on topological vector spaces. The most important of these is the *precompact bornology*, which consists of the precompact subsets. This bornology is useful because many important approximations in analysis are uniform only on precompact subsets.

Definition 2.4. A linear map $f\colon V_1 \to V_2$ between two bornological vector spaces is called *bounded* if it maps bounded sets to bounded sets; a set S of maps $V_1 \to V_2$ is called *equibounded* or *uniformly bounded* if $S(T) \subseteq V_2$ is bounded for all bounded subsets $T \subseteq V_1$. We use similar definitions for *multi*-linear maps.

Let $\mathrm{Hom}(V_1, V_2)$ be the space of bounded linear maps $V_1 \to V_2$ equipped with the uniformly bounded bornology; this is again a complete convex bornological vector space.

Definition 2.5. A *bornological algebra* is a bornological vector space A with a bounded, associative, bilinear multiplication map $A \times A \to A$.

If $V \subseteq W$ is a vector subspace of a bornological vector space, then we call V *closed* if the intersection $V \cap W_S$ is a closed subspace of W_S for each complete bounded disk $S \subseteq W$. Then the subsets $V \cap W_S$ form a complete convex bornology on V, called the *subspace bornology*. The *quotient bornology* on W/V consists of those subsets that are images of bounded subsets of W. This is a complete convex bornology if V is closed (see [66]); W/V is automatically complete because quotients of Banach spaces by closed subspaces are again Banach spaces.

Definition 2.6. A diagram of bornological vector spaces is called a *bornological extension* if it is isomorphic to a diagram of the form $V \to W \to W/V$ for a closed subspace V. A *bornological algebra extension* is a diagram of bornological algebras that is at the same time a bornological extension and an algebra extension.

2.1.1 Spaces of continuous maps

We introduce the space $C_0(X, V)$ of continuous functions $X \to V$ vanishing at ∞ for a locally compact space X and a bornological vector space V and study some of its basic properties.

First let X be a compact topological space. A function $X \to V$ is *continuous* if and only if it is continuous as a function $X \to V_S$ for some bounded complete disk $S \subseteq V$. Let $C(X, V)$ be the space of continuous maps $X \to V$. Since the spaces $C(X, V_S)$ are Banach spaces and $C(X, V)$ is their increasing union, we get a canonical bornology on $C(X, V)$: a subset $S \subseteq C(X, V)$ is bounded if and only if it is von Neumann bounded in $C(X, V_T)$ for some complete bounded disk $T \subseteq V$; equivalently, S is a uniformly bounded set of continuous functions from X to V_T.

More generally, we define $C_0(X, V)$ for a locally compact space X as the kernel of the *evaluation homomorphism*

$$\mathrm{ev}_\infty \colon C(X^+, V) \to V, \qquad f \mapsto f(\infty),$$

where $X^+ := X \cup \{\infty\}$ denotes the one-point compactification of X.

This construction is functorial in X and V. First, a continuous proper map $f \colon X \to Y$ extends to a continuous map $f^+ \colon X^+ \to Y^+$ with $f^+(\infty) = \infty$ and hence induces a map $f^* \colon C_0(Y, V) \to C_0(X, V)$, $g \mapsto g \circ f$. Secondly, a bounded linear map $h \colon V \to W$ induces a map $h_* \colon C_0(X, V) \to C_0(X, W)$, $g \mapsto h \circ g$.

Our functor also has nice exactness properties in both variables. First, if $Y \subseteq X$ is a closed subspace, then we have an extension of bornological vector spaces

$$C_0(X \smallsetminus Y, V) \rightarrowtail C_0(X, V) \twoheadrightarrow C_0(Y, V).$$

Secondly, if $V \rightarrowtail W \twoheadrightarrow W/V$ is an extension of bornological vector spaces, then so is $C_0(X, V) \rightarrowtail C_0(X, W) \twoheadrightarrow C_0(X, W/V)$; the main point is that any uniformly bounded set of maps $X \to W/V$ lifts to a uniformly bounded set of maps $X \to W$. Another useful property is $C_0(X, C_0(Y, V)) \cong C_0(X \times Y, V)$.

All these assertions are well-known facts for Banach spaces. Our definition of $C_0(X, V)$ as $\bigcup C_0(X, V_S)$ makes the extension to bornological vector spaces trivial.

Finally, if A is a bornological algebra, then so is $C_0(X, A)$ with pointwise multiplication. The maps $f^* \colon C_0(Y, A) \to C_0(X, A)$ for a continuous proper map $f \colon X \to Y$ and $h_* \colon C_0(X, A) \to C_0(X, B)$ for a bounded algebra homomorphism $h \colon A \to B$ are always bounded algebra homomorphisms.

2.1.2 Bornological tensor products

Definition 2.7. Let V and W be bornological vector spaces. Their *complete projective bornological tensor product* $V \mathbin{\widehat{\otimes}} W$ is defined by the universal property that bounded linear maps $V \mathbin{\widehat{\otimes}} W \to X$ into a bornological vector space X correspond bijectively to bounded bilinear maps $V \times W \to X$, via composition with a canonical bounded bilinear map $V \times W \to V \mathbin{\widehat{\otimes}} W$.

This universal property determines $V \mathbin{\widehat{\otimes}} W$ up to natural isomorphism. To describe it more explicitly, we assume first that V and W are Banach spaces with closed unit balls S and T. Let $V \otimes W$ be their usual vector space tensor product, and let $S \otimes T \subseteq V \otimes W$ be the convex hull of the set of elementary tensors $v \otimes w$ with $v \in S$, $w \in T$. Then $S \otimes T$ is an absorbing disk in $V \otimes W$ and thus gives rise to a norm on $V \otimes W$. Let $V \mathbin{\widehat{\otimes}} W$ be the completion of $V \otimes W$ with respect to this norm. It is easy to check that this Banach space together with the canonical bilinear map $V \times W \to V \mathbin{\widehat{\otimes}} W$ satisfies the universal property of Definition 2.7. In particular, $V \mathbin{\widehat{\otimes}} W$ satisfies a universal property in the category of Banach spaces. Therefore, $V \mathbin{\widehat{\otimes}} W$ is equal to Alexander Grothendieck's *projective Banach space tensor product* of V and W, which is usually denoted by $V \mathbin{\widehat{\otimes}_\pi} W$ (see [121]).

It follows immediately from the universal property that $(V, W) \mapsto V \widehat{\otimes} W$ is a bifunctor that commutes with arbitrary direct limits in both variables. In particular, it commutes with inductive limits. Since any bornological vector space is an inductive limit of Banach spaces, we get

$$V \widehat{\otimes} W \cong \varinjlim V_S \widehat{\otimes} W_T,$$

where S and T run through the systems of complete bounded disks in V and W, respectively. *Warning*: the natural maps $V_S \widehat{\otimes} W_T \to V_{S'} \widehat{\otimes} W_{T'}$ for $S \subseteq S'$ and $T \subseteq T'$ need not be injective, so that $V_S \widehat{\otimes} W_T$ need not be a subspace of $V \widehat{\otimes} W$; the natural map $V \otimes W \to V \widehat{\otimes} W$ need not be injective. Fortunately, injectivity rarely fails in applications.

Definition 2.8. Let I be a set and let V be a bornological vector space. Let $\ell^1(I, V)$ be the space of all functions $f: I \to V$ for which there are a bounded disk $T \subseteq V$ and $C > 0$ such that $\sum_{i \in I} \|f(i)\|_T \leq C$; a subset of $\ell^1(I, V)$ is bounded if the same T and C work for all its elements.

Tensor product computations can often be reduced to the following case.

Lemma 2.9. $\ell^1(I) \widehat{\otimes} V \cong \ell^1(I, V)$.

Proof. A bilinear map $f: \ell^1(I) \times V \to X$ is bounded if and only if the family of linear maps $f_i: V \to X$, $f_i(v) := f(\delta_i, v)$, is uniformly bounded. Bounded linear maps $\ell^1(I, V) \to X$ also correspond bijectively to such families of linear maps $V \to X$. This implies the assertion using the Yoneda Lemma. $\square$

Example 2.10. We can describe the bornology on $C_c^\infty(M)$ for a smooth manifold M using ℓ^1-estimates on derivatives. Thus we write $C_c^\infty(M)$ as a direct limit of Banach spaces isomorphic to $L^1(M, \mu)$ for some measure μ; the assertion of Lemma 2.9 remains valid in this case, so that we get

$$C_c^\infty(M) \widehat{\otimes} A \cong C_c^\infty(M, A).$$

This isomorphism is related to the *nuclearity* of $C_c^\infty(M)$. If V is nuclear, then there is only one "reasonable" bornology on $V \otimes W$ for any W; thus any "reasonable" completion of $V \otimes W$ is equal to $V \widehat{\otimes} W$.

Another, even simpler case, arises if V carries the *fine* bornology. Since $\widehat{\otimes}$ commutes with direct limits, we get $V \widehat{\otimes} W = \varinjlim V_T \widehat{\otimes} W$, where the spaces V_T are finite-dimensional. Hence $V_T \widehat{\otimes} W \cong W^{\dim V_T}$. It follows that $V \widehat{\otimes} W = V \otimes W$ as a vector space, equipped with a certain canonical bornology. If both V and W carry the fine bornology, then $V \widehat{\otimes} W$ is $V \otimes W$ equipped with the fine bornology.

2.1.3 Local Banach algebras and functional calculus

In this section, we introduce a special class of bornological algebras which have the same holomorphic functional calculus as Banach algebras (see also [65]).

Definition 2.11. A subset S of an algebra is called *submultiplicative* if $S \cdot S \subseteq S$. A bornological algebra is called a *local Banach algebra* if any bounded subset is absorbed by a bounded, submultiplicative, complete disk.

Recall that the complete disks in a vector space V are exactly the unit balls of Banach subspaces of V; similarly, complete submultiplicative disks in an algebra A are exactly the unit balls of Banach subalgebras of A. Let A be a local Banach algebra and let $\mathfrak{S}_{cmd}$ be the set of all bounded submultiplicative complete disks in A; we write $S_1 \prec S_2$ if $S_1 \subseteq C \cdot S_2$ for some $C > 0$ or, equivalently, if $A_{S_1} \subseteq A_{S_2}$ with a bounded inclusion map. If $S_1, S_2 \in \mathfrak{S}_{cmd}$, then $S_1 \cup S_2$ is again bounded and hence absorbed by some $S_3 \in \mathfrak{S}_{cmd}$. Therefore, $(\mathfrak{S}_{cmd}, \prec)$ is a directed partially ordered set. The Banach subalgebras $(A_S)_{S \in \mathfrak{S}_{cmd}}$ form an inductive system indexed by this directed set. All the structure maps in this inductive system are injective. Its inductive limit is naturally isomorphic to A.

Thus any local Banach algebra is an inductive limit of Banach algebras in a canonical way. Conversely, a bornological algebra that is an inductive limit of Banach algebras is a local Banach algebra.

A further analysis of this construction reveals an equivalence of categories between the category of local Banach algebras, with bounded algebra homomorphisms as morphisms, and the category of inductive systems of Banach algebras with injective structure maps, with morphisms of inductive systems as morphisms.

Example 2.12. Let A be a Banach algebra. Then $\mathbb{M}_\infty(A)$ is a local Banach algebra as well: it is defined as the union of the Banach subalgebras $\mathbb{M}_m(A)$ for $m \in \mathbb{N}$. More generally, $\mathbb{M}_\infty(A)$ is a local Banach algebra if A is one.

Example 2.13. If A is a local Banach algebra and X is a locally compact space, then $C_0(X, A)$ is a local Banach algebra because $C_0(X, A_S)$ is a Banach algebra if A_S is one.

Exercise 2.14. An algebra with the fine bornology is a local Banach algebra if and only if it is a union of finite-dimensional subalgebras.

It is easy to see that closed subalgebras and quotients of local Banach algebras are again local Banach algebras. Moreover, being a local Banach algebra is hereditary for inductive limits.

Theorem 2.15. *Let* $I \rightarrowtail E \twoheadrightarrow Q$ *be an extension of bornological algebras. If* I *and* Q *are local Banach algebras, so is* E.

Proof. This is proved in a different notation in [82]. □

But infinite products of local Banach algebras need not be local Banach algebras any more. For example, $\prod_{n \in \mathbb{N}} \mathbb{C}$ is not a local Banach algebra. Thus Fréchet algebras need not be local Banach algebras. We remark here that whether or not an algebra is Fréchet is rather irrelevant for the study of its algebraic K_0. Chris Phillips has extended topological K_0 to (locally multiplicatively convex) Fréchet algebras in [100]. Yet his theory differs from the K_0 constructed above. We will briefly discuss it in §2.3.4.

Definition 2.16. A subset $S \subseteq A$ of a bornological algebra is called *power-bounded* if $S^\infty := \bigcup_{n=1}^\infty S^n$ is bounded.

Notice that S^∞ is the smallest submultiplicative subset containing S. If S^∞ is bounded, then the smallest complete disk containing S^∞ is a submultiplicative bounded complete disk. This yields:

Lemma 2.17. *A bornological algebra is a local Banach algebra if and only if any bounded subset is absorbed by a power-bounded subset.*

Example 2.18. Let M be a smooth manifold and let $C_c^\infty(M)$ be the algebra of smooth functions with compact support on M. A subset S of $C_c^\infty(M)$ is bounded if there is a compact subset $K \subseteq M$ such that all functions in S are supported in K, and for each differential operator D on M there is a constant $c_D \in \mathbb{R}_+$ such that $|D(f)(x)| \leq c_D$ for all $f \in S$, $x \in M$; it suffices to require this for differential operators of the form $D = X_1 \circ \cdots \circ X_n$, where $n \in \mathbb{N}$ and where $X_1, \ldots, X_n$ are vector fields on M, viewed as derivations $X_j \colon C_c^\infty(M) \to C_c^\infty(M)$.

We claim that $C_c^\infty(M)$ is a local Banach algebra. It is easy to check that $C_c^\infty(M)$ is a bornological vector space. By Lemma 2.17, it remains to show: for any bounded subset $S \subseteq C_c^\infty(M)$ there is $r > 0$ such that $r^{-1}S$ is power-bounded.

Let ϱ be the supremum of $|f(x)|$ for $x \in M$, $f \in S$; we claim that $r^{-1}S$ is power-bounded for any $r > \varrho$. Since taking products does not increase the support, we only have to estimate differential operators of the form $X_1 \circ \cdots \circ X_n$ on a product $f_1 \cdots f_k$ with $f_1, \ldots, f_k \in r^{-1}S$. By the *Leibniz rule*, this yields a sum of $n \cdot k$ monomial terms of the form $X_{w_1}(f_1) \cdot X_{w_2}(f_2) \cdots X_{w_k}(f_k)$, where the sets $w_1, \ldots, w_k$ form a partition of $\{1, \ldots, n\}$ and where $X_w := X_{i_1} \circ \cdots \circ X_{i_j}$ if $w = \{i_1, \ldots, i_j\}$ with $i_1 \leq \cdots \leq i_j$; by convention, $X_\emptyset = \mathrm{id}$. Since there are only finitely many possibilities for X_w, the factors $X_w(f)$ are bounded by some constant $C_1 > 0$. The crucial observation is that since there are at most n possible letters, no more than n of the functions f_j are hit by a non-trivial differential operator. Thus we may estimate the supremum norm of the occurring monomials by $C_1^n \cdot (\varrho/r)^{k-n}$. The sum of $n \cdot k$ such monomials is then estimated by $n(C_1 r/\varrho)^n \cdot k(\varrho/r)^k$. This remains bounded for $k \to \infty$ because $\varrho/r < 1$.

Example 2.19. Let $X \subseteq \mathbb{C}$ be a compact subset. For a *compact* neighbourhood $K \supseteq X$, let $A(K)$ be the algebra of continuous functions $K \to \mathbb{C}$ that are holomorphic on the interior of K; this is a Banach algebra because it is a closed subalgebra of $C(K)$. If $K_0 \subseteq K_1$ are two such compact neighbourhoods, then we have a natural bounded restriction homomorphism $A(K_1) \to A(K_0)$. There is a fundamental decreasing sequence (K_n) of such compact neighbourhoods, that is, any compact neighbourhood contains K_n for sufficiently large n. Choosing this sequence carefully, we can achieve that the restriction maps $A(K_n) \to A(K_{n+1})$ are all injective. We let $\mathcal{O}(X) := \varinjlim A(K_n)$; this is the algebra of *germs of holomorphic functions near* X. It is a local Banach algebra by definition.

Definition 2.20. Let A be a unital local Banach algebra over $\mathbb{C}$. The *spectrum* of $x \in A$ is the set $\Sigma_A(x)$ of all $\lambda \in \mathbb{C}$ for which $x - \lambda \cdot 1_A$ is *not* invertible in A.

Definition 2.21. Let A be a local Banach algebra. The *spectral radius* of a bounded subset S is defined by

$$\varrho_A(S) := \inf\{r > 0 \mid r^{-1}S \text{ is power-bounded}\}.$$

The *spectral radius* of an element $x \in A$ is defined by $\varrho_A(x) := \varrho_A(\{x\})$.

These definitions still work for general bornological algebras, but the spectral radius may become ∞ and the spectrum need not have particularly nice properties. It is easy to see that a bornological algebra is a local Banach algebra if and only if $\varrho_A(S) < \infty$ for all bounded subsets $S \subseteq A$ (see [86]).

Exercise 2.22. Check that the spectrum and the spectral radius are local in the following sense. Let A be a unital local Banach algebra, write $A = \bigcup_{S \in \mathfrak{S}} A_S$ for a directed set $\mathfrak{S}$ of unital Banach subalgebras. Then

$$\Sigma_A(x) = \bigcap_{S \in \mathfrak{S}} \Sigma_{A_S}(x), \qquad \varrho_A(T) = \inf_{S \in \mathfrak{S}} \varrho_{A_S}(T)$$

for all $x \in A$ and all bounded subsets $T \subseteq A$.

Theorem 2.23. *Let A be a unital local Banach algebra and let $x \in A$. Its spectrum $\Sigma_A(x) \subseteq \mathbb{C}$ is a non-empty compact subset of $\mathbb{C}$, and we have*

$$\varrho_A(x) = \max\{|\lambda| \mid \lambda \in \Sigma_A(x)\}.$$

There is a unique bounded homomorphism $\mathcal{O}(\Sigma_A(x)) \to A$—called holomorphic functional calculus for x—which sends the identity function in $\mathcal{O}(\Sigma_A(x))$ to x.

Proof. This theorem is well-known if A is a Banach algebra. We take this special case for granted and only explain how to reduce the general case to it, using Exercise 2.22. We write A as a union of unital Banach subalgebras $A_S \subseteq A$ as above. The subsets $\Sigma_{A_T}(x) \subseteq \mathbb{C}$ are non-empty and compact for all T; since we also have $\Sigma_{A_T}(x) \supseteq \Sigma_{A_{T'}}(x)$ if T' absorbs T, the intersection $\Sigma_A(x)$ of these subsets is again non-empty and compact. Moreover, we have

$$\max\{|\lambda| \mid \lambda \in \Sigma_A(x)\} = \lim_T \max\{|\lambda| \mid \lambda \in \Sigma_{A_T}(x)\} = \lim_T \varrho_{A_T}(x) = \varrho_A(x).$$

If K is a compact neighbourhood of $\Sigma_A(x)$, then K is already a compact neighbourhood of $\Sigma_{A_T}(x)$ for sufficiently large T. Therefore, there is a holomorphic functional calculus $A(K) \to A_T$; these maps fit together to a bounded homomorphism $\mathcal{O}(\Sigma_A(x)) \to A$. $\qquad\square$

2.2 Homotopy invariance and exact sequences for local Banach algebras

We prove that the functor K_0 is homotopy invariant for local Banach algebras. Then we define higher K-theory groups $K_n(A)$ for $n \geq 1$ and establish the long exact sequence for extensions of bornological algebras, the Puppe exact sequence, and the Mayer–Vietoris exact sequence. These long exact sequences rely on the homotopy invariance of K_0. Finally, we consider the special case of C^*-algebras, where we can replace idempotents and invertible elements by projections and unitary elements without changing K-theory.

2.2.1 Homotopy invariance of K-theory

Let A be a unital local Banach algebra. We use the spectral radius and the functional calculus to prove the homotopy invariance of K_0 and some related results. We first study invertible elements, then idempotents.

By functional calculus, the exponential function exp yields a map $A \to A$. Using identities of power series, we get $\exp(x)\exp(-x) = \exp(-x)\exp(x) = 1$ for all $x \in A$, that is $\exp(x)$ is invertible with inverse $\exp(-x)$. Thus we get a map $\exp\colon A \to \mathrm{Gl}_1(A)$. The inverse function ln is defined by the power series

$$\ln(x) = \sum_{n=1}^{\infty} \frac{(-1)^{n-1}}{n}(x-1)^n,$$

whose domain of convergence is the circle of radius 1 around 1. Thus we can define $\ln(x)$ for $x \in A$ if $\varrho_A(x-1) < 1$. We have $\exp\big(\ln(x)\big) = x$ for all $x \in A$ with $\varrho_A(x-1) < 1$ because this is an identity of formal power series.

Lemma 2.24. *Two elements* $u_0, u_1 \in \mathrm{Gl}_1(A)$ *are homotopic in* $\mathrm{Gl}_1(A)$ *if and only if* $u_0 \cdot u_1^{-1} = \exp(x_0)\cdots\exp(x_k)$ *for some* $x_0, \ldots, x_k \in A$.

Proof. We may assume without loss of generality that $u_1 = 1$. If $x \in A_S$ for some Banach subalgebra $A_S \subseteq A$, then $\exp(tx)$ for $t \in [0,1]$ is a continuous path from 1 to $\exp(x)$ in A_S and hence in A. Thus any invertible element of the form $\exp(x)$ is homotopic to 1 in $\mathrm{Gl}_1(A)$. Products of such elements are homotopic to 1 as well because we can concatenate homotopies.

Conversely, suppose that u_0 is homotopic to 1 via some homotopy U. This homotopy lies in some Banach subalgebra $A_S \subseteq A$. By continuity, we may find $0 = t_0 \le t_1 \le \cdots \le t_{k+1} = 1$ such that

$$\|U(t_j) - U(t_{j+1})\|_{A_S} < \|U(t_{j+1})^{-1}\|_{A_S}^{-1} \qquad \text{for all } j = 0, \ldots, k.$$

Hence $\varrho_{A_S}(U(t_j)U(t_{j+1})^{-1} - 1) < 1$ for $j = 0, \ldots, k$. This allows us to define $x_j := \ln\big(U(t_j)U(t_{j+1})^{-1}\big)$, so that $U(t_j) = \exp(x_j) \cdot U(t_{j+1})$. By induction, this implies the assertion because $U(0) = u_0$ and $U(1) = 1$. $\qquad\square$

Now let $e_0, e_1 \in \mathrm{Idem}\, A$. Put

$$x := e_0 e_1 + (1-e_0)(1-e_1) = 1 - e_0 - e_1 + 2e_0 e_1 = 1 + (1 - 2e_0)(e_0 - e_1).$$

Then $xe_1 = e_0 e_1 = e_0 x$. Therefore, if x is invertible, then $xe_1 x^{-1} = e_0$, so that e_0 and e_1 are similar. If $\varrho_A\big((1 - 2e_0)(e_0 - e_1)\big) < 1$, then x is invertible. Roughly speaking, *nearby idempotents are similar*.

Proposition 2.25. *Let A be a unital local Banach algebra. If $e_0, e_1 \in \mathrm{Idem}\, A$ are homotopic, then there is a homotopy of invertible elements $u \in \mathrm{Gl}_1\big(C([0,1], A)\big)$ with $u(0) = 1$ and $u(1)e_0 u(1)^{-1} = e_1$. Thus homotopic idempotents are similar. Conversely, equivalent idempotents in A become homotopic in $\mathbb{M}_2(A)$.*

Therefore, homotopy, equivalence, and similarity all provide the same equivalence relation on $\mathrm{Idem}\,\mathbb{M}_\infty(A)$.

Proof. Let $e \in \mathrm{Idem}\, C([0,1], A)$ be a homotopy between e_0 and e_1. We have $e \in \mathrm{Idem}\, C([0,1], A_S)$ for some Banach subalgebra $A_S \subseteq A$. We can find $0 = t_0 \leq t_1 \leq \cdots \leq t_{k+1} = 1$ such that $\|e(t_j) - e(t_{j+1})\|_{A_S} < \|1 - 2e(t_j)\|_{A_S}^{-1}$ for $j = 0, \ldots, k$; thus

$$x_j := e(t_j)e(t_{j+1}) + \big(1 - e(t_j)\big)\big(1 - e(t_{j+1})\big)$$

satisfies $\varrho(x_j) < 1$, so that $\ln(x_j)$ is defined. Moreover, $x_j e(t_{j+1}) x_j^{-1} = e(t_j)$. Now we construct u out of the paths of invertible elements $\exp\big(t \cdot \ln(x_j)\big)$. Thus homotopic idempotents are similar.

Conversely, we claim that there is $u \in \mathrm{Gl}_{2n}\big(C([0,1], A)\big)$ with

$$u_0 = 1, \qquad u_1(e_0 \oplus 0_n)u_1^{-1} = e_1 \oplus 0_n$$

if e_0 and e_1 are equivalent; thus e_0 and e_1 become homotopic in $\mathbb{M}_2(A)$.

Let $v, w \in \mathbb{M}_n(A)$ implement the equivalence $e_0 \sim e_1$ as in (1.9). Let

$$v_t := tv + (1-t), \qquad w_t := tw + (1-t)$$

for $t \in [0,1]$. As in (1.43), define $u \in \mathrm{Gl}_{2n}\big(C([0,1], A)\big)$ by

$$u_t := \begin{pmatrix} v_t & v_t w_t - 1 \\ 1 - w_t v_t & 2w_t - w_t v_t w_t \end{pmatrix}, \qquad u_t^{-1} := \begin{pmatrix} 2w_t - w_t v_t w_t & 1 - w_t v_t \\ v_t w_t - 1 & v_t \end{pmatrix}.$$

It is easy to check that this has the required properties.   □

Corollary 2.26. *Let A be a bornological algebra and B a local Banach algebra, and let $f \colon A \to C([0,1], B)$ be a bounded algebra homomorphism. Then the bounded homomorphisms $\mathrm{ev}_t \circ f \colon A \to B$ induce the same map $\mathrm{K}_0(A) \to \mathrm{K}_0(B)$ for all $t \in [0,1]$. Thus the functor K_0 is homotopy invariant on the category of local Banach algebras. Similarly, the functor $\mathbf{V}$ is homotopy invariant on the category of unital local Banach algebras.*

Proof. It suffices to prove the assertion in the special case where A, B, and f are unital. If $e \in \mathrm{Idem}\,\mathbb{M}_m(A)$, then $\mathrm{ev}_t \circ f(e)$ for $t \in [0,1]$ are homotopic in $\mathrm{Idem}\,\mathbb{M}_m(B)$. Since $\mathbb{M}_m(B)$ is a local Banach algebra, Proposition 2.25 yields that these idempotents are similar. Thus $[\mathrm{ev}_t \circ f(e)]$ does not depend on t.   □

Exercise 2.27. A local Banach algebra A is called *contractible* if its identity map is homotopic to the zero map. Show that $\mathrm{K}_0(A)$ vanishes if A is contractible. Show also that $C_0((-\infty, \infty], A)$ is contractible for any local Banach algebra A.

2.2.2   Higher K-theory

Definition 2.28. Let A be a local Banach algebra. We let $K_1(A)$ be the set of homotopy classes of elements in $\mathrm{Gl}_\infty(A)$.

It is easy to see that homotopy is an equivalence relation on $\mathrm{Gl}_\infty(A)$. If $u_0 \sim u_1$ and $v_0 \sim v_1$, then $u_0 \cdot v_0 \sim u_1 \cdot v_1$ because $C_0([0,1], A)$ is an algebra. Thus $K_1(A)$ is a group.

Lemma 2.29. *Let A be a local Banach algebra. For any $u, v \in \mathrm{Gl}_n(A)$, we have $uv \oplus 1_n \sim u \oplus v \sim vu \oplus 1_n$ in $\mathrm{Gl}_{2n}(A)$. Thus $K_1(A)$ is an Abelian group.*

Proof. Let $R_t \in \mathbb{M}_2(\mathbb{K})$ be the matrix that describes a rotation with angle t, and view R_t as a block matrix in $\mathbb{M}_{2n}(A_{\mathbb{K}}^+)$. Then

$$(u \oplus 1_n) \cdot R_t \cdot (v \oplus 1_n) \cdot R_{-t} \in \mathrm{Gl}_{2n}\big(C_0([0,1], A)\big).$$

For $t \in [0, \pi/2]$, this provides a homotopy between $uv \oplus 1_n$ and $u \oplus v$, that is, $uv \oplus 1_n \sim u \oplus v$. A similar formula yields a homotopy $vu \oplus 1_n \sim u \oplus v$, so that $uv \oplus 1_n \sim vu \oplus 1_n$. It follows that $K_1(A)$ is an Abelian group. $\square$

Lemma 2.30. *Let $I \rightarrowtail E \twoheadrightarrow Q$ be an extension of local Banach algebras. Then there is an exact sequence*

$$K_1(E) \to K_1(Q) \xrightarrow{\ \mathrm{ind}\ } K_0(I) \to K_0(E) \to K_0(Q).$$

Proof. If $u_0, u_1 \in \mathrm{Gl}_\infty(Q)$ are homotopic, then $\mathrm{ind}(u_0)$ and $\mathrm{ind}(u_1)$ are homotopic in $K_0(I)$: apply the index map to the homotopy between u_0 and u_1 to get a homotopy between $\mathrm{ind}(u_0)$ and $\mathrm{ind}(u_1)$. Since we assume homotopy invariance of K_0, we get $\mathrm{ind}(u_0) = \mathrm{ind}(u_1)$. Therefore, the index map descends to the quotient group $K_1(Q)$. Now Theorem 1.47 yields the desired exact sequence. $\square$

Theorem 2.31. *If A is a local Banach algebra, then the index map induces an isomorphism $K_1(A) \cong K_0\big(C_0(\mathbb{R}, A)\big)$.*

Proof. The results of §2.1.1 yield an extension of bornological algebras

$$C_0(\mathbb{R}, A) \rightarrowtail C_0\big((-\infty, \infty], A\big) \twoheadrightarrow A.$$

The algebra in the middle has vanishing K_0 and K_1 because it is *contractible* (Exercise 2.27). Now apply the long exact sequence of Lemma 2.30. $\square$

Definition 2.32. $K_n(A) := K_1\big(C_0(\mathbb{R}^{n-1}, A)\big) \cong K_0\big(C_0(\mathbb{R}^n, A)\big)$.

Roughly speaking, $K_n(A)$ is the $n-1$-th homotopy group of $\mathrm{Gl}_\infty(A)$.

Theorem 2.33. *Let $I \rightarrowtail E \twoheadrightarrow Q$ be an extension of local Banach algebras. Then there is an exact sequence*

$$\cdots \to K_3(I) \to K_3(E) \to K_3(Q) \to K_2(I) \to K_2(E) \to K_2(Q)$$
$$\to K_1(I) \to K_1(E) \to K_1(Q) \to K_0(I) \to K_0(E) \to K_0(Q)$$

that continues indefinitely to the left.

Proof. We use the isomorphism of Theorem 2.31 to rewrite the exact sequence of Lemma 2.30 for the extension $C_0(\mathbb{R}^n, I) \rightarrowtail C_0(\mathbb{R}^n, E) \twoheadrightarrow C_0(\mathbb{R}^n, Q)$ as

$$\mathrm{K}_{n+1}(E) \to \mathrm{K}_{n+1}(Q) \to \mathrm{K}_n(I) \to \mathrm{K}_n(E) \to \mathrm{K}_n(Q).$$

Finally, we put these pieces together for all $n \geq 0$. $\square$

2.2.3 The Puppe exact sequence

Let $f \colon A \to B$ be a bounded algebra homomorphism between two local Banach algebras. We want to measure to what extent f induces an isomorphism on K-theory using *relative K-theory groups* $\mathrm{K}_*^{\mathrm{rel}}(f)$. These should fit into a long exact sequence

$$\ldots \to \mathrm{K}_3^{\mathrm{rel}}(f) \to \mathrm{K}_3(A) \xrightarrow{f_*} \mathrm{K}_3(B) \to \mathrm{K}_2^{\mathrm{rel}}(f) \to \mathrm{K}_2(A) \xrightarrow{f_*} \mathrm{K}_2(B)$$

$$\to \mathrm{K}_1^{\mathrm{rel}}(f) \to \mathrm{K}_1(A) \xrightarrow{f_*} \mathrm{K}_1(B) \to \mathrm{K}_0^{\mathrm{rel}}(f) \to \mathrm{K}_0(A) \xrightarrow{f_*} \mathrm{K}_0(B). \quad (2.34)$$

Definition 2.35. The *mapping cone* $C(f)$ of f is defined by

$$C(f) := \{(a, b) \in A \oplus C_0((0, 1], B) \mid f(a) = \mathrm{ev}_1(b)\};$$

it is again a local Banach algebra.

We define the *relative K-theory with respect to* f by $\mathrm{K}_*^{\mathrm{rel}}(f) := \mathrm{K}_*(C(f))$.

Exercise 2.36. The mapping cone of the identity map on B is naturally isomorphic to the cone $C_0((0, 1], B)$ over B and hence contractible.

Thus $\mathrm{K}_*^{\mathrm{rel}}(f) = 0$ if f is an isomorphism, as it should be.

Exercise 2.37. Let $V \to X$ be a vector bundle over a locally compact space X. Equip V with a metric and let $SV \subseteq V$ be the resulting sphere bundle. Let $f \colon C_0(X) \to C_0(SV)$ be the map induced by the bundle projection $SV \to X$. Then $C(f) \cong C_0(V)$.

Theorem 2.38. *There is a natural exact sequence as in* (2.34), *which is called the Puppe exact sequence.*

Proof. We have natural maps

$$\varepsilon_f \colon C(f) \to A, \qquad\qquad (a, b) \mapsto a,$$
$$\iota_f \colon C_0((0, 1), B) \to C(f), \qquad b \mapsto (0, b),$$

where we omit the obvious inclusion map $C_0((0, 1), B) \to C_0((0, 1], B)$ from our notation. These maps fit into an extension of local Banach algebras

$$C_0((0, 1), B) \xrightarrow{\iota_f} C(f) \xrightarrow{\varepsilon_f} A.$$

Since $(0, 1)$ is homeomorphic to $\mathbb{R}$, we may identify $C_0((0, 1), B)$ with $C_0(\mathbb{R}, B)$. Applying the long exact sequence of Theorem 2.33, we get the exact sequence in (2.34) except for the exactness at $K_0(A)$, which we check directly. It is easy to see that $K_*(f \circ \varepsilon_f)$ vanishes: $f \circ \varepsilon_f$ factors through the contractible algebra $C(\mathrm{id}_B)$ and therefore is smoothly homotopic to 0.

Conversely, choose an element in the kernel of $f_* \colon K_0(A) \to K_0(B)$ and represent it as $[e] - [1_n]$ for $e \in \mathrm{Idem}\,\mathbb{M}_\infty(A^+)$ with $e - 1_n \in \mathbb{M}_\infty(A)$ (Exercise 1.38). Then $f(e)$ and $f(1_n) = 1_n$ are stably equivalent in $\mathbf{V}(B)$; hence they are stably homotopic by Proposition 2.25. Stabilising the pre-images, we can achieve that $f(e)$ and $f(1_n)$ are homotopic. Thus we get $e' \in \mathrm{Idem}\,\mathbb{M}_\infty(C([0,1], B)^+)$ with $e'(1) = f(e)$ and $e'(0) = 1_n$; we can also achieve that $e' - 1_n \in C_0((0,1], B)$. Since $e'(1) = f(e)$, the pair $e'' := (e, e')$ defines an idempotent in $\mathbb{M}_\infty(C(f)^+)$ and $[e''] - [1_n]$ belongs to $K_0(C(f))$. This is the desired pre-image of $[e] - [1_n]$. Thus the sequence (2.34) is exact at $K_0(A)$ as well. $\qquad\square$

Remark 2.39. We can also get the Puppe sequence from the exact sequence that relates the mapping cone and cylinder of f. This has the advantage that we immediately get exactness up to $K_0(A)$.

2.2.4 The Mayer–Vietoris sequence

Let $I \overset{i}{\rightarrowtail} E \overset{p}{\twoheadrightarrow} Q$ be an extension of local Banach algebras. We *pull back* this extension along a bounded algebra homomorphism $f \colon Q' \to Q$. This yields an extension $I \rightarrowtail E' \twoheadrightarrow Q'$ with

$$E' := \{(e, q') \in E \oplus Q' \mid p(e) = f(q')\}$$

together with morphisms $\bar{f} \colon E' \to E$, $i' \colon I \to E'$, $p' \colon E' \to Q'$ defined by

$$\bar{f}(e, q') := e, \qquad i' := (i, 0), \qquad p'(e, q') := q'.$$

It is easy to see that we get a commuting diagram

$$
\begin{array}{ccccc}
I & \overset{i}{\rightarrowtail} & E & \overset{p}{\twoheadrightarrow} & Q \\
\Big\| & & \Big\uparrow{\scriptstyle \bar{f}} & & \Big\uparrow{\scriptstyle f} \\
I & \overset{i'}{\rightarrowtail} & E' & \overset{p'}{\twoheadrightarrow} & Q',
\end{array}
\tag{2.40}
$$

whose rows are extensions of local Banach algebras. Roughly speaking, we form E' by glueing together E and Q' over Q.

Theorem 2.41. *In the above situation, there is a long exact sequence*

$$\ldots \to K_2(E') \to K_2(E) \oplus K_2(Q') \to K_2(Q)$$
$$\to K_1(E') \to K_1(E) \oplus K_1(Q') \to K_1(Q)$$
$$\to K_0(E') \to K_0(E) \oplus K_0(Q') \to K_0(Q)$$

called the Mayer–Vietoris sequence, *where the maps* $K_*(E) \oplus K_*(Q') \to K_*(Q)$ *are* $(-p_*, f_*)$ *and the maps* $K_*(E') \to K_*(E) \oplus K_*(Q')$ *are* $(\bar{f}_*, p'_*)$.

Proof. Our proof follows an idea of Mariusz Wodzicki. By Theorem 2.33, we get long exact sequences for the two rows in (2.40) and natural maps between them:

$$
\begin{array}{ccccccccc}
\cdots \longrightarrow & K_1(E) & \xrightarrow{p_*} & K_1(Q) & \xrightarrow{\delta} & K_0(I) & \xrightarrow{i_*} & K_0(E) & \xrightarrow{p_*} & K_0(Q) \\
& \bar{f}_* \uparrow & & f_* \uparrow & & \| & & \bar{f}_* \uparrow & & f_* \uparrow \\
\cdots \longrightarrow & K_1(E') & \xrightarrow{p'_*} & K_1(Q') & \xrightarrow{\delta'} & K_0(I) & \xrightarrow{i'_*} & K_0(E') & \xrightarrow{p'_*} & K_0(Q').
\end{array}
$$

We consider the rows in this diagram as chain complexes that are exact except at $K_0(Q)$ and $K_0(Q')$, and we view the vertical maps as a chain map between them. Its *mapping cone* is another chain complex of the form

$$
\begin{aligned}
\cdots &\to K_2(I) \oplus K_3(Q) \to K_2(E') \oplus K_2(I) \to K_2(Q') \oplus K_2(E) \\
&\to K_1(I) \oplus K_2(Q) \to K_1(E') \oplus K_1(I) \to K_1(Q') \oplus K_1(E) \\
&\to K_0(I) \oplus K_1(Q) \to K_0(E') \oplus K_0(I) \to K_0(Q') \oplus K_0(E) \to K_0(Q)
\end{aligned}
$$

with boundary maps

$$
\begin{pmatrix} i'_* & 0 \\ \mathrm{id} & -\delta \end{pmatrix} : K_*(I) \oplus K_{*+1}(Q) \to K_*(E') \oplus K_*(I),
$$

$$
\begin{pmatrix} p'_* & 0 \\ \bar{f}_* & -i_* \end{pmatrix} : K_*(E') \oplus K_*(I) \to K_*(Q') \oplus K_*(E),
$$

$$
\begin{pmatrix} \delta' & 0 \\ f_* & -p_* \end{pmatrix} : K_*(Q') \oplus K_*(E) \to K_{*-1}(I) \oplus K_*(Q).
$$

The homology of this mapping cone may be computed by a Puppe exact sequence. It follows that the mapping cone is exact except at $K_0(Q)$ and $K_0(Q') \oplus K_0(E)$.

Let N be the subcomplex of the mapping cone generated by $K_*(I) \subseteq K_*(I) \oplus K_{*+1}(Q)$ and its image under the boundary map. The latter is equal to

$$
\{(x, y) \in K_*(E') \oplus K_*(I) \mid x = i'_*(y)\},
$$

and the boundary map restricts to an isomorphism from $K_*(I)$ onto this space. Thus N is a contractible subcomplex. By the long exact homology sequence for chain maps, dividing by such a contractible subcomplex does not change homology. Thus the quotient complex

$$
\begin{aligned}
\cdots &\to K_3(Q) \to K_2(E') \to K_2(Q') \oplus K_2(E) \to K_2(Q) \to K_1(E') \\
&\to K_1(Q') \oplus K_1(E) \to K_1(Q) \to K_0(E') \to K_0(Q') \oplus K_0(E) \to K_0(Q),
\end{aligned}
$$

is exact except at $K_0(Q)$ and $K_0(Q') \oplus K_0(E)$. Here we use the isomorphisms

$$
K_*(E') \oplus K_*(I)/d\big(K_*(I)\big) \cong K_*(E'), \qquad (x, y) \mapsto x - i'_*(y), \quad (x, 0) \mapsfrom x.
$$

It is straightforward to compute the boundary maps in this new exact chain complex. This yields the desired long exact sequence up to $K_0(E) \oplus K_0(Q')$. We leave it as an exercise to augment it by $K_0(Q)$. This point becomes trivial using Bott periodicity.   □

These long exact sequences are only infinite to the left. In Chapter 4, we shall prove *Bott periodicity*, which allows us to extend them to the right.

2.2.5   Projections and idempotents in C^*-algebras

In this section, we only consider $\mathbb{C}$-algebras. A map $f \colon A \to A$ is called *conjugate-linear* if it is additive and satisfies $f(\lambda \cdot x) = \overline{\lambda} f(x)$ for all $\lambda \in \mathbb{C}$, $x \in A$. An *involution* on a bornological $\mathbb{C}$-algebra is a bounded conjugate-linear map $A \to A$, $x \mapsto x^*$, such that $(x^*)^* = x$ and $(xy)^* = y^* x^*$ for all $x, y \in A$. A *bornological ∗-algebra* or an *involutive bornological algebra* is a bornological algebra equipped with such an involution.

Definition 2.42. A norm ν on a ∗-algebra A is a C^*-*norm* if $\nu(aa^*) = \nu(a)^2$ for all $a \in A$. A C^*-*algebra* is a Banach ∗-algebra A whose norm is a C^*-norm.

For example, the algebra $\mathcal{L}(\mathcal{H})$ of bounded operators on a Hilbert space $\mathcal{H}$ is a C^*-algebra, where x^* denotes the *adjoint* of x. Conversely, any C^*-algebra is isomorphic to a closed involutive subalgebra of $\mathcal{L}(\mathcal{H})$ for some Hilbert space $\mathcal{H}$ [92]. Hence $A_{\mathbb{C}}^+$ is a C^*-algebra for a unique C^*-norm if A is a C^*-algebra.

Definition 2.43. An element x of a ∗-algebra A is called *self-adjoint* if $x = x^*$, and *positive* if $x = y^* y$ for some $y \in A$; there are many other characterisations of positive elements in C^*-algebras.

When we study K-theory for C^*-algebras, we can incorporate some compatibility with the involution into our definition without changing the K-theory groups. We briefly sketch how this works, following [10].

Definition 2.44. Let A be a unital C^*-algebra. An element $v \in A$ is called

- a *projection* if $v^* v = v$;
- *unitary* if $v^{-1} = v^*$, that is, $vv^* = 1 = v^* v$;
- an *isometry* if $v^* v = 1$;
- a *co-isometry* if $vv^* = 1$;
- a *partial isometry* if $vv^* v = v$.

Thus v is unitary if and only if it is both an isometry and a co-isometry. Projections, unitaries, isometries, and co-isometries are partial isometries.

Exercise 2.45. The projections in A are exactly the self-adjoint idempotent elements.

If $v \in A$ is a partial isometry, then $e_0 = vv^*$ and $e_1 = v^*v$ are projections, called the *range and source projections* of v, and they satisfy (1.9) with $w = v^*$. If v^*v or vv^* is idempotent, then v is a partial isometry.

Proposition 2.46. *Let A be a unital C^*-algebra.*

(1) *The set of projections in A is a deformation retract of the set of idempotents in A; thus any idempotent is homotopic to a projection, and two projections that are homotopic among idempotents are homotopic among projections.*

(2) *Two projections p, q in A are similar if and only if they are unitarily equivalent, that is, $upu^{-1} = q$ for some unitary $u \in A$.*

(3) *Two projections p, q in A are equivalent if and only if they are Murray–von-Neumann equivalent, that is, there is a partial isometry v with $v^*v = p$ and $vv^* = q$.*

(4) *The set of unitary elements in A is a deformation retract of the set of invertible elements in A.*

More generally, these assertions still hold if A is a unital local Banach $*$-algebra with the additional property that $\Sigma_A(x^*x) \subseteq \mathbb{R}_+$ for all $x \in A$.

Proof. First we prove (1). Let $e \in \mathrm{Idem}\, A$. Then $z := 1 + (e^* - e)(e^* - e)^*$ satisfies $z - 1 \geq 0$, so that z is invertible in A. Let $p := ee^* \cdot z^{-1}$. We have $ez = ze = ee^*e$. This implies that z^{-1} commutes with e and e^*. Thus $p = p^*$ and $p^2 = z^{-1}ee^*ee^*z^{-1} = z^{-1}zee^*z^{-1} = p$. Moreover, $pe = e$ and $ep = p$, so that $(e-p)^2 = 0$. Therefore, $1 - t(e - p)$ is invertible with inverse $1 + t(e - p)$ for all $t \in \mathbb{R}$. Thus $t \mapsto (1 + t(e - p)) \cdot e \cdot (1 - t(e - p))$ is a continuous path in $\mathrm{Idem}\, A$ from e to p. Since this depends continuously on e, we get the desired deformation retraction.

Now we turn to (4). If $y \in \mathrm{Gl}_1(A)$ is invertible, so is y^*y. Since $y^*y \geq 0$, the spectrum of y^*y is contained in $\mathbb{R}_{>0}$. Being compact, it must be contained in $[\varepsilon, \varepsilon^{-1}]$ for some $\varepsilon > 0$. Thus $(t, z) \mapsto z^{\pm t/2}$ defines elements of $C([0, 1], \mathcal{O}(\Sigma_A(y)))$. By holomorphic functional calculus, we get $(y^*y)^{\pm t/2} \in \mathrm{Gl}_1(C([0, 1], A))$. Thus $t \mapsto y(y^*y)^{-t/2}$ is a continuous path of invertible elements that joins y to the unitary $y(y^*y)^{-1/2}$. Thus we get a deformation retraction as in (4).

Next we prove (2). Let p, q be projections in A and let $z \in A$ be invertible with $zpz^{-1} = q$. Then $zp = qz$, hence $pz^* = z^*q$ and $pz^*z = z^*zp$, that is, p commutes with z^*z. The square root of z^*z still commutes with p. Let $u := z(z^*z)^{-1/2}$, then u is unitary with $upu^{-1} = q$.

Finally, we show (3). Let p, q be equivalent projections in A. Then $p' := p \oplus 0$ and $q' := q \oplus 0$ are similar in $\mathbb{M}_2(A)$ by Lemma 1.18. By (2), p' and q' are unitarily equivalent via some unitary $u \in \mathbb{M}_2(A)$ with $up'u^* = q'$. Now let $w := q'u = up'$, then $w^*w = p'$ and $ww^* = q'$, so that w is a partial isometry. Moreover, we have $w = v \oplus 0$, and $v \in A$ is a partial isometry with $v^*v = p$ and $vv^* = q$.

These arguments only use the functional calculus for power series—which works for any local Banach $*$-algebra—and $\Sigma_A(x^*x) \subseteq \mathbb{R}_+$ for all $x \in A$. $\qquad\square$

2.3   Invariance of K-theory for isoradial subalgebras

We show that K-theory is invariant under passage to certain nice dense subalgebras. Many of the results and definitions in this section come from [84]. See also [86] for a more detailed account.

2.3.1   Isoradial homomorphisms

Definition 2.47. A subset X of a bornological vector space V is called *locally dense* if for any bounded subset $S \subseteq V$ there is a bounded disk $T \subseteq V$ such that the norm closure of $X \cap T$ in the normed space V_T contains S.

Definition 2.48. Let A and B be two bornological algebras and let $f \colon A \to B$ be a bounded homomorphism. Suppose that B is a local Banach algebra and that $f(A)$ is locally dense in B. We call f *isoradial* if $\varrho_A(S) = \varrho_B(f(S))$ for all bounded subsets $S \subseteq A$.

Roughly speaking, isoradiality means that noncommutative power series in A have the same radius of convergence in B and A. In the situation of Definition 2.48, A is necessarily a local Banach algebra because $\varrho_A(S) < \infty$ for all bounded subsets $S \subseteq A$. Often, this is a convenient method for checking that a given bornological algebra is a local Banach algebra.

Lemma 2.49. *Let A be a bornological algebra, B a local Banach algebra, and $f \colon A \to B$ a bounded algebra homomorphism with locally dense range. Suppose that any bounded subset $S \subseteq A$ with $\varrho_B(f(S)) < 1$ satisfies $\varrho_A(S^n) \leq 1$ for some $n \in \mathbb{N}_{\geq 1}$. Then f is isoradial.*

Proof. We have $\varrho_A(S^n) = \varrho_A(S)^n$ for all $n \in \mathbb{N}_{\geq 1}$ by [84, Lemma 6.3]. Using $\varrho(\lambda S) = \lambda \, \varrho(S)$ for $\lambda > 0$, we get $\varrho_A(S) \leq \varrho_B(f(S))$ for all bounded subsets $S \subseteq A$. The reverse inequality holds for any bounded algebra homomorphism. $\qquad \square$

The homomorphism f in Definition 2.48 need not be injective, although this happens in all examples that we meet here. It is crucial to require f to have dense range for the following lemma:

Lemma 2.50. *Let A and B be unital local Banach algebras and let $f \colon A \to B$ be an isoradial bounded unital algebra homomorphism (with locally dense range). Then f preserves spectra of elements, that is, $\Sigma_A(x) = \Sigma_B(f(x))$ for all $x \in A$.*

Proof. We only have to prove that $x \in A$ is invertible if and only if $f(x)$ is invertible in B. It is clear that invertibility of x implies invertibility of $f(x)$. Conversely, suppose that $f(x)$ is invertible. By local density, there is a sequence (y_n) in A such that $f(y_n)$ converges towards $f(x)^{-1}$. This implies

$$\lim_{n \to \infty} \varrho_B(f(y_n) \cdot f(x) - 1) = 0, \qquad \lim_{n \to \infty} \varrho_B(f(x) \cdot f(y_n) - 1) = 0.$$

Since f is isoradial, we also get

$$\lim_{n\to\infty} \varrho_A(y_n \cdot x - 1) = 0, \qquad \lim_{n\to\infty} \varrho_A(x \cdot y_n - 1) = 0.$$

Therefore, $y_n \cdot x$ and $x \cdot y_n$ are both invertible for sufficiently large n. Hence x is invertible. □

Example 2.51. Let M be a smooth manifold. The spectral radius computations in Example 2.18 show that the embedding $C_c^\infty(M) \to C_0(M)$ is isoradial.

More generally, if A is any local Banach algebra, we may consider the embedding $C_c^\infty(M, A) \to C_0(M, A)$. Here we define $C_c^\infty(M, A)$ in the usual way if A is a Banach space; the general case is reduced to this one by defining $C_c^\infty(M, A)$ as the increasing union of the subspaces $C_c^\infty(M, A_S)$.

Let B be $C_0(M, A)$ or $C_c^\infty(M, A)$. Let $\mathrm{ev}_x \colon B \to A$ for $x \in X$ be the point evaluation homomorphism. It is shown in [84, Proposition 6.9] that we have

$$\varrho_B(S) = \sup_{x \in M} \varrho_A\big(\mathrm{ev}_x(S)\big) \tag{2.52}$$

for all bounded subsets $S \subseteq B$. Hence the embedding $C_c^\infty(M, A) \to C_0(M, A)$ is isoradial. If $A = \mathbb{C}$, then (2.52) follows from the computations in Example 2.18. The general case requires an additional step, which we leave as an exercise.

Example 2.53. Let G be a compact Lie group and let A be a local Banach algebra equipped with a continuous action of G by algebra automorphisms. That is, we are given a bounded algebra homomorphism $\alpha \colon A \to C(G, A)$, sending $a \in A$ to the function $g \mapsto g \cdot a$. This is a bornological embedding because the algebra homomorphism $f \mapsto f(1_G)$ provides a section.

We call $a \in A$ a *smooth element* for the action of G if $\alpha(a) \in C^\infty(G, A)$; a set S of smooth elements is called *uniformly smooth* if $\alpha(S)$ is bounded in $C^\infty(G, A)$. The smooth elements form a bornological algebra whose bounded subsets are the uniformly smooth subsets. The embedding $\mathrm{SE}(\alpha) \to A$ is a bounded algebra homomorphism with locally dense range.

We have seen in Example 2.51 that the embedding $C^\infty(G, A) \to C(G, A)$ is isoradial. Hence so is the map $\mathrm{SE}(\alpha) \to A$.

The construction of $\mathrm{SE}(\alpha)$ is generalised to arbitrary locally compact groups in [85]. The natural embedding $\mathrm{SE}(\alpha) \to A$ remains isoradial in this generality by [84, Proposition 6.12].

Lemma 2.54. *Let A and B be local Banach algebras. If $f \colon A \to B$ is isoradial, so is its unital extension $f_{\mathbb{K}}^+ \colon A_{\mathbb{K}}^+ \to B_{\mathbb{K}}^+$.*

Proof. We write $A^+ = A_{\mathbb{K}}^+$ and $B^+ = B_{\mathbb{K}}^+$ to avoid clutter. Let $S \subseteq A^+$ be a bounded subset such that $\varrho_{B^+}\big(f(S)\big) < 1$; we must show that $\varrho_{A^+}(S) \le 1$. There are $r < 1$ and a Banach subalgebra $B_T \subseteq B$ such that $r^{-1} f^+(S)$ is power-bounded in B_T^+. Equip B_T^+ with the norm

$$\|(x, t)\| := \max\big\{\|x\|_{B_T}, |t|\big\} \qquad \text{for all } x \in B_T,\, t \in \mathbb{K}.$$

Since this norm remains bounded on $\bigcup r^{-n} f(S^n)$, there is $n \in \mathbb{N}_{\geq 1}$ with $\|f(x)^n\| \leq r/2$ for all $x \in S$.

By Lemma 2.49, we may replace S by S^n, so that we may assume $n = 1$. Decomposing x as $(x_A, x_\mathbb{K}) \in A \oplus \mathbb{K}$, we get $\|f(x_A)\|_{B_T} = \|f^+(x)_B\|_{B_T} \leq r/2$ and $|f^+(x)_\mathbb{K}| = |x_\mathbb{K}| \leq r/2$ for all $x \in S$. Since f is isoradial, this implies $\varrho_A(S_A) \leq 1/2$. Then $\varrho_{A^+}(S) \leq 1$ because S is contained in the convex hull of $2 \cdot S_A \cup 2 \cdot S_\mathbb{K}$. $\qquad\square$

Lemma 2.55. *Let A and B be local Banach algebras and let $f \colon A \to B$ be isoradial. Then $\mathbb{M}_m(f) \colon \mathbb{M}_m(A) \to \mathbb{M}_m(B)$ is isoradial for all $m \in \mathbb{N} \cup \{\infty\}$.*

Proof. Since any bounded subset of $\mathbb{M}_\infty(A)$ is already contained in and bounded in $\mathbb{M}_m(A)$ for some $m \in \mathbb{N}$, it suffices to prove the assertion for finite m. Let $S \subseteq \mathbb{M}_m(A)$ be bounded and suppose that $\varrho_{\mathbb{M}_m(B)}(S) < 1$. Then there are $r > 1$, $C > 0$ and a Banach subalgebra $B_T \subseteq B$ such that

$$\sup_{1 \leq i,j \leq m} \|x_{ij}\|_{B_T} \leq C \qquad \text{for all } x \in \bigcup (rf(S))^n = \bigcup r^n f(S^n).$$

As in the proof of Lemma 2.54, we can find $n \in \mathbb{N}_{\geq 1}$ such that

$$\sup_{1 \leq i,j \leq m} \|x_{ij}\|_{B_T} < 1/rm \qquad \text{for all } x \in f(S^n)$$

and may assume $n = 1$. Now let $S_{ij} \subseteq A$ be the set of all (i, j)th entries of matrices in S. Using that f is isoradial, we get $\varrho_A(\bigcup_{1 \leq i,j \leq m} S_{ij}) \leq 1/rm$. Hence there is a Banach subalgebra $A_{T'} \subseteq A$ such that

$$\sup_{1 \leq i,j \leq m} \|x_{ij}\|_{A_{T'}} \leq 1/, \qquad \text{for all } x \in S.$$

This inequality defines a submultiplicative disk in $\mathbb{M}_m(A)$ that contains S; hence S is power-bounded. $\qquad\square$

2.3.2 Nearly idempotent elements

The set of invertible elements in a unital Banach algebra is open, that is, an element that is sufficiently close to an invertible element is itself invertible. The corresponding assertion for idempotents is obviously false. We can, however, use the functional calculus to replace an element that is nearly idempotent by a nearby one that is exactly idempotent.

Definition 2.56. *Let A be a local Banach algebra and let $x \in A$. We call x nearly idempotent if $\varrho_A(x - x^2) < 1/4$.*

Lemma 2.57. *Let A be a unital local Banach algebra, let $x \in A$ be nearly idempotent, and let $\varepsilon > 0$. Then there are a commutative Banach subalgebra $A_S \subseteq A$ with $x \in A_S$ and $e \in \mathrm{Idem}\, A_S$ such that:*

$$\|x^2 - x\|_{A_S} < \varrho_A(x^2 - x) + \varepsilon,$$
$$\|x - e\|_{A_S} < 1/2 - \sqrt{1/4 - \varrho_A(x^2 - x)} + \varepsilon,$$
$$\|1 - 2e\|_{A_S} = 1.$$

Proof. The idea of the proof of Theorem 1.54 yields an explicit formula for a nearby idempotent. Let $y := x - x^2$, then we may take

$$e := x + (2x - 1) \sum_{n=1}^{\infty} \binom{2n-1}{n} y^n = x + (x - 1/2)\big((1 - 4y)^{-1/2} - 1\big).$$

The power series that we need has radius of convergence $1/4$. Hence the hypothesis $\varrho_A(x - x^2) < 1/4$ ensures that this formula works. By construction, e and x commute. The second statement means that the set

$$S_0 := \{\lambda^{-1} \cdot (x^2 - x), \mu^{-1} \cdot (x - e), 1 - 2e\}$$

is power-bounded for all $\lambda > \varrho_A(x^2 - x)$, $\mu > 1/2 - \sqrt{1/4 - \varrho_A(x^2 - x)}$; then we can take S to be the complete disked hull of S_0^{∞}. Since we are now working in a commutative subalgebra of A, we have

$$S_0^{\infty} = \{\lambda^{-1}(x^2 - x)\}^{\infty} \cdot \{\mu^{-1}(x - e)\}^{\infty} \cdot \{1 - 2e\}^{\infty}.$$

The boundedness of the first factor follows from $\lambda > \varrho_A(x^2 - x)$ and the definition of the spectral radius. The boundedness of the last factor is trivial because $(1 - 2e)^2 = 1$. It remains to show that the spectral radius of $x - e$ is at most $1/2 - \sqrt{1/4 - \varrho_A(x^2 - x)}$. A straightforward computation shows that $(x - e)^2 = z^2$ with $z := 1/2 - \sqrt{1/4 - y}$. Hence $x - e$ and z have the same spectral radius. Using the functional calculus homomorphism for y, we may replace z by the corresponding element of the bornological algebra $\mathcal{O}(B_\varrho)$, where $B_\varrho \subseteq \mathbb{C}$ is the closed disk of radius $\varrho_A(y)$ around 0. Computing the spectral radius of the latter, we get the desired estimate. $\qquad\square$

2.3.3 The invariance results

Proposition 2.58. *Let A and B be local Banach algebras, let $f \colon A \to B$ be an isoradial bounded homomorphism, and let $m \in \mathbb{N} \cup \{\infty\}$.*

Any $e \in \mathrm{Idem}\, \mathbb{M}_m(B)$ is homotopic to $f(e')$ for some $e' \in \mathrm{Idem}\, \mathbb{M}_m(A)$; if $e_0, e_1 \in \mathrm{Idem}\, \mathbb{M}_m(A)$ have the property that $f(e_0)$ and $f(e_1)$ are homotopic, similar, or equivalent in $\mathrm{Idem}\, \mathbb{M}_m(B)$, then e_0 and e_1 are homotopic, similar, or equivalent in $\mathrm{Idem}\, \mathbb{M}_m(A)$, respectively.

Any $u \in \mathrm{Gl}_m(B)$ is homotopic to $f(u')$ for some $u' \in \mathrm{Gl}_m(A)$. If $u_0, u_1 \in$ $\mathrm{Gl}_m(A)$ become homotopic in $\mathrm{Gl}_m(B)$, then u_0 and u_1 are already homotopic in $\mathrm{Gl}_m(A)$.

Proof. Lemma 2.55 allows us to reduce to the case $m = 1$, and Lemma 2.54 allows us to assume that A, B, and f are unital.

Let $e \in \mathrm{Idem}\,B$. Since $f(A)$ is locally dense, we can find a sequence (x_n) in A with $\lim f(x_n) = e$. Therefore, $\lim f(x_n^2 - x_n) = 0$. Since f is isoradial, this implies $\lim \varrho_A(x_n^2 - x_n) = 0$. Lemma 2.57 yields $e'_n \in \mathrm{Idem}\,A$ that are close to the nearly idempotent elements x_n by applying functional calculus to x_n. The idempotents $f(e'_n)$ are obtained by the same recipe from the nearly idempotent elements $f(x_n) \approx e$. Therefore, $\lim f(e'_n) = e$ in B. This implies that $f(e'_n)$ is homotopic to e for sufficiently large n.

Now let $e_0, e_1 \in A$ be idempotents such that $f(e_0)$ and $f(e_1)$ are homotopic via an idempotent $H \in C([0,1], B)$. The embedding $C([0,1], A) \to C([0,1], B)$ again has locally dense range; therefore, we can find (non-idempotent) elements $H'_n \in C([0,1], A)$ such that $\lim f(H'_n) = H$. We may modify H'_n by a partition of unity such that $H'_n(t) = e_t$ for $t = 0, 1$.

Equation (2.52) yields that the embedding $C([0,1], A) \to C([0,1], B)$ is isoradial. Therefore, H'_n is nearly idempotent for sufficiently large n. Using Lemma 2.57, we replace H'_n by an idempotent homotopy; since this construction uses only functional calculus, the endpoints of the homotopy remain equal to e_0 and e_1. Thus e_0 and e_1 are homotopic idempotents in A.

We leave it as an exercise to prove that equivalence or similarity of $f(e_0)$ and $f(e_1)$ implies the same relation for e_0 and e_1.

For any $u \in \mathrm{Gl}_1(B)$, there is a sequence (x_n) in A with $\lim f(x_n) = u$. This convergence already happens in B_S for some Banach subalgebra $B_S \subseteq B$; since $\mathrm{Gl}_1(B_S)$ is open in B_S, it follows that $f(x_n)$ is invertible and homotopic to u in $\mathrm{Gl}_1(B)$ for sufficiently large n. Hence x_n is invertible as well by Lemma 2.50.

If $u_0, u_1 \in \mathrm{Gl}_1(A)$ become homotopic in $\mathrm{Gl}_1(B)$, then we can find a sequence of paths (H_n) in $C([0,1], A)$ with $H_n(t) = u_t$ for $t = 0, 1$ such that $f(H_n)$ converges towards an *invertible* homotopy between $f(u_0)$ and $f(u_1)$ in $\mathrm{Gl}_1\big(C([0,1], B)\big)$. As above, this convergence already occurs in $C([0,1], B_S)$ for some Banach subalgebra $B_S \subseteq B$. Hence $f(H_n)$ is invertible for sufficiently large n. Since f is isoradial, Lemma 2.50 shows that H_n is invertible for sufficiently large n, so that u_0 and u_1 are homotopic in $\mathrm{Gl}_1(A)$. $\square$

Corollary 2.59. *Let A and B be unital local Banach algebras. Let $f\colon A \to B$ be an isoradial bounded unital algebra homomorphism. Then $f_*\colon \mathbf{V}(A) \to \mathbf{V}(B)$ is bijective.*

When we apply this to the isoradial embedding $C^\infty(M) \to C(M)$ for a smooth compact manifold M, we get that any vector bundle on M admits an essentially unique smooth structure.

Theorem 2.60. *Let A and B be two local Banach algebras and let $f : A \to B$ be isoradial. Then $f_* : K_n(A) \to K_n(B)$ is an isomorphism for all $n \in \mathbb{N}$.*

Proof. Using Lemma 2.54 and the split-exactness of K-theory, we reduce to the case where A, B, and f are unital. Recall that we can describe $K_0(A)$ and $K_1(A)$ using homotopy classes of idempotents and invertibles in $\mathbb{M}_m(A)$. Thus Proposition 2.58 yields the assertion for K_0 and K_1. This extends to K_n for $n \geq 2$ because the induced maps $C_0(\mathbb{R}^n, A) \to C_0(\mathbb{R}^n, B)$ are isoradial for all $n \in \mathbb{N}$. □

2.3.4 Continuity and stability

We apply Theorem 2.60 to study how K-theory behaves for *completed* inductive limits and stabilisations (compare Exercise 1.15).

Let $(A_i)_{i \in I}$ be an inductive system of bornological algebras with injective structure maps $A_i \to A_j$ for $i \leq j$. Let A be the direct limit of this system, equipped with its natural bornology. This is the increasing union of the subalgebras A_i for $i \in I$; a subset in A is bounded if and only if it is bounded in A_i for some $i \in I$.

Definition 2.61. A bornological algebra A' equipped with an algebra homomorphism $f : A \to A'$ is called a *completed direct limit* of (A_i) if $f(A)$ is locally dense in A' and the restriction of f to A_i is a bornological embedding for each $i \in I$; that is, $S \subseteq A_i$ is bounded in A_i if and only if $f(S)$ is bounded in A'.

Theorem 2.62. *Let A' be a local Banach algebra that is a completed direct limit of an inductive system of local Banach algebras (A_i). Then the canonical map*

$$\varinjlim K_*(A_i) \to K_*(A')$$

is an isomorphism. In particular, this holds for direct limits in the category of C^*-algebras.

Proof. Exercise 1.15 implies $K_*(\varinjlim A_i) \cong K_*(A)$ (first for unital A, then in general). The embedding $f : A \to A'$ is isoradial: any bounded subset S of A is already bounded in A_i for some $i \in I$, and $\bigcup S^n \subseteq A_i$ is bounded in A if and only if it is bounded in A_j for some $j \in I$, if and only if $f(\bigcup S^n) = \bigcup f(S)^n$ is bounded in A'. Hence $K_*(A) \cong K_*(A')$ by Theorem 2.60.

Let (A_i) be an inductive system of C^*-algebras with injective structure maps. Its C^*-algebraic direct limit A' is the completion of A with respect to the unique C^*-norm that extends the given C^*-norms on the subalgebras A_i. Clearly, this is a completed direct limit. □

The case of the direct system $\big(\mathbb{M}_m(A)\big)_{m \in \mathbb{N}}$ for a bornological algebra A is particularly important. Its direct limit is $\mathbb{M}_\infty(A)$.

Exercise 2.63. Check that the embeddings $A \cong \mathbb{M}_1(A) \to \mathbb{M}_m(A)$ induce isomorphisms $K_*(A) \cong K_*\big(\mathbb{M}_m(A)\big)$ for all $m \in \mathbb{N} \cup \{\infty\}$ for any ring A.

Definition 2.64. A completed limit A' of the direct system $\big(\mathbb{M}_m(A)\big)_{m \in \mathbb{N}}$ is called a *stabilisation* of A.

Theorem 2.65. *If A and A' are local Banach algebras and A' is a stabilisation of A, then the standard embedding $A \cong \mathbb{M}_1(A) \subseteq \mathbb{M}_\infty(A) \subseteq A'$ induces an isomorphism $\mathrm{K}_*(A) \xrightarrow{\cong} \mathrm{K}_*(A')$.*

Thus K-*theory for local Banach algebras is stable* for any stabilisation.

Proof. Use Theorem 2.62 and Exercise 2.63.   □

Now we consider some useful stabilisations.

The C^*-stabilisation

The C^*-*stabilisation* $\mathcal{K}_{C^*}(A)$ of a C^*-algebra A is defined as the completion of $\mathbb{M}_\infty(A)$ with respect to the unique norm that satisfies the C^*-condition $\|a^*a\| = \|a\|^2$ for all $a \in \mathbb{M}_\infty(A)$. We may also describe $\mathcal{K}_{C^*}(A)$ as the minimal or maximal C^*-tensor product of $\mathcal{K}_{C^*} := \mathcal{K}\big(\ell^2(\mathbb{N})\big)$ and A; both tensor products agree, that is, $\mathcal{K}\big(\ell^2(\mathbb{N})\big)$ is a nuclear C^*-algebra.

This stabilisation plays an important role because it is the only one that is again a C^*-algebra; we will use it in §8.5.

The smooth stabilisation

We need the *Schwartz space* $\mathscr{S}(\mathbb{N}^2, A)$. A scalar-valued function $(x_{ij})_{i,j \in \mathbb{N}}$ on $\mathbb{N}^2$ is *rapidly decreasing* if the function $x_{ij} \cdot (1 + i + j)^k$ on $\mathbb{N}^2$ remains bounded for all $k \in \mathbb{N}$; this is equivalent to $x_{ij} \cdot (1 + i + j)^k \in \ell^p(\mathbb{N}^2)$ for all $k \in \mathbb{N}$ for any $p \in [1, \infty]$.

Definition 2.66. Let V be a bornological vector space. A function $f \colon \mathbb{N}^2 \to V$ is *rapidly decreasing* if there are a rapidly decreasing sequence of scalars (ε_{ij}) and a bounded subset $T \subseteq V$ such that $f(i, j) \in \varepsilon_{ij} T$ for all $i, j \in \mathbb{N}^2$. A set of such functions is *uniformly rapidly decreasing* if the same ε_{ij} and T work for all its elements.

We let $\mathscr{S}(\mathbb{N}^2, V)$ be the bornological vector space of rapidly decreasing functions $\mathbb{N}^2 \to V$ with the bornology of uniformly rapidly decreasing subsets.

By definition, $\mathscr{S}(\mathbb{N}^2, V)$ is the direct limit of the subspaces $\mathscr{S}(\mathbb{N}^2, V_T)$. Since we can define $\mathscr{S}(\mathbb{N}^2)$ by ℓ^1-estimates, we can write it as an increasing union of Banach spaces isomorphic to $\ell^1(\mathbb{N}^2)$. If we use this and Lemma 2.9 to compute $\mathscr{S}(\mathbb{N}^2) \widehat{\otimes} A$, we get

$$\mathscr{S}(\mathbb{N}^2) \widehat{\otimes} A \cong \mathscr{S}(\mathbb{N}^2, A).$$

Thus $\mathcal{K}_{\mathscr{S}}(A) \cong \mathscr{S}(\mathbb{N}^2) \widehat{\otimes} A$ as a bornological vector space.

Let A be a bornological algebra. We may identify $\mathbb{M}_\infty(A)$ with the subspace of $\mathscr{S}(\mathbb{N}^2, A)$ of functions $\mathbb{N}^2 \to A$ with finite support. The multiplication on $\mathbb{M}_\infty(A)$ is given by

$$f_1 * f_2(i,j) = \sum_{k \in \mathbb{N}} f_1(i,k) f_2(k,j).$$

The same formula still works on $\mathscr{S}(\mathbb{N}^2, A)$ and turns it into a bornological algebra, which we denote by $\mathcal{K}_{\mathscr{S}}(A)$ and call the *smooth stabilisation* of A.

One checks easily that $\mathcal{K}_{\mathscr{S}}(A)$ is a local Banach algebra if A is one. The proof reduces immediately to the case where A is $\mathbb{C}$ and hence a C^*-algebra. In this case, we have the following stronger result:

Lemma 2.67. *Let A be a C^*-algebra. Then the smooth stabilisation $\mathcal{K}_{\mathscr{S}}(A)$ is an isoradial subalgebra of the C^*-stabilisation $\mathcal{K}_{C^*}(A)$.*

Proof. Let $N \colon \mathscr{S}(\mathbb{N}) \to \mathscr{S}(\mathbb{N})$ be the *number operator*

$$N\varphi(n) := (1+n) \cdot \varphi(n). \tag{2.68}$$

This is an unbounded multiplier of $\mathcal{K}_{C^*}(A)$ and a bounded multiplier of $\mathcal{K}_{\mathscr{S}}(A)$ via

$$Nx := \big((i+1) \cdot x_{ij}\big)_{i,j \in \mathbb{N}}, \qquad xN := \big((j+1) \cdot x_{ij}\big)_{i,j \in \mathbb{N}}$$

for $x \in \mathcal{K}_{\mathscr{S}}(A)$. It is easy to see that

$$\mathcal{K}_{\mathscr{S}}(A) = \{x \in \mathcal{K}_{C^*}(A) \mid N^m x N^n \in \mathcal{K}_{C^*}(A) \text{ for all } m,n \in \mathbb{N}\}$$

as bornological algebras; that is, a subset S of $\mathcal{K}_{C^*}(A)$ is bounded in $\mathcal{K}_{\mathscr{S}}(A)$ if and only if $N^m S N^n$ is bounded in $\mathcal{K}_{C^*}(A)$ for all $m,n \in \mathbb{N}$.

If S is bounded in $\mathcal{K}_{\mathscr{S}}(A)$ and power-bounded in $\mathcal{K}_{C^*}(A)$, then $\bigcup_{n=1}^\infty N^k S^n N^l$ is bounded in $\mathcal{K}_{C^*}(A)$ for all $k,l \in \mathbb{N}$. Hence S is power-bounded in $\mathcal{K}_{\mathscr{S}}(A)$. Thus the embedding $\mathcal{K}_{\mathscr{S}}(A) \to \mathcal{K}_{C^*}(A)$ is isoradial. $\qquad\square$

As we shall see, the smooth stabilisation is the smallest one that works for the proof of Bott periodicity and the Pimsner–Voiculescu Theorem. This is why we use it in connection with bivariant K-theory (see §7.1). In addition, the smooth stabilisation is a good choice for problems in cyclic cohomology, and it is used in Phillips' definition of topological K-theory for locally multiplicatively convex topological algebras in [100]: his definition is equivalent to $K_0\big(\mathcal{K}_{\mathscr{S}}(A)\big)$. This may differ from $K_0(A)$ because the stability of K_0 with respect to $\mathcal{K}_{\mathscr{S}}$ only works for local Banach algebras.

Stabilisation by Schatten ideals

For some purposes, $\mathcal{K}_{\mathscr{S}}$ is too small and $\mathcal{K}_{C^*}$ is too large. A good intermediate choice is to stabilise by a *Schatten ideal*

$$\mathscr{L}^p = \mathscr{L}^p\big(\ell^2(\mathbb{N})\big)$$

for some $1 \leq p < \infty$ (see [115]). The Schatten ideals $\mathscr{L}^p(\mathcal{H})$ are Banach algebras and ideals in $\mathcal{L}(\mathcal{H})$. They are dense in $\mathcal{K}(\mathcal{H})$. The easiest one to describe is $\mathscr{L}^2(\mathcal{H})$, the ideal of *Hilbert–Schmidt operators*: we simply have

$$\mathscr{L}^2\big(\ell^2(\mathbb{N})\big) \cong \ell^2(\mathbb{N} \times \mathbb{N})$$

equipped with matrix multiplication.

The ideal $\mathscr{L}^1$ is also called the *trace class* because it is the natural domain of definition for the trace on infinite matrices. We have $T \in \mathscr{L}^p$ if and only if $|T|^p \in \mathscr{L}^1$, that is, $\mathrm{tr}(|T|^p) < \infty$, if and only if $|T|^{p/2} \in \mathscr{L}^2$. Another equivalent characterisation is that the sequence of singular values of T belongs to $\ell^p(\mathbb{N})$.

The space $\mathscr{L}^p$ is a Banach algebra with respect to the norm $\|x\|_p^p := \mathrm{tr}\,|x|^p$. The multiplication in $\mathcal{K}\big(\ell^2(\mathbb{N})\big)$ restricts to bounded linear maps

$$\mathscr{L}^p \cdot \mathscr{L}^q \to \mathscr{L}^r \qquad \text{for all } p,q,r \geq 1 \text{ with } 1/p + 1/q \geq 1/r.$$

The *stabilisation of A by the Schatten ideal $\mathscr{L}^p$* is defined by

$$\mathcal{K}_{\mathscr{L}^p}(A) := \mathscr{L}^p \mathbin{\widehat{\otimes}} A.$$

Schatten ideals are useful in connection with pseudo-differential operator extensions, which we shall discuss in §12.1.1. The ideal of pseudo-differential operators of order $-\infty$ is isomorphic to $\mathscr{S}(\mathbb{N}^2)$. Often, we need the larger ideal of pseudo-differential operators of order -1, which is contained in a Schatten ideal.

A problem with the definition of $\mathcal{K}_{\mathscr{L}^p}(A)$ for $p > 1$ is that the projective bornological tensor product $\widehat{\otimes}$ is hard to compute unless we have ℓ^1-spaces and usually rather small. For instance, an A-valued diagonal matrix whose entries satisfy $\sum_{i \in \mathbb{N}} \|a_i\|_S^p < \infty$ for some bounded disk $S \subseteq A$ need not belong to $\mathcal{K}_{\mathscr{L}^p}(A)$. It would be desirable to use another tensor product here that works better for ℓ^p-spaces. Such tensor products are considered in [41], but we have not checked whether they work for our purposes.

Chapter 3

Homotopy invariance of stabilised algebraic K-theory

There are many interesting algebras that are not local Banach algebras (see Exercise 2.14), so that the results of Chapter 2 do not apply to them. Problems with homotopy invariance already occur in a purely algebraic context: the evaluation homomorphism

$$\mathrm{ev}_0 \colon A[t] := A \otimes_{\mathbb{Z}} \mathbb{Z}[t] \to A$$

for a ring A need not induce an isomorphism on K_0 although it is a homotopy equivalence. Since ev_0 is a split-surjection, the induced map $\mathrm{K}_0(A[t]) \to \mathrm{K}_0(A)$ is always surjective. Its kernel is denoted $\mathrm{NK}_0(A)$ (see [109, Definition 3.2.14]) and may be non-trivial. An example for this is $A = \mathbb{C}[t^2, t^3]$ (see [109, Exercise 3.2.24]).

We can upgrade K_0 to a homology theory with good properties for general bornological algebras by stabilising it. Here we prove the homotopy invariance result that lies at the heart of this. We do not study the resulting long exact sequences here because the proofs would mainly be repetitions of arguments in §2.2 and because we will later get them from general properties of bivariant K-theories.

The amount of homotopy invariance that we get depends on the stabilisation we choose. If we stabilise by the algebra of all compact operators on a Hilbert space or similar algebras, then we get homotopy invariance for all continuous homotopies. We will explain later why it is desirable to use smaller stabilisations. For them, we still get homotopy invariance for Hölder continuous homotopies.

The proof that we present here is new and was found by Ralf Meyer while preparing this book. It is much simpler than another proof due to Joachim Cuntz and Andreas Thom in [39], which is based on an idea of Nigel Higson [60]. Like earlier proofs, it applies to any functor defined on the category of bornological algebras that is split-exact and has suitable stability properties.

Thus we are dealing with a rigidity property of the category of noncommutative algebras, not with a special feature of K_0. Later, we will meet more such results like Bott periodicity and the Pimsner–Voiculescu exact sequence. Their proofs will also depend on the methods that we introduce here.

As in the previous chapter, it makes no difference whether we work with real or complex bornological algebras.

3.1   Ingredients in the proof

Here we introduce the tools needed in our argument:

- split-exact functors and quasi-homomorphisms,

- the relationship between stable functors and inner endomorphisms,

- a carefully chosen stabilisation, and

- Hölder continuous functions.

3.1.1   Split-exact functors and quasi-homomorphisms

First we generalise the definition of split-exact functor to allow functors with values in an *additive* category $\mathfrak{C}$; we will need this later when we study bivariant K-theory.

A sequence $A \to B \to C$ in $\mathfrak{C}$ is called *split-exact* if it is isomorphic to the sequence $A \to A \oplus C \to C$, where the maps $A \to A \oplus C \to C$ are the obvious ones. Let BAlg be the category of bornological algebras with bounded algebra homomorphisms as morphisms. We call a functor $F \colon \mathrm{BAlg} \to \mathfrak{C}$ *split-exact* if it maps any split extension of bornological algebras to a split-exact sequence in $\mathfrak{C}$.

If $I \overset{i}{\rightarrowtail} E \overset{p}{\twoheadrightarrow} Q$ is a split extension of bornological algebras, then the map $F(E) \to F(Q)$ automatically has a section $F(Q) \to F(E)$; hence the sequence $F(I) \to F(E) \to F(Q)$ is split-exact if and only if $F(i)$ is a kernel for $F(p)$. Therefore, if $\mathfrak{C}$ is Abelian then $F(I) \to F(E) \to F(Q)$ is split-exact if and only if it is a short exact sequence.

When we apply split-exactness to the trivial extension $A \to A \oplus B \to B$, we get that the coordinate embeddings induce an isomorphism $F(A) \oplus F(B) \cong F(A \oplus B)$ for any split-exact functor F. That is, split-exact functors are *additive*.

Definition 3.1. Let B and D be bornological algebras and let $i \colon B \to D$ be an injective bounded algebra homomorphism. We call B a *(generalised) ideal* in D if the multiplication on D restricts to bounded bilinear maps $B \times D \to B$ and $D \times B \to B$; here we view $B \subseteq D$ using i. The ideal is called *closed* if i is a bornological embedding, that is, B carries the subspace bornology from D.

Definition 3.2. Let A, B, and D be bornological algebras and suppose that B is an ideal in D.

A *quasi-homomorphism* $A \rightrightarrows D \rhd B$ is a pair of bounded homomorphisms $f_\pm \colon A \to D$ such that $f_+(a) - f_-(a) \in B$ for all $a \in A$ and the resulting linear map $f_+ - f_-: A \to B$ is bounded. It is called *special* if the map $A \oplus B \to D$, $(a, b) \mapsto f_+(a) + b$ is a bornological isomorphism.

If $f_\pm: A \rightrightarrows D \triangleright B$ is a special quasi-homomorphism, then B is a closed ideal in D and $D/B \cong A$ via $f_\pm^{-1}$, so that we get an extension of bornological algebras $B \rightarrowtail D \twoheadrightarrow A$. The bounded homomorphisms f_+ and f_- are sections for this extension. Since F is split-exact, we get a split-exact sequence

$$F(B) \to F(D) \to F(A)$$

with two sections $F(f_\pm)$ (we write $F(f_\pm)$ if an assertion holds for both $F(f_+)$ and $F(f_-)$). Thus we get a map

$$\tilde{F}(f_\pm) := F(f_+) - F(f_-): F(A) \to F(B) \subseteq F(D).$$

We write $\tilde{F}(f_\pm)$ to avoid confusion with the notation $F(f_\pm)$.

If the quasi-homomorphism is not special, we are going to define an associated special quasi-homomorphism $f'_\pm: A \rightrightarrows D' \triangleright B$. Then we let $\tilde{F}(f_\pm) := \tilde{F}(f'_\pm): F(A) \to F(B)$.

We drop the map i from our notation and assume $B \subseteq D$. We let $D' := B \oplus A$ as a bornological vector space, with multiplication

$$(b_1, a_1) \cdot (b_2, a_2) := (b_1 b_2 + f_+(a_1) \cdot b_2 + b_1 \cdot f_+(a_2), a_1 \cdot a_2).$$

It is easy to check that this is bounded and associative. We get homomorphisms $B \to D' \to A$ by $b \mapsto (b, 0)$ and $(b, a) \mapsto a$. These yield an extension of bornological algebras $B \rightarrowtail D' \twoheadrightarrow A$, which has two sections

$$f'_\pm: A \to D', \qquad f'_+(a) := (0, a), \quad f'_-(a) := (f_-(a) - f_+(a), a).$$

The maps f'_+ and f'_- are bounded homomorphisms and form a special quasi-homomorphism $f': A \rightrightarrows D' \triangleright B$. If we forget about bornologies, we can get the extension $B \rightarrowtail D' \twoheadrightarrow A$ by pulling back the extension $B \rightarrowtail D \twoheadrightarrow D/B$ (which need not be bornological) along $f_+: A \to D/B$.

Proposition 3.3. *The construction of $\tilde{F}(f_\pm)$ has the following properties:*

(a) *Consider a commuting diagram*

$$
\begin{array}{ccc}
A \xrightarrow[f_-]{f_+} & D_1 & \triangleright & B_1 \\
& \downarrow{\psi_D} & & \downarrow{\psi_B} \\
& D_2 & \triangleright & B_2
\end{array}
$$

whose first row is a quasi-homomorphism. Then $(\psi_D \circ f_\pm): A \rightrightarrows D_2 \triangleright B_2$ is a quasi-homomorphism, and $\tilde{F}(\psi_D \circ f_\pm) = F(\psi_B) \circ \tilde{F}(f_\pm)$.

(b) *We have*
$$\tilde{F}(f, f) = 0; \tag{3.4}$$

such *quasi-homomorphisms are called* degenerate.

(c) *If* (f_+, f_-) *is a quasi-homomorphism, so is* (f_-, f_+), *and*
$$\tilde{F}(f_+, f_-) = -\tilde{F}(f_-, f_+). \tag{3.5}$$

(d) *If* (f_+, f_-) *and* (f_+, f_0) *are quasi-homomorphisms, so is* (f_-, f_0), *and*
$$\tilde{F}(f_+, f_0) + \tilde{F}(f_0, f_-) = \tilde{F}(f_+, f_-). \tag{3.6}$$

(e) *If* $(f_\pm)$ *is a pair of bounded homomorphisms* $A \to B$, *then*
$$\tilde{F}(f_\pm) = F(f_+) - F(f_-). \tag{3.7}$$

(f) *Two quasi-homomorphisms* $(f_+^1, f_-^1), (f_+^2, f_-^2): A \rightrightarrows D \triangleright B$ *are called* orthogonal *if* $f_+^1(x) \cdot f_+^2(y) = 0$ *and* $f_-^1(x) \cdot f_-^2(y) = 0$ *for all* $x, y \in A$. *In this case,*
$f_+^1 + f_+^2$ *and* $f_-^1 + f_-^2$ *are homomorphisms and we get a quasi-homomorphism*
$$(f_+^1, f_-^1) + (f_+^2, f_-^2) := (f_+^1 + f_+^2, f_-^1 + f_-^2): A \rightrightarrows D \triangleright B.$$

We have
$$\tilde{F}\big((f_+^1, f_-^1) + (f_+^2, f_-^2)\big) = \tilde{F}(f_+^1, f_-^1) + \tilde{F}(f_+^2, f_-^2). \tag{3.8}$$

Proof. Statement (a) formalises the naturality of the construction of $\tilde{F}(f_\pm)$ and
is, therefore, trivial. Statement (b) is trivial as well.

We prove (c). If we exchange the roles of f_+ and f_-, then we get an isomorphic
split extension D' via $(b, a) \leftrightarrow (b + f_+(a) - f_-(a), a)$. We get (c) because this
isomorphism exchanges the roles of f_+ and f_-.

We prove (d). Clearly, all combinations of f_+, f_-, f_0 are quasi-homomorphisms; as in the proof of (c), they yield canonically isomorphic split extensions.
Now (3.6) reduces to the trivial computation
$$F(f_+) - F(f_-) = F(f_+) - F(f_0) + F(f_0) - F(f_-).$$

We prove (e). Let $f_+, f_-: A \to B$ be a pair of bounded homomorphisms,
viewed as a quasi-homomorphism $A \rightrightarrows B \triangleright B$. If $f_+ = 0$, then the associated split
extension is the direct sum extension, and we immediately get $\tilde{F}(0, f_-) = -F(f_-)$;
the general case of (3.7) reduces to this situation using (3.6) and (3.5).

Finally, we prove (f). The orthogonality assumption yields that there is a
quasi-homomorphism $(\tilde{f}_+, \tilde{f}_-): A \oplus A \rightrightarrows D \triangleright B$ whose restriction to the kth
summand A is (f_+^k, f_-^k) for $k = 1, 2$. We get the sum quasi-homomorphism by
composing $(\tilde{f}_+, \tilde{f}_-)$ with the *diagonal embedding*
$$\Delta: A \to A \oplus A, \qquad a \mapsto (a, a).$$

Hence $\tilde{F}\big((f^1_+, f^1_-) + (f^2_+, f^2_-)\big) = \tilde{F}(\tilde{f}_+, \tilde{f}_-) \circ F(\Delta)$ by naturality.

Since F is additive, $F(A \oplus A) \cong F(A) \oplus F(A)$. Hence $F(\Delta) = F(i_1) + F(i_2)$, where $i_1, i_2 \colon A \to A \oplus A$ are the coordinate embeddings. Naturality of $\tilde{F}$ yields $\tilde{F}(\tilde{f}_+, \tilde{f}_-) \circ F(i_k) = \tilde{F}(f^k_+, f^k_-)$. Now the assertion follows. □

3.1.2 Inner automorphisms and stability

For our proof, we need a reason for two maps $A \to B$ to induce the same map $F(A) \to F(B)$. Our sufficient condition is purely algebraic and hence works for arbitrary rings.

Definition 3.9. Let R be a ring and let $\iota^n_R \colon R \to \mathbb{M}_n(R)$ for $n \in \mathbb{N} \cup \{\infty\}$ be the upper left corner embeddings. A functor F is called $\mathbb{M}_n$-*stable* if ι^n_R induces an isomorphism $F(R) \cong F\big(\mathbb{M}_n(R)\big)$ for all R.

It follows from Exercise 2.63 that K_0 is $\mathbb{M}_n$-stable for all $n \in \mathbb{N} \cup \{\infty\}$.

Lemma 3.10. *Suppose that F is $\mathbb{M}_m$-stable for some $m \in \mathbb{N} \cup \{\infty\}$. Then F is $\mathbb{M}_n$-stable for all $n \in \mathbb{N}$, and $\mathbb{M}_\infty$-stable if F commutes with inductive limits.*

Proof. Let $F_k := F\big(\mathbb{M}_k(R)\big)$. Suppose first that n is finite. It is clear that $\mathbb{M}_m$-stability implies $\mathbb{M}_{m^l}$-stability for all $l \in \mathbb{N}$. Therefore, we may assume $m \geq n$. We have upper left corner embeddings $R \to \mathbb{M}_n(R) \to \mathbb{M}_m(R) \to \mathbb{M}_{mn}(R)$, which induce maps $F_1 \to F_n \to F_m \to F_{mn}$; here we declare $\infty \cdot n = \infty$ if $m = \infty$. The maps $F_1 \to F_m$ and $F_n \to F_{mn}$ are invertible because F is $\mathbb{M}_m$-stable and they are induced by upper left corner embeddings. It follows first that the map $F_n \to F_m$ is invertible because it has both a left and a right inverse, then that $F_1 \to F_n$ is invertible because it has an invertible one-sided inverse.

Finally, we get $F_\infty = \varinjlim F_n \cong F_1$ if F commutes with inductive limits because $\mathbb{M}_\infty(R) = \varinjlim \mathbb{M}_n(R)$. □

Now we want to show that $\mathbb{M}_2$-stable functors are invariant under inner automorphisms and endomorphisms. We first define these notions if R is unital. Any invertible element $u \in R$ gives rise to an automorphism

$$\mathrm{Ad}_u \colon R \to R, \qquad x \mapsto uxu^{-1}.$$

Such automorphisms are called *inner*. They form a normal subgroup in the automorphism group of R.

More generally, if $v, w \in R$ satisfy $wv = 1$, then we get a ring endomorphism

$$\mathrm{Ad}_{v,w} \colon R \to R, \qquad x \mapsto vxw.$$

We need $wv = 1$ for $\mathrm{Ad}_{v,w}(x) \cdot \mathrm{Ad}_{v,w}(y) = \mathrm{Ad}_{v,w}(xy)$. We also define

$$\hat{v} := \begin{pmatrix} 1 & 0 \\ 0 & v \end{pmatrix}, \qquad \hat{w} := \begin{pmatrix} 1 & 0 \\ 0 & w \end{pmatrix}.$$

Then $\hat{v}, \hat{w} \in \mathbb{M}_2(R)$ satisfy $\hat{w}\hat{v} = 1$ as well and hence generate an inner endomorphism of $\mathbb{M}_2(R)$. We compute

$$\mathrm{Ad}_{\hat{v},\hat{w}} \begin{pmatrix} x_{11} & x_{12} \\ x_{21} & x_{22} \end{pmatrix} = \begin{pmatrix} x_{11} & x_{12} \cdot w \\ v \cdot x_{21} & v \cdot x_{22} \cdot w \end{pmatrix}.$$

Definition 3.11. Let R be a ring, possibly without unit. An endomorphism (or automorphism) $\alpha \colon R \to R$ is called *inner* if there is an endomorphism (or automorphism) $\hat{\alpha} \colon \mathbb{M}_2(R) \to \mathbb{M}_2(R)$ of the form

$$\hat{\alpha} \begin{pmatrix} x_{11} & x_{12} \\ x_{21} & x_{22} \end{pmatrix} = \begin{pmatrix} x_{11} & \hat{\alpha}_{12}(x_{12}) \\ \hat{\alpha}_{21}(x_{21}) & \alpha(x_{22}) \end{pmatrix}.$$

Exercise 3.12. Any inner endomorphism of a unital ring R is of the form $\mathrm{Ad}_{v,w}$ for some $v, w \in R$ with $wv = 1$.

In the non-unital case, we get many inner endomorphisms using *multipliers*.

Definition 3.13. A *multiplier* of a ring R is a pair (l, r) consisting of a left and a right module homomorphism $l, r \colon R \to R$ such that $x \cdot r(y) = l(x) \cdot y$ for all $x, y \in R$. Multipliers are added in the obvious fashion and multiplied by the rule

$$(l_1, r_1) \cdot (l_2, r_2) = (l_2 \circ l_1, r_1 \circ r_2).$$

With these operations, the multipliers of R form a unital ring, which we denote by $\mathcal{M}(R)$; the unit element is $(\mathrm{id}_R, \mathrm{id}_R)$. We have a natural ring homomorphism $R \to \mathcal{M}(R)$, sending $x \in R$ to (l_x, r_x) with $l_x(y) := y \cdot x$ and $r_x(y) := x \cdot y$.

Exercise 3.14. This map $R \to \mathcal{M}(R)$ is an isomorphism if and only if R is unital.

If $m = (l, r)$ is a multiplier of R, then we also write $l(x) = x \cdot m$ and $r(x) = m \cdot x$. This turns R into a left and a right $\mathcal{M}(R)$-module. If the map $R \to \mathcal{M}(R)$ is injective, then we always have $(m_1 \cdot x) \cdot m_2 = m_1 \cdot (x \cdot m_2)$ because $\mathcal{M}(R)$ is associative; thus R is a $\mathcal{M}(R)$-*bi*module. In general, there may be elements $x \in R$ with $x \cdot R = 0 = R \cdot x$. Then the associativity condition for bimodules may fail.

Exercise 3.15. Let Z be an Abelian group equipped with the zero multiplication map. Compute $\mathcal{M}(Z)$ and check that Z is not a $\mathcal{M}(Z)$-bimodule.

Let $v, w \in \mathcal{M}(R)$ satisfy $wv = 1$, and suppose that $\alpha(x) := (v \cdot x) \cdot w = v \cdot (x \cdot w)$ holds in R for all $x \in R$. Then α is an inner ring endomorphism because the formula for $\mathrm{Ad}_{\hat{v},\hat{w}}$ above makes sense and defines the required ring endomorphism of $\mathbb{M}_2(R)$. We will only use inner endomorphisms of this form.

Proposition 3.16. *Let F be $\mathbb{M}_2$-stable and let $\varrho \colon R \to R$ be an inner endomorphism. Then $F(\varrho) \colon F(R) \to F(R)$ is the identity map.*

Proof. Let $\hat{\varrho} \colon \mathbb{M}_2(R) \to \mathbb{M}_2(R)$ be as in the definition of an inner endomorphism. Let $j_1, j_2 \colon R \to \mathbb{M}_2(R)$ be the two embeddings in the upper left and lower right corner. Then $F(j_1) \colon F(R) \to F(\mathbb{M}_2(R))$ is invertible by assumption. Notice that conjugation by the matrix $\left(\begin{smallmatrix} 0 & 1 \\ -1 & 0 \end{smallmatrix}\right)$ defines an (inner) automorphism σ of $\mathrm{M}_2(R)$ such that $\sigma \circ j_1 = j_2$. Hence $F(j_2) = F(\sigma) \circ F(j_1)$ is invertible as well.

Since $\hat{\varrho} \circ j_1 = j_1$ and $F(j_1)$ is invertible, we conclude that $F(\hat{\varrho})$ is the identity map on $F(\mathrm{M}_2(R))$. Since $\hat{\varrho} \circ j_2 = j_2 \circ \varrho$ and $F(j_2)$ is invertible, this implies $F(\varrho) = \mathrm{id}_{F(R)}$. $\qquad\square$

Equation (3.8) allows us to add orthogonal quasi-homomorphisms. Using stability, we can add arbitrary (quasi)-homomorphisms $f^1_\pm \colon A \rightrightarrows D \rhd B$ and $f^2_\pm \colon A \rightrightarrows D \rhd B$. Let $\iota_1, \iota_2 \colon D \to \mathrm{M}_2(D)$ be the upper left and lower right corner embeddings. Then $\iota_1 \circ f^1_\pm$ and $\iota_2 \circ f^2_\pm$ are orthogonal, so that we may add them. This yields a quasi-homomorphism

$$f^1_\pm \oplus f^2_\pm \colon A \rightrightarrows \mathrm{M}_2(D) \rhd \mathrm{M}_2(B).$$

If F is split-exact and M_2-stable, then this induces a map

$$F(A) \xrightarrow{\tilde{F}(f^1_\pm \oplus f^2_\pm)} F(\mathrm{M}_2(B)) \xrightarrow{\cong} F(B).$$

Exercise 3.17. Check that (3.8) extends to this situation, that is,

$$\tilde{F}(f^1_\pm) + \tilde{F}(f^2_\pm) = \tilde{F}(f^1_\pm \oplus f^2_\pm) \colon F(A) \to F(B).$$

Using (3.5) as well, we see that the space of maps $F(A) \to F(B)$ that can be constructed from quasi-homomorphisms $A \rightrightarrows \mathrm{M}_\infty(D) \rhd \mathrm{M}_\infty(B)$ is a subgroup of $\mathrm{Hom}(F(A), F(B))$.

3.1.3 A convenient stabilisation

Now we construct some stabilisations that we use in our homotopy invariance result. They are chosen rather carefully to fulfil various conditions, some of which will only become apparent later. We work in the spaces $\ell^1(\mathbb{N})$ and $C_0(\mathbb{N})$ instead of $\ell^2(\mathbb{N})$ because this yields a slightly smaller stabilisation—which means a stronger homotopy invariance result—and because this simplifies estimates.

First we describe the Banach algebras $\mathcal{L}(\ell^1(\mathbb{N}))$ and $\mathcal{K}(\ell^1(\mathbb{N}))$ of bounded and compact operators on the Banach space $\ell^1(\mathbb{N})$. Let $\delta_i \in \ell^1(\mathbb{N})$ be the characteristic function of $i \in \mathbb{N}$. A bounded operator T on $\ell^1(\mathbb{N})$ yields a bounded sequence $(T(\delta_i))_{i \in \mathbb{N}}$ in $\ell^1(\mathbb{N})$; conversely, any such sequence comes from a unique bounded operator on $\ell^1(\mathbb{N})$. Thus

$$\mathcal{L}(\ell^1(\mathbb{N})) \cong \ell^\infty(\ell^1(\mathbb{N})).$$

We usually represent such an operator by the matrix $T_{ij} := T(\delta_j)_i$, so that T acts on $\ell^1(\mathbb{N})$ by matrix multiplication. The operator norm is $\sup_{j \in \mathbb{N}} \sum_{i \in \mathbb{N}} |T_{ij}|$.

The closure of the subalgebra M_∞ of finite matrices is

$$\mathcal{K}(\ell^1(\mathbb{N})) = \{x \in \mathcal{L}(\ell^1(\mathbb{N})) \mid \lim_{j \to \infty} x(\delta_j) = 0\} \cong C_0(\ell^1(\mathbb{N})).$$

This algebra is not invariant under the transposition of matrices. To repair this, we consider

$$\mathcal{K}^* := \big\{(x_{ij}) \mid (x_{ij}), (x_{ji}) \in C_0\big(\ell^1(\mathbb{N})\big)\big\}.$$

This is a Banach algebra for the norm

$$\|x\| := \sup_{i\in\mathbb{N}} \sum_{j\in\mathbb{N}} |x_{ij}| + \sup_{j\in\mathbb{N}} \sum_{i\in\mathbb{N}} |x_{ij}| = \sup_{i\in\mathbb{N}} \sum_{j\in\mathbb{N}} |x_{ij}| + |x_{ji}|.$$

Exercise 3.18. If $x \in \mathcal{K}^*$, then matrix multiplication by x defines compact operators $\ell^1(\mathbb{N}) \to \ell^1(\mathbb{N})$ and $\ell^\infty(\mathbb{N}) \to C_0(\mathbb{N})$. By interpolation, it defines compact operators $\ell^p(\mathbb{N}) \to \ell^p(\mathbb{N})$ for all $p \in [1, \infty]$. Thus $\mathcal{K}^*$ embeds in $\mathcal{K}\big(\ell^p(\mathbb{N})\big)$ for all $p \in [1, \infty]$ and, in particular, for $p = 2$.

We let $\mathcal{K}^*(A)$ for a bornological algebra A consist of matrices $(x_{ij})_{ij\in\mathbb{N}}$ with entries in A for which there is a bounded disk $S \subseteq A$ with $\big(\|x_{ij}\|_S\big)_{ij\in\mathbb{N}} \in \mathcal{K}^*$; a subset $T \subseteq \mathcal{K}^*(A)$ is bounded if there is a bounded disk $S \subseteq A$ such that the matrices $\big(\|x_{ij}\|_S\big)$ for $(x_{ij}) \in T$ form a von Neumann bounded subset of $\mathcal{K}^*$. The multiplication on $\mathbb{M}_\infty(A)$ extends uniquely to a bounded multiplication on $\mathcal{K}^*(A)$.

Recall that the number operator in (2.68) is defined by

$$N: \mathscr{S}(\mathbb{N}) \to \mathscr{S}(\mathbb{N}), \qquad N\varphi(i) := (1 + i) \cdot \varphi(i)$$

We also view N as an unbounded operator on the spaces $\ell^p(\mathbb{N})$. We have $\varphi \in \mathscr{S}(\mathbb{N})$ if and only if $N^k(\varphi) \in \ell^p(\mathbb{N})$ for all $k \in \mathbb{N}$; here $p \in \mathbb{R}_{>0}$ is arbitrary.

Definition 3.19. Let $r \in \mathbb{R}$. We define

$$\mathcal{CK}^r := \{T \mid \forall a, k, l \in \mathbb{R} \; \forall b \in \mathbb{R}_{\le r-a}\colon N^a(1 + \ln N)^k \, T \, N^b(1 + \ln N)^l \in \mathcal{K}^*\}.$$

Given a bornological vector space V, we define $\mathcal{CK}^r(V)$ for $r \in \mathbb{R}$ as the space of all matrices (x_{ij}) with entries in V such that there is a bounded disk $S \subseteq V$ such that $\big(\|x_{ij}\|_S\big)$ belongs to $\mathcal{CK}^r$. The bornology is defined by requiring this estimate uniformly.

More concretely, a matrix $(v_{ij})_{ij\in\mathbb{N}}$ with entries in V belongs to $\mathcal{CK}^r(V)$ if and only if there is a bounded subset $S \subseteq V$ with

$$\sup_{i\in\mathbb{N}} \sum_{j\in\mathbb{N}} \big(\|v_{ij}\|_S + \|v_{ji}\|_S\big)(1+i)^a(1+j)^b\big(1+\ln(1+i)\big)^k\big(1+\ln(1+j)\big)^l < \infty \quad (3.20)$$

for all $a, b \in \mathbb{R}$ with $a + b \le r \in \mathbb{R}$ and all $k, l \in \mathbb{R}$. Here it suffices to consider $b = r - a$ and $k = l \in \mathbb{N}$. To simplify manipulations with such expressions, we shall replace $\mathbb{N}$ by $\mathbb{N}_{\ge 1}$ in the following. Then (3.20) becomes

$$\sup_{i\in\mathbb{N}_{\ge 1}} \sum_{j\in\mathbb{N}_{\ge 1}} \big(\|v_{ij}\|_S + \|v_{ji}\|_S\big)i^a \, j^{r-a}\, (1+\ln i)^k(1+\ln j)^k < \infty. \quad (3.21)$$

We have $\mathcal{CK}^r(V) \subseteq \mathcal{CK}^s(V)$ for $r \geq s$ and

$$\bigcap_{r \geq 0} \mathcal{CK}^r = \mathscr{S}(\mathbb{N}^2)$$

if V is trivial. This need not be true for general V (the issue here is bornological metrisability, see [84]).

Our homotopy invariance proof mainly depends on certain diagonal matrices. A diagonal matrix with entries $(x_i)_{i \in \mathbb{N}}$ belongs to $\mathcal{CK}^r(V)$ if and only if it satisfies (3.21), if and only if there is a bounded disk S such that

$$\lim_{i \to \infty} i^r (1 + \ln i)^k \|x_i\|_S = 0$$

for all $k \in \mathbb{N}$. Notice that the parameter a is gone. The same cancellation happens in the following more general situation:

Lemma 3.22. *Let (T_{ij}) be a matrix with values in A for which there are $C > 0$ and $k \in \mathbb{N}$ with $T_{ij} = 0$ whenever $i > Cj \cdot (1 + \ln j)^k$ or $j > Ci \cdot (1 + \ln i)^k$. Then $(T_{ij}) \in \mathcal{CK}^r(A)$ if and only if*

$$\sup_{i \in \mathbb{N}} \sum_{j \in \mathbb{N}} (\|v_{ij}\|_S + \|v_{ji}\|_S) i^r (1 + \ln i)^k < \infty.$$

Proof. We write $i^a j^{r-a} = i^r (i/j)^{a-r}$ and notice that $(i/j)^{a-r}$ is controlled by $(1+\ln i)^{k(a-r)} + (1+\ln j)^{k(a-r)}$. Furthermore, $1 + \ln i = O((1+\ln j)^2)$ and $1 + \ln j = O((1 + \ln i)^2)$, so that it makes no difference whether we use powers of $1 + \ln i$ or $1 + \ln j$. $\square$

If A is a bornological algebra, then so is $\mathcal{CK}^r(A)$ for all $r \geq 0$ via matrix multiplication; even more, we have bounded bilinear maps

$$m \colon \mathcal{CK}^r(A) \times \mathcal{CK}^s(A) \to \mathcal{CK}^{r+s}(A), \qquad x, y \mapsto x \circ y, \qquad \forall r, s \in \mathbb{R}. \qquad (3.23)$$

The bornological algebras $\mathcal{CK}^r(A)$ and $\mathcal{K}^*(A)$ are stabilisations of A in the sense of Definition 2.64, that is, they contain $\mathbb{M}_\infty(A)$ as a dense subalgebra and their bornologies restrict to the usual one on $\mathbb{M}_n(A)$ for all $n \in \mathbb{N}$.

If A is a local Banach algebra, then so are $\mathcal{K}^*(A)$ and $\mathcal{CK}^r(A)$, and the embedding $\mathcal{CK}^r(A) \to \mathcal{K}^*(A)$ is isoradial. Since our goal here is to treat bornological algebras that are not local Banach algebras, we will not use this fact.

3.1.4 Hölder continuity

Definition 3.24. Let V be a bornological vector space, let X be a compact metric space, and let $\alpha \in \mathbb{R}_{>0}$. A function $f \colon X \to V$ is called α-*Hölder continuous* if there is a bounded subset $S \subseteq V$ such that $f(x) - f(y) \in d(x,y)^\alpha \cdot S$ for all $x, y \in X$. A set T of functions is called *uniformly α-Hölder continuous* if the same set S works for all $f \in T$ and, in addition, $f(x) \in S$ for all $f \in T$, $x \in X$. We let $HC^\alpha(X, V)$ be the space of Hölder continuous functions $X \to V$, equipped with the bornology of uniform Hölder continuity.

If A is a bornological algebra, then $HC^\alpha(X, A)$ is a bornological algebra with respect to the pointwise product.

Definition 3.25. An α-*Hölder continuous homotopy* between $f_0, f_1 \colon A \to B$ is a bounded algebra homomorphism $\bar{f} \colon A \to HC^\alpha([0,1], B)$ with $\mathrm{ev}_t \circ \bar{f} = f_t$ for $t = 0, 1$; here $[0,1]$ carries the standard distance $d(x,y) := |x - y|$.

We call f_0 and f_1 HC^α-*homotopic* if such a homotopy exists.

3.2 The homotopy invariance result

3.2.1 A key lemma

We formulate and establish a key lemma for our homotopy invariance proof.

Lemma 3.26. *Let F be a split-exact, $\mathbb{M}_2$-stable functor on* BAlg *and let A and B be bornological algebras. Let ι be the stabilisation homomorphism $A \to \mathcal{K}^*(A)$ or $A \to \mathcal{CK}^r(A)$ for some $r \in \mathbb{R}_{\geq 0}$.*

If $f_0, f_1 \colon A \to B$ are homotopic, then $\iota \circ f_0$ and $\iota \circ f_1$ induce the same map $F(A) \to F(\mathcal{K}^*(B))$.

If $f_0, f_1 \colon A \to B$ are HC^α-homotopic for some $\alpha \in (0, 1]$, then $\iota \circ f_0$ and $\iota \circ f_1$ induce the same map $F(A) \to F(\mathcal{CK}^r(B))$ for any $r \in [0, \alpha)$.

Proof. We abbreviate $\tilde{A} := C([0,1], B)$ and $D := \mathcal{K}^*(B)$ in the first case and $\tilde{A} := HC^\alpha([0,1], B)$ and $D := \mathcal{CK}^r(B)$ in the second case. Thus our homotopy is a map $\bar{f} \colon A \to \tilde{A}$. Since $f_t = \mathrm{ev}_t \circ \bar{f}$, we are done if we show that ev_0 and ev_1 induce the same map $F(\tilde{A}) \to F(D)$. To simplify notation, we assume from now on that $A = \tilde{A}$, replacing f_t by ev_t. Furthermore, we often write $\mathrm{ev}(t) = \mathrm{ev}_t$ and $F(\mathrm{ev}_t) = \mathrm{ev}(t)_*$ to improve readability.

We prepare for the proof with some heuristic considerations. The difference $F(\mathrm{ev}_0) - F(\mathrm{ev}_1)$ is associated to the quasi-homomorphisms $(\mathrm{ev}_0, \mathrm{ev}_1) \colon A \rightrightarrows B$ by (3.7). The direct sum quasi-homomorphism

$$\bigoplus_{k=0}^{2^l - 1} \left(\mathrm{ev}(k\, 2^{-l}), \mathrm{ev}((k+1)2^{-l}) \right) \colon A \rightrightarrows \mathbb{M}_{2^l}(B)$$

induces the same map $F(A) \to F(B)$ by Exercise 3.17 and the computation

$$\sum_{k=0}^{2^l - 1} \tilde{F}\left(\mathrm{ev}(k\, 2^{-l}), \mathrm{ev}((k+1)2^{-l}) \right)$$

$$= \sum_{k=0}^{2^l - 1} \mathrm{ev}(k\, 2^{-l})_* - \mathrm{ev}((k+1)2^{-l})_* = F(\mathrm{ev}_0) - F(\mathrm{ev}_1).$$

Although these quasi-homomorphisms take values in the subalgebra B^{2^l} of diagonal matrices, it is important to use $\mathbb{M}_{2^l}(B)$ because $F(B^{2^l}) \cong F(B)^{2^l}$ is quite different from $F(\mathbb{M}_{2^l}(B)) \cong F(B)$.

The idea of the proof is to consider the infinite sum

$$\bigoplus_{l=0}^{\infty} \bigoplus_{k=0}^{2^l-1} \Big(\mathrm{ev}(k\,2^{-l}), \mathrm{ev}\big((k+1)2^{-l}\big) \Big).$$

We will see that this defines a quasi-homomorphism from A to D. Our computations suggest that it should induce the map $\sum_{l=0}^{\infty} F(\mathrm{ev}_0) - F(\mathrm{ev}_1)$, which indicates that $F(\mathrm{ev}_0) - F(\mathrm{ev}_1) = 0$. Now we supply the formal argument that makes this heuristic idea work.

Let X be the set of pairs (l,k) with $l \in \mathbb{N}$ and $k \in \{0, \ldots, 2^l - 1\}$. We identify $X \cong \mathbb{N}$ using the bijection $(l,k) \mapsto k + 2^l - 1$. This allows us to index matrices in $\mathcal{CK}^r(A)$ by X instead of $\mathbb{N}$. The number operator becomes multiplication by $k + 2^l$ on X.

Let $\ell^\infty(X;B)$ be the bornological algebra of bounded sequences in B with the pointwise product and the bornology of uniform boundedness. We view elements of $\ell^\infty(X;B)$ as diagonal matrices. This embeds $\ell^\infty(X;B) \to \mathcal{M}(D)$, that is, D is closed under multiplication on the left or right by bounded diagonal matrices.

We need three homomorphisms $\varphi_+, \varphi_0, \varphi_0 \colon A \to \ell^\infty(X;B)$ defined by

$$\varphi_+(l,k) := \mathrm{ev}(k\,2^{-l}), \qquad \varphi_0(l,k) := \begin{cases} \mathrm{ev}(k\,2^{-l}) & k \text{ even}, \\ \mathrm{ev}\big((k-1)\,2^{-l}\big) & k \text{ odd}, \end{cases}$$

$$\varphi_-(l,k) := \mathrm{ev}\big((k+1)\,2^{-l}\big),$$

for $(l,k) \in X$; our notation means that $\varphi_+(f)(l,k) := f(k\,2^{-l})$ for all $f \in A$.

If $f \in C([0,1], B)$ is continuous, then f is automatically *uniformly* continuous because $[0,1]$ is compact and continuity with values in bornological vector spaces is defined by reduction to Banach space valued functions. Hence we get $\lim_{(l,k)\to\infty} \varphi_\pm(f)(l,k) - \varphi_0(f)(l,k) = 0$, so that $\varphi_\pm(f) - \varphi_0(f) \in C_0(X;B) \subseteq \ell^\infty(X;B)$. Hence we may view $\varphi_\pm - \varphi_0$ as a bounded map $A \to \mathcal{K}^*(B) = D$.

If $f \in HC^\alpha([0,1], B)$, the maps

$$X \to B, \qquad (l,k) \mapsto 2^{l\alpha} \cdot \big(\varphi_\pm(f)(l,k) - \varphi_0(f)(l,k) \big)$$

are bounded by Hölder continuity. This remains so if we replace $2^{l\alpha}$ by $(2^l + k)^\alpha$ because $k < 2^l$ for all $(l,k) \in X$, and it holds uniformly for f in a bounded subset of $HC^\alpha([0,1], B)$. Using the specialisation of (3.21) for diagonal matrices, we get $\varphi_\pm(f) - \varphi_0(f) \in \mathcal{CK}^r(A)$ for $r < \alpha$.

Hence (φ_+, φ_0) and (φ_-, φ_0) are quasi-homomorphisms $A \rightrightarrows \mathcal{M}(D) \triangleright D$ in all cases we consider. Equation (3.6) shows that

$$\tilde{F}(\varphi_+, \varphi_-) = \tilde{F}(\varphi_+, \varphi_0) + \tilde{F}(\varphi_0, \varphi_-).$$

Next we compute the right-hand side in a different way using (3.8). We split $X = X^1 \sqcup X^2 \sqcup X^3$ with

$$X^1 := \{(0,0)\},$$
$$X^2 := \{(l,k) \in X \mid k \text{ even and } l \neq 0\},$$
$$X^3 := \{(l,k) \in X \mid k \text{ odd}\}.$$

We split $\ell^\infty(X; B)$ accordingly into sequences supported in the subspaces X^j. This yields decompositions of φ_+, φ_0, φ_- into orthogonal pieces φ_+^j, φ_0^j, φ_-^j. We use (3.8) to write

$$\tilde{F}(\varphi_+, \varphi_0) + \tilde{F}(\varphi_0, \varphi_-) = \sum_{j=1}^{3} \tilde{F}(\varphi_+^j, \varphi_0^j) + \tilde{F}(\varphi_0^j, \varphi_-^j). \tag{3.27}$$

Now we examine the summands in (3.27). We have $(\varphi_0^1, \varphi_-^1) = (\mathrm{ev}_0, \mathrm{ev}_1)$, so that

$$\tilde{F}(\varphi_0^1, \varphi_-^1) = F(\iota \circ \mathrm{ev}_0) - F(\iota \circ \mathrm{ev}_1)$$

by (3.7). Since $\varphi_+(f)(l,k) = \varphi_0(f)(l,k)$ for even k, the quasi-homomorphisms $(\varphi_+^j, \varphi_0^j)$ are degenerate for $j = 1, 2$, so that $\tilde{F}$ annihilates them by (3.4).

The quasi-homomorphisms $(\varphi_+^3, \varphi_0^3)$ and $(\varphi_0^2, \varphi_-^2)$ both involve the same evaluation homomorphisms:

$$\varphi_+^3(l, 2k+1) = \varphi_-^2(l, 2k) = \mathrm{ev}((2k+1)\,2^{-l}),$$
$$\varphi_0^3(l, 2k+1) = \varphi_0^2(l, 2k) = \mathrm{ev}((2k)\,2^{-l})$$

for all $l \in \mathbb{N}_{\geq 1}$, $k \in \{0, \ldots, 2^{l-1} - 1\}$. There are two differences: first, the order is reversed, which generates a sign by (3.5); secondly, these two maps live in orthogonal subalgebras of $\ell^\infty(X; B)$. We claim that these parts are related by inner endomorphisms of D, so that this has no effect.

Let $V \colon X \to X$ be the involution that fixes $(0,0)$ and exchanges $(l, 2k+1) \leftrightarrow (l, 2k)$ for all $l \in \mathbb{N}_{\geq 1}$ and all $k \in \{0, \ldots, 2^{l-1} - 1\}$. Since V does not move elements of $X \cong \mathbb{N}$ by more than 1, it defines a multiplier of D with $V^2 = \mathrm{id}$. Conjugation by this multiplier exchanges the roles of X^2 and X^3, so that we have

$$(\varphi_+^3, \varphi_0^3) = \mathrm{Ad}_V \circ (\varphi_-^2, \varphi_0^2).$$

Since F is $\mathbb{M}_2$-stable, inner endomorphisms act identically on $F(D)$. This yields $\tilde{F}(\varphi_+^3, \varphi_0^3) = \tilde{F}(\varphi_-^2, \varphi_0^2)$, so that the resulting two summands in (3.27) cancel.

Similarly, the quasi-homomorphisms $(\varphi_0^3, \varphi_-^3)$ and (φ_+, φ_-) both involve exactly the same evaluation homomorphisms:

$$\varphi_0^3(l+1, 2k+1) = \varphi_+(l, k) = \mathrm{ev}(k\,2^{-l}),$$
$$\varphi_-^3(l+1, 2k+1) = \varphi_-(l, k) = \mathrm{ev}((k+1)\,2^{-l})$$

for all $l \in \mathbb{N}_{\geq 0}$, $k \in \{0, \ldots, 2^l - 1\}$. The only difference is that they occur in different points of X. Again both are related by an inner endomorphism. This time, we use the embedding

$$\bar{V} \colon X \xrightarrow{\cong} X^3 \subseteq X, \qquad (l, k) \mapsto (l+1, 2k+1).$$

If we identify $X \cong \mathbb{N}$ as above, then $\bar{V}(n) = 2n+2$. We define associated operators V and W on $\ell^1(X)$ by $W(f) := f \circ \bar{V}$ and $V f(x) = f(y)$ if $\bar{V}(y) = x$ and $V f(x) = 0$ if there is no such y. Notice that the associated matrices are transpose to each other and satisfy $WV = \mathrm{id}$. Since $\bar{V}(n) = 2n + 2$, these operators V and W yield multipliers of D (compare Lemma 3.22). The associated inner endomorphism satisfies

$$(\varphi_0^3, \varphi_-^3) = \mathrm{Ad}_{V,W} \circ (\varphi_+, \varphi_-),$$

so that $\tilde{F}(\varphi_0^3, \varphi_-^3) = \tilde{F}(\varphi_+, \varphi_-)$ as above.

Finally, plugging all this into (3.27), almost everything cancels and we remain with the identity $0 = F(\iota \circ \mathrm{ev}_0) - F(\iota \circ \mathrm{ev}_1)$. This finishes the proof. $\qquad \square$

There is a C^*-algebraic version of this lemma as well:

Lemma 3.28. *Let F be a split-exact, $\mathbb{M}_2$-stable functor on the category of C^*-algebras, and let A and B be C^*-algebras. Let ι be the stabilisation homomorphism $A \to \mathcal{K}_{C^*}(A)$.*

If $f_0, f_1 \colon A \to B$ are homotopic $*$-homomorphisms, then $\iota \circ f_0$ and $\iota \circ f_1$ induce the same map $F(A) \to F(\mathcal{K}_{C^*}(B))$.

Proof. Copy the proof of Lemma 3.26 for the stabilisation $\mathcal{K}^*(A)$ and observe that all the relevant homomorphisms are $*$-homomorphisms between C^*-algebras. $\qquad \square$

3.2.2 The main results

We shall need the following weakening of the stability of a functor (compare Definition 3.9):

Definition 3.29. A functor F on BAlg is called *weakly stable* (*stable*) with respect to a stabilisation $A \mapsto \mathcal{K}_?(A)$ if the stabilisation homomorphism $A \to \mathcal{K}_?(A)$ induces an injective (or bijective) map $F(A) \to F(\mathcal{K}_?(A))$ for all A.

If a stabilisation $\mathcal{K}_1$ dominates another stabilisation $\mathcal{K}_2$ in the sense that the identity map on $\mathbb{M}_\infty(A)$ extends to a bounded map $\mathcal{K}_2(A) \to \mathcal{K}_1(A)$ for all A, then weak $\mathcal{K}_1$-stability implies weak $\mathcal{K}_2$-stability. The corresponding assertion for strong stability holds in some special cases but not in general.

Proposition 3.30. *Let F be a functor from BAlg to an additive category that is split-exact and $\mathbb{M}_2$-stable. If F is weakly $\mathcal{K}^*$-stable, then F is homotopy invariant (with respect to continuous homotopies). If F is weakly $\mathcal{CK}^r$-stable for some $0 \leq r < 1$, then F is homotopy invariant with respect to α-Hölder continuous homotopies for all $\alpha \in (r, 1]$.*

Proof. Lemma 3.26 yields

$$F(\iota) \circ F(f_1) = F(\iota \circ f_1) = F(\iota \circ f_0) = F(\iota) \circ F(f_0).$$

The weak stability hypothesis allows us to cancel by $F(\iota)$. $\square$

This begs the question: how do we get weakly stable functors?

Proposition 3.31. *If a functor F on* BAlg *is $\mathbb{M}_2$-stable, then $A \mapsto F\big(\mathcal{K}^*(A)\big)$ is weakly $\mathcal{K}^*$-stable and $A \mapsto F\big(\mathcal{CK}^r(A)\big)$ for $r \geq 0$ is weakly $\mathcal{CK}^r$-stable. The functor $A \mapsto F\big(\mathbb{M}_\infty(A)\big)$ is $\mathbb{M}_\infty$-stable.*

If a functor F on the category of C^*-algebras is $\mathbb{M}_2$-stable, then the functor $A \mapsto F\big(\mathcal{K}_{C^*}(A)\big)$ is $\mathcal{K}_{C^*}$-stable.

Before we prove this proposition, we formulate its main consequences.

Theorem 3.32. *Let F be a functor from* BAlg *to an additive category that is split-exact and $\mathbb{M}_2$-stable. Then the functor $A \mapsto F\big(\mathcal{K}^*(A)\big)$ is homotopy invariant for continuous homotopies, $\mathcal{K}^*$-stable, split-exact, and $\mathbb{M}_2$-stable; similarly, the functor $A \mapsto F\big(\mathcal{CK}^r(A)\big)$ is homotopy invariant for α-Hölder continuous homotopies with $\alpha \in (r, 1]$, $\mathcal{CK}^r$-stable, split-exact, and $\mathbb{M}_2$-stable.*

Proof. The homotopy invariance assertions follow immediately from Propositions 3.30 and 3.31. It is also clear that the stabilised functors remain split-exact and $\mathbb{M}_2$-stable because the functors that we stabilise with preserve split extensions and tensor products with $\mathbb{M}_2$. The remaining stability assertions will be proved later in this section. $\square$

Specialising to K_0, which we know is split-exact and $\mathbb{M}_2$-stable, we get the desired homotopy invariance result for stabilised algebraic K-theory:

Corollary 3.33. *The functor $A \mapsto K_0\big(\mathcal{K}^*(A)\big)$ is homotopy invariant for continuous homotopies, $\mathcal{K}^*$-stable, split-exact, and $\mathbb{M}_2$-stable.*

The functor $A \mapsto K_0\big(\mathcal{CK}^r(A)\big)$ is homotopy invariant for α-Hölder-continuous homotopies with $\alpha \in (r, 1]$, $\mathcal{CK}^r$-stable, split-exact, and $\mathbb{M}_2$-stable.

If $r \geq s$, then $\mathcal{CK}^r(A)$ is a (generalised) ideal in $\mathcal{CK}^s(A)$ and the quotient ring $Q := \mathcal{CK}^s(A)/\mathcal{CK}^r(A)$ is nilpotent by (3.23). Whereas we have very good excision results for extensions with nilpotent *kernel*, we cannot say much about extensions with nilpotent *quotient*, so that we have little control over the kernel of the map $K_0\big(\mathcal{CK}^r(A)\big) \to K_0\big(\mathcal{CK}^s(A)\big)$. Nevertheless, an idea of Guillermo Cortiñas and Andreas Thom [32] allows us to infer the homotopy invariance (with respect to smooth homotopies) of $K_0\big(\mathcal{CK}^r(A)\big)$ from that of $K_0\big(\mathcal{CK}^s(A)\big)$. Using the machinery of bivariant K-theory, we can then infer that $K_0\big(\mathcal{CK}^r(A)\big) \cong K_0\big(\mathcal{CK}^s(A)\big)$ for all r, s; hence $K_0\big(\mathcal{CK}^r(A)\big)$ is homotopy invariant with respect to α-Hölder continuous homotopies for arbitrary pairs r, α.

If we let $r \to \infty$, we get arbitrarily close to $\mathcal{K}_{\mathscr{S}}(A)$, but we do not quite reach it. We remark that $\mathcal{K}_{\mathscr{S}}(A)$ is also an ideal in $\mathcal{CK}^s(A)$ and that the quotient

$\mathcal{CK}^s(A)/\mathcal{K}_\mathscr{S}(A)$ is a projective limit of nilpotent rings. It is not clear whether this suffices for the argument of Cortiñas and Thom.

Proof of Proposition 3.31. We first prove the assertion about $\mathbb{M}_\infty$. Any bijection $\mathbb{N}^2 \cong \mathbb{N}$ induces an isomorphism $\mathbb{M}_\infty(\mathbb{M}_\infty(A)) \cong \mathbb{M}_\infty(A)$. Composition with the stabilisation homomorphism $\mathbb{M}_\infty(A) \to \mathbb{M}_\infty(\mathbb{M}_\infty(A))$ on either side yields *inner* endomorphisms of $\mathbb{M}_\infty(A)$ and $\mathbb{M}_\infty(\mathbb{M}_\infty(A))$, respectively, because any injective map $\mathbb{N} \to \mathbb{N}$ yields a pair V, W of multipliers of $\mathbb{M}_\infty(A)$ satisfying $WV = 1$. Since inner endomorphisms act trivially on F by Proposition 3.16, the functor $A \mapsto F(\mathbb{M}_\infty(A))$ is $\mathbb{M}_\infty$-stable.

To treat more interesting stabilisations, we seek a bijection $\mathbb{N} \cong \mathbb{N}^2$ for which the induced isomorphism $\mathbb{M}_\infty(\mathbb{M}_\infty(A)) \cong \mathbb{M}_\infty(A)$ extends to a bounded map $\mathcal{K}_?(\mathcal{K}_?(A)) \to \mathcal{K}_?(A)$ and the multipliers V, W on $\mathbb{M}_\infty(A)$ extend to multipliers on $\mathcal{K}_?(A)$. If we can achieve this, we get the *weak* $\mathcal{K}_?$-stability of $F \circ \mathcal{K}_?$. If we also get the corresponding assertion for the induced multipliers on $\mathcal{K}_?(\mathcal{K}_?(A))$, we get full $\mathcal{K}_?$-stability.

For $\mathcal{K}_{C^*}(A)$, we can carry this out easily because

$$\mathcal{K}_{C^*}(\mathcal{K}_{C^*}(A)) \cong \mathcal{K}(\ell^2(\mathbb{N} \times \mathbb{N})) \,\widehat{\otimes}_{C^*}\, A$$

and all separable Hilbert spaces are isomorphic. It does not matter which bijection $\mathbb{N} \to \mathbb{N}^2$ we choose, and we get the full stability of $F \circ \mathcal{K}_{C^*}$.

Similarly, the bijection $\mathbb{N}^2 \to \mathbb{N}$ does not matter for $\mathcal{K}^*(A)$ because it treats all elements of $\mathbb{N}$ equally. It is straightforward to check that the above strategy works in this case and yields the weak $\mathcal{K}^*$-stability of $F \circ \mathcal{K}^*$. We only get a bounded homomorphism $\mathcal{K}^*(\mathcal{K}^*(A)) \to \mathcal{K}^*(A)$, not an algebra isomorphism, because ℓ^1- and ℓ^∞-estimates do not commute; but this does not matter. It seems that we only get weak stability because the resulting endomorphism of $\mathcal{K}^*(\mathcal{K}^*(A))$ is not inner.

Finally, we consider the more difficult case of $\mathcal{CK}^r$. We replace $\mathbb{N}$ by $\mathbb{N}_{\geq 1}$ to simplify the norm estimates. A bijection $\mathbb{N}_{\geq 1} \cong \mathbb{N}_{\geq 1}^2$ may be specified by a well-ordering on $\mathbb{N}_{\geq 1}^2$: there is a unique bijection $\varphi \colon \mathbb{N}_{\geq 1}^2 \to \mathbb{N}_{\geq 1}$ that satisfies $x \leq y \iff \varphi(x) \leq \varphi(y)$. We define a well-ordering by

$$(i_1, i_2) \leq (k_1, k_2) \iff \text{either } i_1 \cdot i_2 < k_1 \cdot k_2 \text{ or } (i_1 \cdot i_2 = k_1 \cdot k_2 \text{ and } i_1 \leq k_1).$$

This yields the enumeration of $\mathbb{N}_{\geq 1}^2$ that begins with:

$$(1,1), \ (1,2), \ (2,1), \ (1,3), \ (3,1), \ (1,4), \ (2,2), \ (4,1), \ (1,5), \ (5,1),$$
$$(1,6), \ (2,3), \ (3,2), \ (6,1), \ (1,7), \ (7,1), \ (1,8), \ (2,4), \ (4,2), \ (8,1), \ \dots \,.$$

Explicitly, we get the bijection

$$\varphi \colon \mathbb{N}_{\geq 1}^2 \to \mathbb{N}_{\geq 1}, \qquad \varphi(i_1, i_2) = \sum_{j=1}^{i_1 i_2 - 1} \left\lfloor \frac{i_1 i_2}{j} \right\rfloor + \#\{d \leq i_1 \mid d \mid i_1 i_2\},$$

where $\lfloor a \rfloor \in \mathbb{N}$ is the integral part of $a \in \mathbb{R}_+$. We shall only need the resulting estimate

$$\varphi(i_1, i_2) = i_1 \cdot i_2 \cdot \ln(i_1 \cdot i_2) + O(i_1 \cdot i_2).$$

The logarithmic term that occurs here is the reason why we included the factor $(1 + \ln i)^k$ in the definition of $\mathcal{CK}^r(A)$. Now straightforward computations show:

- φ defines a bounded algebra homomorphism $\mathcal{CK}^r\big(\mathcal{CK}^r(A)\big) \to \mathcal{CK}^r(A)$;

- the embedding $\mathbb{N}_{\geq 1} \rightarrowtail \mathbb{N}_{\geq 1}^2 \xrightarrow{\varphi} \mathbb{N}_{\geq 1}$, $i \mapsto \varphi(i, 1)$, yields bounded multipliers of $\mathcal{CK}^r(A)$, so that the composite endomorphism $\mathcal{CK}^r(A) \to \mathcal{CK}^r(A)$ is inner.

For the first assertion, we use (3.21) twice to explicitly describe $\mathcal{CK}^r\big(\mathcal{CK}^r(A)\big)$; inspection shows that the growth estimates in $\mathcal{CK}^r\big(\mathcal{CK}^r(A)\big)$ for the matrix coefficient at $(i_1, j_1), (i_2, j_2) \in \mathbb{N}_{\geq 1}^4$ are stronger than those in $\mathcal{CK}^r(A)$ for the matrix coefficient at $\varphi(i_1, j_1), \varphi(i_2, j_2) \in \mathbb{N}_{\geq 1}^2$. For the second assertion, we use the estimate $\varphi(i, 1) = i \cdot \ln(i) + O(i)$ and argue as in the proof of Lemma 3.22. Further details are left to the reader. $\qquad\square$

This finishes the proof of homotopy invariance for stabilised algebraic K-theory. We can now proceed as in §2.2 and define higher stabilised K-theory groups and construct various long exact sequences for them. If we stabilise by $\mathcal{CK}^r$, then we have to modify our treatment of homotopy invariance, using spaces of Hölder continuous functions on the usual subspaces of $[0, 1]$. We will explain how this works for smooth homotopies in §6.1.

Finally, we formulate the C^*-algebraic version of our result. Since functors of the form $A \mapsto F\big(\mathcal{K}_{C^*}(A)\big)$ are automatically strongly stable by Proposition 3.31, there is no need to consider weakly stable functors. A similar argument yields:

Exercise 3.34. A $\mathcal{K}_{C^*}$-stable functor is automatically $\mathbb{M}_2$-stable.

Using these simplifications, we arrive at the following theorem of Nigel Higson [60]:

Theorem 3.35. *Any split-exact, $\mathcal{K}_{C^*}$-stable functor on the category of C^*-algebras is homotopy invariant.*

3.2.3   Weak versus full stability

The notion of a weakly stable functor is only an auxiliary concept. In many cases, weakly stable functors are automatically stable. To prove this rather technical result, we will use some ideas from §7.1. The following lemma is an instance of this:

Lemma 3.36. *If the functor F is smoothly homotopy invariant, then $F \circ \mathcal{CK}^r$ is $\mathcal{CK}^r$-stable. The same conclusion holds if F is $\mathbb{M}_2$-stable and $F \circ \mathcal{CK}^r$ is smoothly homotopy invariant.*

Proof. Let $\iota\colon C\mathcal{K}^r(A) \to C\mathcal{K}^r C\mathcal{K}^r(A)$ be the stabilisation homomorphism. We have to invert $F(\iota)$. First we recall the proof of Proposition 3.31, which shows that $F(\iota)$ is injective provided F is $\mathbb{M}_2$-stable. There we have constructed a bijection $\varphi\colon \mathbb{N}_{\geq 1}^2 \to \mathbb{N}_{\geq 1}$ with good growth properties, which induces a bounded algebra homomorphism $\alpha\colon C\mathcal{K}^r C\mathcal{K}^r(A) \to C\mathcal{K}^r(A)$; moreover, the composite map $\alpha \circ \iota$ is an inner endomorphism of $C\mathcal{K}^r(A)$.

We claim that the composite maps $\alpha \circ \iota$ and $\iota \circ \alpha$ are smoothly homotopic to the identity maps on $C\mathcal{K}^r(A)$ and $C\mathcal{K}^r C\mathcal{K}^r(A)$, respectively. This yields the assertion if F is smoothly homotopy invariant or if F is $\mathbb{M}_2$-stable and $F \circ C\mathcal{K}^r$ is smoothly homotopy invariant because $\alpha \circ \iota$ is both an inner endomorphism and smoothly homotopic to the identity map.

We first discuss the notion of a *rotation homotopy*. Let $V_0, V_1\colon \mathbb{N} \to \mathbb{N}$ be two injective maps. They induce isometric bornological embeddings

$$\hat{V}_0, \hat{V}_1\colon \mathbb{C}[\mathbb{N}] \to \mathbb{C}[\mathbb{N}], \qquad \hat{V}_t(\delta_n) := \delta_{V_t(n)},$$

which in turn induce inner endomorphisms of $\mathbb{M}_\infty(A)$. Now assume $V_0(i) \neq V_1(j)$ for all $i, j \in \mathbb{N}$ with $i \neq j$ (we allow $V_0(i) = V_1(i)$). We define $V_t\colon \mathbb{C}[\mathbb{N}] \to \mathbb{C}[\mathbb{N}]$ by

$$V_t(\delta_i) := \begin{cases} \delta_{V_0(i)} & \text{if } V_0(i) = V_1(i), \\ \sqrt{1-t^2}\delta_{V_0(i)} + t\delta_{V_1(i)} & \text{otherwise.} \end{cases}$$

The hypothesis on V_0 and V_1 ensures that these maps are again isometric, so that they define a smooth homotopy of standard homomorphisms.

Now we write down a sequence of maps $V_n\colon \mathbb{N}_{\geq 1} \to \mathbb{N}_{\geq 1}$, $n \in \mathbb{N}$, with the above properties, which will eventually lead to a smooth homotopy between id and $\alpha \circ \iota$ on $C\mathcal{K}^r(A)$. We begin with $V_0(i) := \varphi(i,1)$; the associated standard homomorphism is $\alpha \circ \iota$. We define V_n for $n \geq 1$ by

$$V_n(i) = \begin{cases} i & \text{if } i \leq 2^{n-1}, \\ \varphi(i,2) & \text{if } 2^{n-1} < i \leq 2^n, \\ \varphi(i,1) & \text{if } 2^n < i. \end{cases}$$

This map is injective because φ is injective and satisfies $\varphi(i,j) \geq ij$ for all $i,j \in \mathbb{N}_{\geq 1}^2$. Moreover, $V_n(\mathbb{N}_{\geq 1})$ is disjoint from $\{2^{n-1}+1, \ldots, 2^n\}$ because $\varphi(i,j) \geq i \cdot j$ for all $i,j \in \mathbb{N}_{\geq 1}$. Therefore, there is a rotation homotopy between V_n and V_{n+1}. We have $V_n(i)/i = O(1 + \ln i)$ because $\varphi(i,j) = O\big(i \cdot j \cdot (1 + \ln(ij))\big)$. Hence the maps $V_t\colon \mathbb{N}_{\geq 1} \to \mathbb{N}_{\geq 1}$ induce inner endomorphisms of $C\mathcal{K}^r(A)$.

Now we reparametrise the homotopy from V_{n-1} to V_n to occur on the interval $[1/n, 1/(n+1)]$. The lengths of these intervals decrease like n^{-2}, so that the kth derivative of the resulting rotation homotopy grows like n^{2k}. However, we only get contributions to this derivative in the region $i > 2^{n-1}$ because $V_n(i) = V_{n+1}(i)$ for $i \leq 2^{n-1}$. Hence the growth of the derivatives is $O\big((1+\ln i)^{2k}\big)$. Since such factors are absorbed by the norms that define $\mathcal{CK}^r(A)$, we conclude that we get a smooth homotopy parametrised by $[0,1]$.

Next we construct a sequence of injective maps $W_n\colon \mathbb{N}^2_{\geq 1} \to \mathbb{N}^2_{\geq 1}$, starting with $W_0(i,j) := (\varphi(i,j),1)$, so that the associated standard homomorphism is $\iota\circ\alpha$. We let

$$W_n(i,j) := \begin{cases} (i,j) & \text{if } j \leq 2^n, \\ (\varphi(i,j-2^n), 2^{n+1}+1) & \text{if } j > 2^n. \end{cases}$$

You may check that each of these maps is injective and that $W_n(i,j) = W_{n+1}(i',j')$ if and only if $j = j' \leq 2^n$ and $i = i'$. Hence there are again rotation homotopies between consecutive W_n. As above, we define a homotopy of isometries $W_t\colon \mathbb{N}^2_{\geq 1} \times [0,1] \to \mathbb{N}^2_{\geq 1}$ by rotating between W_{n-1} and W_n on the interval $[1/n, 1/(n+1)]$.

Now a crucial point about our construction is that $W_n(i,j) = W_{n+1}(i,j)$ for $j \leq 2^n$ and that $W_{n+1}(i,j)$ and $W_n(i,j)$ do not differ by more than a constant factor in each coordinate because $\varphi(i,j-2^n) \approx i\cdot(j-2^n)\cdot\ln(i\cdot(j-2^n))$. The resulting estimates show that our rotation homotopy is a smooth homotopy of endomorphisms of $\mathcal{CK}^r\mathcal{CK}^r(A)$. $\qquad\square$

Corollary 3.37. *If $0 < r < 1$, then the functor $\mathrm{K}_0 \circ \mathcal{CK}^r$ is $\mathcal{CK}^r$-stable.*

Proof. The functor K_0 is $\mathbb{M}_2$-stable, and $\mathrm{K}_0 \circ \mathcal{CK}^r$ is smoothly homotopy invariant by Corollary 3.33. $\qquad\square$

Recall that $\bigcap_{r>0}\mathcal{CK}^r = \mathcal{K}_{\mathscr{S}}$. This no longer holds for general coefficient algebras. Nevertheless, if an assertion holds for $\mathcal{CK}^r(A)$ for all $r \geq 0$, then the proof often carries over to $\mathcal{K}_{\mathscr{S}}(A)$ as well. Lemma 3.36 is an instance of this:

Lemma 3.38. *If the functor F is smoothly homotopy invariant, then the functor $A \mapsto F\bigl(\mathcal{K}_{\mathscr{S}}(A)\bigr)$ is $\mathcal{K}_{\mathscr{S}}$-stable and $\mathbb{M}_n$-stable for all $n \in \mathbb{N} \cup \{\infty\}$.*

Proof. The proof of $\mathcal{K}_{\mathscr{S}}$-stability is literally the same as for Lemma 3.36. A similar argument, replacing $\mathbb{N}^2$ by $\mathbb{N} \times \{1,\ldots,n\}$, yields $\mathbb{M}_n$-stability for $n \in \mathbb{N}$. We omit the proof of $\mathbb{M}_\infty$-stability because we are not going to use it, anyway. $\qquad\square$

Chapter 4

Bott periodicity

Bott periodicity asserts that $K_{*+2}(A) \cong K_*(A)$ if A is a local Banach algebra over $\mathbb{C}$. It is crucial to work with algebras over $\mathbb{C}$ here. We shall follow the proof of Joachim Cuntz based on the *Toeplitz extension* [33]. Like the homotopy invariance proof in Chapter 3, it uses only formal properties of K-theory and therefore works for all functors with certain properties. We will consider this generalisation in §7.3.

Bott periodicity is crucial for most K-theory computations. To highlight this, we end this section with some simple computations.

4.1 Toeplitz algebras

We recall the definition of the Toeplitz C^*-algebra and then introduce some dense subalgebras. Joachim Cuntz originally formulated his proof for C^*-algebras (see [33, 92]). When dealing with local Banach algebras, it is more convenient to work with suitable dense subalgebras of the Toeplitz C^*-algebra.

Definition 4.1. The *Toeplitz C^*-algebra* $\mathcal{T}_{C^*}$ is the universal unital C^*-algebra generated by an isometry; that is, it has one generator v that is subject to the single relation $v^*v = 1$.

This means that there are natural bijections between unital $*$-homomorphisms $\mathcal{T}_{C^*} \to A$ and isometries in A for all unital C^*-algebras A.

Let $(e_n)_{n \in \mathbb{N}}$ denote the standard basis of $\ell^2(\mathbb{N})$. The *unilateral shift operator* $S \colon \ell^2(\mathbb{N}) \to \ell^2(\mathbb{N})$ is the isometry defined by $S(e_n) := e_{n+1}$:

The following theorem identifies $\mathcal{T}_{C^*}$ with the concrete C^*-algebra generated by S:

Theorem 4.2 (Coburn's Theorem). *The representation of $\mathcal{T}_{C^*}$ on $\ell^2(\mathbb{N})$ generated by the isometry* S *is faithful, that is, it identifies $\mathcal{T}_{C^*}$ with the C^*-subalgebra of $\mathcal{L}(\ell^2(\mathbb{N}))$ generated by* S. *The latter fits into a C^*-algebra extension*

$$\mathcal{K}(\ell^2(\mathbb{N})) \rightarrowtail C^*(S) \twoheadrightarrow C(\mathbb{S}^1).$$

Proof. The first assertion $\mathcal{T}_{C^*} \cong C^*(S)$ is proved in [92]. We denote the matrix units in $\mathcal{K}_{C^*} := \mathcal{K}(\ell^2(\mathbb{N}))$ by E_{mn} for $m, n \in \mathbb{N}$. One checks easily that $1 - SS^* = E_{00}$, so that

$$S^m(1 - SS^*)(S^*)^n = E_{mn}$$

for all $m, n \in \mathbb{N}$. Thus $C^*(S)$ contains $\mathcal{K}_{C^*}$. Since $1 - SS^* \in \mathcal{K}_{C^*}$ and $S^*S = 1$, the image of S in the Calkin algebra $\mathcal{L}(\ell^2(\mathbb{N}))/\mathcal{K}_{C^*}$ is unitary. By functional calculus, the C^*-subalgebra of the Calkin algebra that it generates is $C(X)$, where $X \subseteq \mathbb{S}^1$ is the essential spectrum of S. Since there is no $\lambda \in \mathbb{S}^1$ for which $\lambda - S$ is a Fredholm operator, we get $C^*(S)/\mathcal{K}_{C^*} \cong C(\mathbb{S}^1)$. $\square$

Therefore, we get an extension of C^*-algebras $\mathcal{K}_{C^*} \rightarrowtail \mathcal{T}_{C^*} \twoheadrightarrow C(\mathbb{S}^1)$; however, it is conceptually better to think of $C(\mathbb{S}^1)$ as the group algebra of $\mathbb{Z}$. If we work with real C^*-algebras, then $C^*(\mathbb{Z})$ and $C(\mathbb{S}^1)$ become different; this is why our proof of Bott periodicity fails for real K-theory.

Let $\mathcal{T}_{\mathrm{alg}}$ be the $*$-subalgebra of $\mathcal{T}_{C^*}$ generated by S without any completion. As above, we get an algebra extension $\mathbb{M}_\infty \rightarrowtail \mathcal{T}_{\mathrm{alg}} \twoheadrightarrow \mathbb{C}[\mathbb{Z}]$, where $\mathbb{C}[\mathbb{Z}]$ is the group algebra of $\mathbb{Z}$ or, equivalently, the algebra of Laurent polynomials. Using fine bornologies, we turn this into an extension of bornological algebras.

Given a bornological algebra A we get an extension of bornological algebras

$$\mathbb{M}_\infty(A) \rightarrowtail \mathcal{T}_{\mathrm{alg}}(A) \twoheadrightarrow \mathbb{C}[\mathbb{Z}] \,\widehat{\otimes}\, A,$$

tensoring the above extension with A. We may identify $\mathbb{M}_\infty(A)$ and $\mathbb{C}[\mathbb{Z}] \,\widehat{\otimes}\, A$ with the spaces of all functions $\mathbb{N}^2 \to A$ or $\mathbb{Z} \to A$ with finite support. We view a pair of such functions $(f_{\mathbb{N}^2}, g_{\mathbb{Z}})$ as the sum of the (finite) series

$$\sum_{i,j \in \mathbb{N}} f_{\mathbb{N}^2}(i,j) \cdot E_{ij} + \sum_{n=0}^{\infty} g_{\mathbb{Z}}(n) S^n + \sum_{n=1}^{\infty} g_{\mathbb{Z}}(-n)(S^*)^n,$$

where $E_{ij} = S^i(1 - SS^*)(S^*)^j$ as above. This yields an explicit isomorphism $\mathcal{T}_{\mathrm{alg}}(A) \cong \mathbb{M}_\infty(A) \oplus \mathbb{C}[\mathbb{Z}] \,\widehat{\otimes}\, A$. In this description, the multiplication in $\mathcal{T}_{\mathrm{alg}}(A)$ looks as follows: $(f_{\mathbb{N}^2}^1, g_{\mathbb{Z}}^1) \cdot (f_{\mathbb{N}^2}^2, g_{\mathbb{Z}}^2) = (f_{\mathbb{N}^2}, g_{\mathbb{Z}})$ with

$$f_{\mathbb{N}^2}(i,j) = \sum_{k=0}^{\infty} f_{\mathbb{N}^2}^1(i,k) f_{\mathbb{N}^2}^2(k,j) + \sum_{k=0}^{\infty} g_{\mathbb{Z}}^1(i-k) f_{\mathbb{N}^2}^2(k,j)$$

$$+ \sum_{k=0}^{\infty} f_{\mathbb{N}^2}^1(i,k) g_{\mathbb{Z}}^2(k-j) - \sum_{k=1}^{\infty} g_{\mathbb{Z}}^1(i+k) g_{\mathbb{Z}}^2(-j-k),$$

$$g_{\mathbb{Z}}(n) = \sum_{k \in \mathbb{Z}} g_{\mathbb{Z}}^1(k) g_{\mathbb{Z}}^2(n-k).$$

Now we enlarge $\mathcal{T}_{\mathrm{alg}}$ to the *smooth Toeplitz algebra*

$$\mathcal{T}_{\mathscr{S}}(A) := \mathscr{S}(\mathbb{N}^2, A) \oplus \mathscr{S}(\mathbb{Z}, A) \cong \mathcal{K}_{\mathscr{S}}(A) \oplus C^{\infty}(\mathbb{S}^1, A).$$

Here we use the smooth stabilisation (see §2.3.4) and the isomorphism $\mathscr{S}(\mathbb{Z}, A) \cong C^{\infty}(\mathbb{S}^1, A)$ induced by the Fourier transform on $\mathbb{Z}$. The multiplication on $\mathcal{T}_{\mathrm{alg}}(A)$ extends to a bounded bilinear map on $\mathcal{T}_{\mathscr{S}}(A)$. Thus $\mathcal{T}_{\mathscr{S}}(A)$ becomes a bornological algebra as well, and it is part of an extension of bornological algebras

$$\mathcal{K}_{\mathscr{S}}(A) \rightarrowtail \mathcal{T}_{\mathscr{S}}(A) \twoheadrightarrow C^{\infty}(\mathbb{S}^1, A). \tag{4.3}$$

If A is a local Banach algebra, then so are $\mathcal{K}_{\mathscr{S}}(A)$ and $C^{\infty}(\mathbb{S}^1, A)$. The same holds for $\mathcal{T}_{\mathscr{S}}(A)$ by Theorem 2.15. We observe that

$$\mathcal{T}_{\mathrm{alg}}(A) \cong \mathcal{T}_{\mathrm{alg}} \widehat{\otimes} A, \qquad \mathcal{T}_{\mathscr{S}}(A) \cong \mathcal{T}_{\mathscr{S}} \widehat{\otimes} A$$

for all bornological algebras A, where $\mathcal{T}_{\mathscr{S}} = \mathcal{T}_{\mathscr{S}}(\mathbb{C})$. Therefore, we can often reduce computations with these algebras to the special case $A = \mathbb{C}$.

Exercise 4.4. The unital bornological algebra $\mathcal{T}_{\mathrm{alg}}$ is the universal one that is generated by two elements v, w with $wv = 1$; that is, bounded unital algebra homomorphisms $\mathcal{T}_{\mathrm{alg}} \to A$ for a unital bornological algebra A correspond bijectively to pairs (v, w) in A with $wv = 1$.

There is a similar universal property for the smooth Toeplitz algebra. Let (v, w) satisfy $wv = 1$. We say that (v, w) has *polynomial growth* if $\{\varepsilon_n v^n, \varepsilon_n w^n \mid n \in \mathbb{N}\}$ is bounded in A for any rapidly decreasing sequence of scalars $(\varepsilon_n)_{n \in \mathbb{N}}$.

Lemma 4.5. *The smooth Toeplitz algebra $\mathcal{T}_{\mathscr{S}}$ is the universal unital algebra generated by (v, w) satisfying the relation $wv = 1$ and having polynomial growth.*

Proof. Since (S, S^*) clearly has polynomial growth in $\mathcal{T}_{\mathscr{S}}$, a pair (v, w) can only generate a bounded representation of $\mathcal{T}_{\mathscr{S}}$ if it has polynomial growth. Conversely, let (v, w) in A satisfy $wv = 1$ and have polynomial growth. Then the induced map on $\mathbb{C}[\mathbb{Z}] \subseteq \mathcal{T}_{\mathscr{S}}$ extends to a bounded linear map $\mathscr{S}(\mathbb{Z}) \to A$.

The matrix units $E_{mn} \in \mathcal{T}_{\mathrm{alg}}$ are represented by $v^m(1 - vw)w^n \in A$. Since the multiplication in A is bounded, $\varepsilon_m \varepsilon_n v^m(1 - vw)w^n$ remains bounded in A for any $(\varepsilon_m) \in \mathscr{S}(\mathbb{N})$; equivalently, $\varepsilon_{mn} v^m(1 - vw)w^n$ remains bounded in A for any $(\varepsilon_{mn}) \in \mathscr{S}(\mathbb{N}^2)$. Thus we can extend the homomorphism $\mathcal{T}_{\mathrm{alg}} \to A$ to a bounded homomorphism on $\mathcal{T}_{\mathscr{S}}$. $\square$

4.2 The proof of Bott periodicity

First we need a slight variant of the Toeplitz extension. Let A be a local Banach algebra. Let $C_0^{\infty}(\mathbb{S}^1 \smallsetminus \{1\}, A) \subseteq C^{\infty}(\mathbb{S}^1, A)$ be the ideal of all A-valued functions that vanish at 1. Let $\mathcal{T}_{\mathscr{S}}^0(A) \subseteq \mathcal{T}_{\mathscr{S}}(A)$ be the pre-image of this ideal, equipped with the subspace bornology. Then we get an extension of bornological algebras

$$\mathcal{K}_{\mathscr{S}}(A) \rightarrowtail \mathcal{T}_{\mathscr{S}}^0(A) \twoheadrightarrow C_0^{\infty}(\mathbb{S}^1 \smallsetminus \{1\}, A). \tag{4.6}$$

Theorem 2.65 yields $K_m(\mathcal{K}_{\mathscr{S}}(A)) \cong K_m(A)$ for all $m \in \mathbb{N}$. Since $C^\infty(\mathbb{S}^1, A)$ is an isoradial subalgebra of $C(\mathbb{S}^1, A)$, $C_0^\infty(\mathbb{S}^1 \smallsetminus \{1\}, A)$ is an isoradial subalgebra of $C_0(\mathbb{S}^1 \smallsetminus \{1\}, A)$. We may use the Möbius transformation $\Phi \colon x \mapsto \frac{x+i}{x-i}$ to identify $\mathbb{R} \cong \mathbb{S}^1 \smallsetminus \{1\}$ and hence $C_0(\mathbb{S}^1 \smallsetminus \{1\}, A) \cong C_0(\mathbb{R}, A)$. As a result,

$$K_m(C_0^\infty(\mathbb{S}^1 \smallsetminus \{1\}, A)) \cong K_m(C_0(\mathbb{S}^1 \smallsetminus \{1\}, A)) \cong K_m(C_0(\mathbb{R}, A)) \cong K_{1+m}(A)$$

for all $m \geq 0$. Hence the K-theory boundary maps of the extension (4.6) become maps

$$\beta \colon K_{2+m}(A) \cong K_{1+m}(C_0^\infty(\mathbb{S}^1 \smallsetminus \{1\}, A)) \xrightarrow{\text{ind}} K_m(\mathcal{K}_{\mathscr{S}}(A)) \cong K_m(A).$$

Theorem 4.7. *The maps* $\beta \colon K_{2+m}(A) \to K_m(A)$ *are isomorphisms for all local Banach algebras A and all $m \in \mathbb{N}$, so that topological K-theory for local Banach algebras is 2-periodic.*

Proof. The index map for (4.6) is part of a long exact sequence

$$\cdots \to K_1(\mathcal{K}_{\mathscr{S}}(A)) \to K_1(\mathcal{T}_{\mathscr{S}}^0(A)) \to K_1(C_0^\infty(\mathbb{S}^1 \smallsetminus \{1\}, A))$$
$$\to K_0(\mathcal{K}_{\mathscr{S}}(A)) \to K_0(\mathcal{T}_{\mathscr{S}}^0(A)) \to K_0(C_0^\infty(\mathbb{S}^1 \smallsetminus \{1\}, A))$$

by Theorem 2.33. Hence β is an isomorphism if $K_*(\mathcal{T}_{\mathscr{S}}^0(A)) = 0$; this is what we are going to prove.

We need an auxiliary algebra $\widehat{\mathcal{T}}_{\mathscr{S}} \subseteq \mathcal{T}_{\mathscr{S}} \mathbin{\widehat{\otimes}} \mathcal{T}_{\mathscr{S}}$. Notice that $\mathcal{K}_{\mathscr{S}} \mathbin{\widehat{\otimes}} \mathcal{T}_{\mathscr{S}}^0$ is a closed ideal in $\mathcal{T}_{\mathscr{S}} \mathbin{\widehat{\otimes}} \mathcal{T}_{\mathscr{S}}$ and that it has trivial intersection with $\mathcal{T}_{\mathscr{S}} \mathbin{\widehat{\otimes}} 1$. We let

$$\widehat{\mathcal{T}}_{\mathscr{S}} := \mathcal{K}_{\mathscr{S}} \mathbin{\widehat{\otimes}} \mathcal{T}_{\mathscr{S}}^0 + \mathcal{T}_{\mathscr{S}} \mathbin{\widehat{\otimes}} 1 \subseteq \mathcal{T}_{\mathscr{S}} \mathbin{\widehat{\otimes}} \mathcal{T}_{\mathscr{S}},$$

equipped with the subspace bornology. We also let $\widehat{\mathcal{T}}_{\mathscr{S}}(A) := \widehat{\mathcal{T}}_{\mathscr{S}} \mathbin{\widehat{\otimes}} A$.

We get an extension of bornological algebras

$$\mathcal{K}_{\mathscr{S}} \mathbin{\widehat{\otimes}} \mathcal{T}_{\mathscr{S}}^0(A) \rightarrowtail \widehat{\mathcal{T}}_{\mathscr{S}}(A) \twoheadrightarrow \mathcal{T}_{\mathscr{S}}(A), \tag{4.8}$$

which splits by the bounded homomorphism $k \colon x \otimes a \mapsto x \otimes 1 \otimes a$. The algebra $\widehat{\mathcal{T}}_{\mathscr{S}}(A)$ is a local Banach algebra by Theorem 2.15. Since K-theory is stable and split-exact, the embedding

$$j \colon \mathcal{T}_{\mathscr{S}}^0(A) \to \mathcal{K}_{\mathscr{S}} \mathbin{\widehat{\otimes}} \mathcal{T}_{\mathscr{S}}^0(A) \subseteq \widehat{\mathcal{T}}_{\mathscr{S}}(A), \qquad x \mapsto E_{00} \mathbin{\widehat{\otimes}} x,$$

induces an *injective* map on K-theory. We will finish the proof by showing that j induces the *zero* map on K-theory.

Conjugation by the isometry $S \otimes 1 \otimes 1$ in $\widehat{\mathcal{T}}_{\mathscr{S}}(A^+)$ defines an inner endomorphism of $\widehat{\mathcal{T}}_{\mathscr{S}}(A)$. It is orthogonal to j, that is,

$$j(x) \cdot (S \otimes 1 \otimes a) = 0 = (S^* \otimes 1 \otimes a) \cdot j(x)$$

for all $x \in T^0_{\mathscr{S}}(A)$. Composing this endomorphism with $k \colon T_{\mathscr{S}}(A) \to \widehat{T}_{\mathscr{S}}(A)$, we get a homomorphism $\varphi^0_1 \colon T^0_{\mathscr{S}}(A) \to \widehat{T}_{\mathscr{S}}(A)$, which is orthogonal to j. Hence $\varphi^0_0 := j + \varphi^0_1$ is a homomorphism as well, and (3.8) specialises to $K_*(\varphi^0_0) = K_*(\varphi^0_0) + K_*(j)$.

We will show that φ^0_1 and φ^0_0 are smoothly homotopic, so that $K_*(\varphi^0_1) = K_*(\varphi^0_0)$ by Corollary 2.26. This implies $K_*(j) = 0$ and finishes the proof. It suffices to construct a smooth homotopy between φ^0_1 and φ^0_0 for $A = \mathbb{C}$ because

$$C^\infty\big([0,1], \widehat{T}_{\mathscr{S}}(A)\big) \cong C^\infty\big([0,1], \widehat{T}_{\mathscr{S}}(\mathbb{C})\big) \mathbin{\widehat{\otimes}} A.$$

Thus we assume $A = \mathbb{C}$ from now on.

Before we construct the required homotopy, we visualise the homomorphisms φ^0_0 and φ^0_1. We may represent $\widehat{T}_{\mathscr{S}} \subseteq T_{\mathscr{S}} \mathbin{\widehat{\otimes}} T_{\mathscr{S}}$ faithfully as algebras of bounded linear operators on $\ell^2(\mathbb{N}^2)$ or $\mathscr{S}(\mathbb{N}^2)$; this representation is generated by the two isometries $S \otimes 1$ and $1 \otimes S$, which are illustrated in Figure 4.1 on page 68. A bounded unital $*$-homomorphism $T_{\mathscr{S}} \to \widehat{T}_{\mathscr{S}}$ is uniquely determined by the image of S, which may be any isometry of polynomial growth by Lemma 4.5. We may extend φ^0_0 and φ^0_1 uniquely to such bounded unital $*$-homomorphisms; they are associated to the isometries $\hat{S}_0$ and $\hat{S}_1$ illustrated in Figure 4.1. Now we define U_0, U_1 as in Figure 4.1. It is easy to check that they are self-adjoint unitaries in $\widehat{T}_{\mathrm{alg}}$, so that they solve the polynomial equation $x^2 = 1$. We connect U_0 and U_j to 1 by the smooth paths of unitaries

$$\tfrac{1}{2}(1 + U_j) - \tfrac{1}{2}\exp(\pi \mathrm{i} t)(1 - U_j)$$

for $t \in [0,1]$. Hence there is a smooth path (U_t) of unitaries in $\widehat{T}$ that connects U_0 and U_1. (There is a technical problem with the concatenation of *smooth* homotopies because we need the derivatives at the end points to match; we will address this in §6.1.)

We have $\hat{S}_0 = U_0 \circ (S \otimes 1)$ and $\hat{S}_1 = U_1 \circ (S \otimes 1)$. Hence the homotopy between U_0 and U_1 generates an isometry $\hat{S}_t := U_t \circ (S \otimes 1)$ in $C^\infty([0,1], \widehat{T})$ connecting $\hat{S}_0$ and $\hat{S}_1$. A tedious computation shows that $(\hat{S}_t, \hat{S}_t^*)$ has polynomial growth in $C^\infty([0,1], \widehat{T})$ (see [36]). Hence we get a $*$-homomorphism $\varphi \colon T_{\mathscr{S}} \to C^\infty([0,1], \widehat{T}_{\mathscr{S}})$ by Lemma 4.5. Its restriction to $T^0_{\mathscr{S}}$ is the desired homotopy between φ^0_0 and φ^0_1. This finishes the proof of Bott periodicity. $\qquad\square$

Bott periodicity tells us that the long exact sequences in Theorems 2.33, 2.38, and 2.41 become periodic with only six different entries. In the situation of Theorem 2.33, this looks as follows:

$$
\begin{array}{ccccc}
K_0(I) & \xrightarrow{K_0(i)} & K_0(E) & \xrightarrow{K_0(p)} & K_0(Q) \\[4pt]
\uparrow{\scriptstyle\mathrm{ind}} & & & & \downarrow{\scriptstyle\mathrm{ind}\circ\beta^{-1}} \\[4pt]
K_1(Q) & \xleftarrow[K_1(q)]{} & K_1(E) & \xleftarrow[K_1(i)]{} & K_1(I)
\end{array}
$$

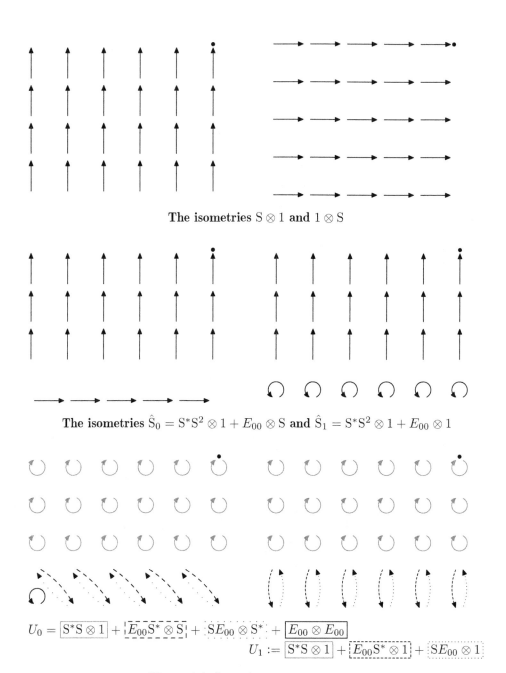

Figure 4.1: Some important operators

From now on, we use Bott periodicity to view $K_*(A)$ for $* \in \mathbb{Z}/2$ as a $\mathbb{Z}/2$-graded Abelian group.

The vanishing of K-theory for Toeplitz algebras depends on functional analysis. For general rings, we only have the following weaker statement, which does not suffice to get periodicity:

Exercise 4.9. For any ring R, the embedding $\mathbb{M}_\infty(R) \to \mathcal{T}_{\mathrm{alg}}(R)$ induces the zero map $K_0(\mathbb{M}_\infty(R)) \to K_0(\mathcal{T}_{\mathrm{alg}}(R))$ (with a suitable definition of $\mathcal{T}_{\mathrm{alg}}(R)$).

4.3 Some K-theory computations

We begin by computing the K-theory of some simple compact spaces. We write $K^*(X) := K_*(C_0(X))$ for a locally compact space X to reduce clutter.

Example 4.10. Consider the one-point-space $\star$ or, equivalently, $C(\star) = \mathbb{C}$. We have $K^0(\star) \cong \mathbb{Z}$ because $\mathbb{C}$ is a field, and $K^1(\star) \cong 0$ because $\mathrm{Gl}_m(\mathbb{C})$ is connected for all $m \in \mathbb{N}_{\geq 1}$. Bott periodicity yields

$$K^n(\star) = K_n(\mathbb{C}) \cong \begin{cases} \mathbb{Z} & n \text{ even,} \\ 0 & n \text{ odd.} \end{cases}$$

Example 4.11. It follows from Example 4.10 that

$$K^n(\mathbb{R}^m) \cong K^{m+n}(\star) \cong \begin{cases} \mathbb{Z} & m+n \text{ even,} \\ 0 & m+n \text{ odd.} \end{cases}$$

Adjoining a unit to $C_0(\mathbb{R}^m)$, we get $C_0(\mathbb{R}^m)_{\mathbb{C}}^+ \cong C(\mathbb{S}^m)$. Since K-theory is split-exact and $C_0(\mathbb{R}^m) \rightarrowtail C(\mathbb{S}^m) \twoheadrightarrow \mathbb{C}$ is a split extension, we get

$$K^n(\mathbb{S}^m) \cong K^n(\mathbb{R}^m) \oplus K^n(\star) \cong \begin{cases} \mathbb{Z}^2 & m \text{ even, } n \text{ even,} \\ 0 & m \text{ even, } n \text{ odd,} \\ \mathbb{Z} & m \text{ odd, } n \text{ arbitrary.} \end{cases}$$

In particular, $K^n(\mathbb{S}^1) \cong \mathbb{Z}$ for $n = 0$ and $n = 1$.

Example 4.12. For the n-torus $\mathbb{T}^n := \mathbb{R}^n/\mathbb{Z}^n$, we have

$$K^m(\mathbb{T}^n) \cong \mathbb{Z}^{2^{n-1}}$$

for all $m \in \mathbb{Z}/2$, $n \in \mathbb{N}_{\geq 1}$. We prove this by induction on n. The assertion for $n = 1$ is a special case of Example 4.11 because $\mathbb{T}^1 = \mathbb{S}^1$. Suppose the assertion holds for $\mathbb{T}^n$. Since $\mathbb{T}^{n+1} = \mathbb{T}^n \times \mathbb{S}^1$ and since we have a split extension $C_0(\mathbb{R}) \rightarrowtail C(\mathbb{S}^1) \twoheadrightarrow \mathbb{C}$, we get a split extension $C_0(\mathbb{T}^n \times \mathbb{R}) \rightarrowtail C(\mathbb{T}^{n+1}) \twoheadrightarrow C(\mathbb{T}^n)$. Thus

$$K^*(\mathbb{T}^{n+1}) \cong K^{*+1}(\mathbb{T}^n) \oplus K^*(\mathbb{T}^n) \cong \mathbb{Z}^{2^{n-1}} \oplus \mathbb{Z}^{2^{n-1}} \cong \mathbb{Z}^{2^n}.$$

Exercise 4.13. Describe the 2^n generators for $K^*(\mathbb{T}^n)$. It is useful to index these generators by subsets of $\{1, \ldots, n\}$.

Example 4.14. Let $\varphi \colon \mathbb{S}^1 \to \mathbb{S}^2$ be a simple closed loop, that is, φ is an injective continuous map; then φ is a homeomorphism onto its image. Let X be the complement of $\varphi(\mathbb{S}^1)$ in $\mathbb{S}^2$. By *Jordan's Curve Theorem*, X is a union of two open disks, that is, it is homeomorphic to $\mathbb{R}^2 \sqcup \mathbb{R}^2$. Therefore,

$$K^n(X) \cong K^n(\mathbb{R}^2) \oplus K^n(\mathbb{R}^2) \cong \begin{cases} \mathbb{Z}^2 & n \text{ even,} \\ 0 & n \text{ odd.} \end{cases}$$

We can also compute this directly, using the C^*-algebra extension

$$C_0(X) \rightarrowtail C(\mathbb{S}^2) \twoheadrightarrow C(\mathbb{S}^1).$$

By Example 4.11, the associated long exact sequence in K-theory is

$$\begin{array}{ccc}
K^0(X) \longrightarrow \mathbb{Z}^2 \longrightarrow \mathbb{Z} \\
\uparrow \qquad\qquad\qquad\qquad \downarrow \\
\mathbb{Z} \longleftarrow 0 \longleftarrow K^1(X).
\end{array} \qquad (4.15)$$

Recall that $K^0(\mathbb{S}^1) \cong \mathbb{Z}$ is generated by the class of the unit element, which lifts to an element of $C(\mathbb{S}^2)$. Hence the map $\mathbb{Z}^2 \to \mathbb{Z}$ in the top row of (4.15) is surjective. Its kernel is generated by the image of $K^0(\mathbb{R}^2) \cong \mathbb{Z}$ in $K^0(\mathbb{S}^2)$. Hence we get

$$K^0(X) \cong \mathbb{Z}^2, \qquad K^1(X) = 0.$$

Thus we can compute the K-theory of X without Jordan's Curve Theorem.

Now we turn to a mildly noncommutative example. Before we discuss it, we mention a general fact. Let G be a compact topological group and let X be a locally compact space equipped with a continuous action of G. Then one can define a G-equivariant version of K-theory $K_G^*(X)$ using G-equivariant vector bundles on the one-point-compactification of X (see [114]). We can also describe equivariant K-theory as the K-theory of the crossed product by [10, Theorem 11.7.1]:

$$K_G^*(X) \cong K_*\big(G \ltimes C_0(X)\big).$$

This is a special case of the *Green–Julg Theorem* [53, 68]. We will define crossed products by locally compact groups and study their K-theory in §5.3.

For our immediate purposes, the following description is most useful. Let $L^2 G$ be the Hilbert space of square-integrable functions on G with respect to the Haar measure on G. The group G acts on $L^2 G$ by the regular representation ϱ. This unitary representation induces a continuous action of G on the C^*-algebra of compact operators $\mathcal{K}(L^2 G)$ by $g \cdot T := \varrho_g \circ T \circ \varrho_g^{-1}$ for all $g \in G$, $T \in \mathcal{K}(L^2 G)$. Let G act on the C^*-algebra $C_0\big(X, \mathcal{K}(L^2 G)\big)$ by $(g \cdot f)(x) := g \cdot \big(f(g^{-1}x)\big)$ for all $g \in G$, $f \in C_0\big(X, \mathcal{K}(L^2 G)\big)$, $x \in X$. Then $G \ltimes C_0(X)$ is naturally isomorphic to the C^*-subalgebra of G-invariant elements in $C_0\big(X, \mathcal{K}(L^2 G)\big)$.

Example 4.16. Let $\mathbb{Z}/2$ act on $C_0(\mathbb{R})$ by reflection at the origin. We are interested in the crossed product $\mathbb{Z}/2 \ltimes C_0(\mathbb{R})$. In our case, $\mathcal{K}(L^2 G) \cong M_2(\mathbb{C})$. If we choose the two characters of $\mathbb{Z}/2$ as our basis in $L^2 G$, then the generator g of $\mathbb{Z}/2$ acts by

$$g \cdot \begin{pmatrix} a_{11} & a_{12} \\ a_{21} & a_{22} \end{pmatrix} = \begin{pmatrix} a_{11} & -a_{12} \\ -a_{21} & a_{22} \end{pmatrix}.$$

A $\mathbb{Z}/2$-invariant function $\mathbb{R} \to M_2(\mathbb{C})$ is already determined by its restriction to $\mathbb{R}_+$, which may be any function $\mathbb{R}_+ \to M_2(\mathbb{C})$ whose value at 0 is diagonal. Thus we get an isomorphism

$$\mathbb{Z}/2 \ltimes C_0(\mathbb{R}) \cong \{f \colon \mathbb{R}_+ \to M_2(\mathbb{C}) \mid f(0) \in M_2(\mathbb{C}) \text{ is diagonal}\}.$$

This algebra fits into a C^*-algebra extension

$$M_2\big(C_0(\mathbb{R}_{>0})\big) \rightarrowtail \mathbb{Z}/2 \ltimes C_0(\mathbb{R}) \twoheadrightarrow \mathbb{C} \oplus \mathbb{C}. \tag{4.17}$$

We have $K_*\big(M_2\big(C_0(\mathbb{R}_{>0})\big)\big) \cong K_{*+1}(\mathbb{C})$ because K-theory is Morita invariant and $\mathbb{R}_{>0} \cong \mathbb{R}$. Hence the K-theory long exact sequence for the extension (4.17) is

$$\begin{array}{ccc}
0 \longrightarrow K_0\big(\mathbb{Z}/2 \ltimes C_0(\mathbb{R})\big) & \longrightarrow & \mathbb{Z}^2 \\
\big\uparrow & & \big\downarrow \\
0 \longleftarrow K_1\big(\mathbb{Z}/2 \ltimes C_0(\mathbb{R})\big) & \longleftarrow & \mathbb{Z}.
\end{array} \tag{4.18}$$

It remains to compute the vertical map $\mathbb{Z}^2 \to \mathbb{Z}$, which is the boundary map of the extension (4.17). The easiest way is to use the naturality of the boundary map. Consider the subalgebra of $\mathbb{Z}/2 \ltimes C_0(\mathbb{R})$ of all functions into $M_1(\mathbb{C}) \subseteq M_2(\mathbb{C})$. This subalgebra is isomorphic to $C_0(\mathbb{R}_{\geq 0})$ and fits into the *cone extension*

$$C_0(\mathbb{R}_{>0}) \rightarrowtail C_0(\mathbb{R}_{\geq 0}) \twoheadrightarrow \mathbb{C}.$$

Since $C_0(\mathbb{R}_{\geq 0})$ is contractible, it has vanishing K-theory. Therefore, the boundary map for the cone extension is invertible. Using the naturality of the boundary map, we find that the vertical map $\mathbb{Z}^2 \to \mathbb{Z}$ in (4.18) sends the first basis vector to $\pm 1 \in \mathbb{Z}$. (A similar argument shows that it sends the second basis vector to ± 1 as well.) Thus the map $\mathbb{Z}^2 \to \mathbb{Z}$ is surjective and has kernel isomorphic to $\mathbb{Z}$. We conclude that

$$K_0\big(\mathbb{Z}/2 \ltimes C_0(\mathbb{R})\big) \cong \mathbb{Z}, \qquad K_1\big(\mathbb{Z}/2 \ltimes C_0(\mathbb{R})\big) \cong 0.$$

We do not discuss the equivariant generalisation of Bott periodicity here in any detail. We should mention, however, that $\mathbb{Z}/2 \ltimes C_0(\mathbb{R})$ is an example where equivariant Bott periodicity *fails*. We have $\mathbb{Z}/2 \ltimes \mathbb{C} \cong C^*(\mathbb{Z}/2) \cong \mathbb{C} \oplus \mathbb{C}$. Thus

$$K_n^{\mathbb{Z}/2}(\mathbb{C}) \cong K_n(\mathbb{C} \oplus \mathbb{C}) \cong K_n(\mathbb{C}) \oplus K_n(\mathbb{C}) \cong \begin{cases} \mathbb{Z}^2 & n \text{ even}, \\ 0 & n \text{ odd}, \end{cases}$$

which is quite different from $\mathrm{K}_{n+1}^{\mathbb{Z}/2}(C_0(\mathbb{R}))$. Bott periodicity fails here because the reflection at 0 reverses orientation.

Example 4.19. Given $p, q \in \mathbb{N}_{\geq 1}$, we consider the C^*-algebra

$$I(p,q) := \{f \colon [0,1] \to \mathbb{M}_p \otimes \mathbb{M}_q \mid f(0) \in \mathbb{M}_p \otimes 1_q, \ f(1) \in 1_p \otimes \mathbb{M}_q\}.$$

Here $1_q \in \mathbb{M}_q$ and $1_p \in \mathbb{M}_p$ denote the unit elements. Since $\mathbb{M}_p \otimes \mathbb{M}_q \cong \mathbb{M}_{pq}$, this fits into a C^*-algebra extension

$$C_0\big((0,1), \mathbb{M}_{pq}\big) \rightarrowtail I(p,q) \twoheadrightarrow \mathbb{M}_p \oplus \mathbb{M}_q.$$

The associated long exact sequence is

$$
\begin{array}{ccc}
0 \longrightarrow \mathrm{K}_0\big(I(p,q)\big) \longrightarrow \mathbb{Z}^2 \\
\uparrow \qquad\qquad\qquad\qquad\qquad\quad \downarrow \\
0 \longleftarrow \mathrm{K}_1\big(I(p,q)\big) \longleftarrow \mathbb{Z}.
\end{array}
\tag{4.20}
$$

To compute the value of the boundary map $\mathbb{Z}^2 \to \mathbb{Z}$ on the first generator, we compare our extension to the subextension

$$C_0\big((0,1), \mathbb{M}_p \otimes 1_q\big) \rightarrowtail C_0\big((0,1], \mathbb{M}_p \otimes 1_q\big) \twoheadrightarrow \mathbb{M}_p \otimes 1_q.$$

This is a cone extension, so that its boundary map is bijective. Thus the image of the first basis vector of $\mathbb{Z}^2$ under the boundary map in (4.20) is, up to a sign, the image of $e \otimes 1_q \in \mathbb{M}_{pq}$ in $\mathrm{K}_0(\mathbb{M}_{pq})$, where $e \in \mathbb{M}_p$ is a rank-one-projection. Under the canonical isomorphism $\mathrm{K}_0(\mathbb{M}_{pq}) \cong \mathbb{Z}$, the class $[e \otimes 1_q]$ is mapped to q. Thus the boundary map $\mathbb{Z}^2 \to \mathbb{Z}$ in (4.20) sends $(1,0) \mapsto \pm q$. Similarly, we compute that $(0,1) \mapsto \pm p$. Thus the range of the boundary map is the ideal in $\mathbb{Z}$ generated by p and q or, equivalently, by their greatest common divisor (p,q); the kernel of the boundary map is isomorphic to $\mathbb{Z}$. As a result,

$$\mathrm{K}_0\big(I(p,q)\big) \cong \mathbb{Z}, \qquad \mathrm{K}_1\big(I(p,q)\big) \cong \mathbb{Z}/(p,q).$$

Notice that $\mathbb{Z}/(p,q) = 0 \iff (p,q) = 1 \iff p$ and q are coprime.

4.3.1   The Atiyah–Hirzebruch spectral sequence

Let X be a compact CW-complex. By definition, this means that X has an increasing filtration $(X^{(n)})_{n \in \mathbb{N}}$ by closed subsets called *skeleta*, such that $X^{(n)} = X$ for sufficiently large n and

$$X^{(n)} \setminus X^{(n-1)} \cong C_n X \times \mathbb{R}^n \qquad \text{for all } n \in \mathbb{N},$$

where the $C_n X$ are certain finite sets. By convention, $X^{(k)} = \emptyset$ for $k < 0$, so that $X^{(0)} \cong C_0 X$ is finite. The components $\gamma \times \mathbb{R}^n$ of $X^{(n)} \setminus X^{(n-1)}$ for $\gamma \in C_n X$ are also called *open n-cells* in X.

The inclusion $X^{(n-1)} \subseteq X^{(n)}$ for $n \in \mathbb{N}$ gives rise to a C^*-algebra extension

$$C_0(X^{(n)} \smallsetminus X^{(n-1)}) \rightarrowtail C(X^{(n)}) \twoheadrightarrow C(X^{(n-1)}),$$

which yields a long exact sequence

$$
\begin{array}{ccc}
\mathrm{K}^0(X^{(n)} \smallsetminus X^{(n-1)}) \longrightarrow \mathrm{K}^0(X^{(n)}) \longrightarrow \mathrm{K}^0(X^{(n-1)}) \\
\Big\uparrow \qquad\qquad\qquad\qquad\qquad\qquad \Big\downarrow \\
\mathrm{K}^1(X^{(n-1)}) \longleftarrow \mathrm{K}^1(X^{(n)}) \longleftarrow \mathrm{K}^1(X^{(n)} \smallsetminus X^{(n-1)}).
\end{array}
\tag{4.21}
$$

Bott periodicity implies

$$
\mathrm{K}^m(X^{(n)} \smallsetminus X^{(n-1)}) \cong \begin{cases} \mathbb{Z}[C_n X] & n+m \text{ even}, \\ 0 & n+m \text{ odd}. \end{cases}
$$

This yields a recipe for computing the K-theory of $X^{(n)}$ by induction on n, and thus also the K-theory of X. This iterative computation may be difficult to carry out. Spectral sequences are designed to help in the bookkeeping.

The most elegant way to get spectral sequences is via *exact couples* (see [78]). Our filtration produces an exact couple in the following fashion. We define bigraded Abelian groups by

$$
D := \sum_{p=1}^{\infty} \sum_{q \in \mathbb{Z}/2} D_{pq}, \qquad D_{pq} := \mathrm{K}^{q+p-1}(X^{(p-1)}),
$$

$$
E := \sum_{p=0}^{\infty} \sum_{q \in \mathbb{Z}/2} E_{pq}, \qquad E_{pq} := \mathrm{K}^{q+p}(X^{(p)} \smallsetminus X^{(p-1)}).
$$

The maps in (4.21) yield homogeneous group homomorphisms

$$
\begin{array}{c}
D \xrightarrow{\ i\ } D \\
{}^{k}\diagdown \quad \diagup{}^{j} \\
E
\end{array}
\qquad
\begin{array}{l}
\deg i = (-1,1), \\
\deg j = (0,0), \\
\deg k = (+1,0).
\end{array}
$$

The exactness of (4.21) means that (D, E, i, j, k) is an *exact couple*. This yields a spectral sequence as in [78, p. 336–337]. By design,

$$
E^1_{pq} = E_{pq} \cong \begin{cases} \mathbb{Z}[C_p X] \cong \mathrm{Hom}(\mathbb{Z}[C_p X], \mathbb{Z}) & q \text{ even}, \\ 0 & q \text{ odd}. \end{cases}
$$

A computation shows that the boundary map $d^1 = jk \colon E^1 \to E^1$ corresponds to the usual cellular coboundary map that computes the cohomology of X. Thus

$$
E^2_{pq} \cong H^p(X, \mathrm{K}_q(\mathbb{C})).
$$

All the even boundary maps d_{pq}^{2n} vanish because the spectral sequence is supported in the rows with even q. Thus $E_{pq}^2 = E_{pq}^3$ and the first non-trivial boundary map is d^3, which maps $H^p(X, \mathrm{K}_q(\mathbb{C})) \to H^{p+3}(X, \mathrm{K}_q(\mathbb{C}))$. This is a natural map on the cohomology of X. Such cohomology operations can be classified, and this allows us to describe d^3: it is the Steenrod operation Sq^3 (see also Exercise 9.20 and [4, 107]).

Chapter 5

The K-theory of crossed products

Crossed products for group actions yield many interesting C^*-algebras. First we consider the case of crossed products by $\mathbb{Z}$. Their K-theory is computed by the *Pimsner–Voiculescu exact sequence* [101]. We use *crossed Toeplitz algebras* to get it, following [33]. For crossed products by more general groups, there is a good guess for what the K-theory ought to be: this is the celebrated *Baum–Connes conjecture*. We discuss an alternative formulation of this conjecture, which is once again based on Toeplitz algebras.

5.1   Crossed products for a single automorphism

Definition 5.1. Let A be a C^*-algebra and $\alpha \in \mathrm{Aut}(A)$. The C^*-*algebraic crossed product* $U_{C^*}(A, \alpha)$ is the universal C^*-algebra with a unitary multiplier u and an essential $*$-homomorphism $j_U \colon A \to U_{C^*}(A, \alpha)$ such that

$$u j_U(a) u^* = j_U\big(\alpha(a)\big) \qquad \text{for all } a \in A.$$

We remark that $u \in U_{C^*}(A, \alpha)$ if and only if A is unital. In general, we only have $j_U(a) u^n \in U_{C^*}(A, \alpha)$ for all $n \in \mathbb{Z}$, but $u^n \notin U_{C^*}(A, \alpha)$. We use the unusual notation $U_{C^*}(A, \alpha)$ in order to distinguish between smooth and C^*-algebraic crossed products and Toeplitz algebras.

The elements $a \cdot u^m$ for $a \in A$, $m \in \mathbb{Z}$ span a dense $*$-subalgebra of $U_{C^*}(A, \alpha)$, which we call the *algebraic crossed product* $U_{\mathrm{alg}}(A, \alpha)$. To define it, we do not need A to be a C^*-algebra:

Definition 5.2. Let A be a bornological algebra and $\alpha \in \mathrm{Aut}(A)$, that is, α is a bounded algebra isomorphism whose inverse is also bounded. The *algebraic crossed product* $U_{\mathrm{alg}}(A, \alpha)$ is the bornological vector space

$$\mathbb{C}[\mathbb{Z}] \,\widehat{\otimes}\, A \cong \bigoplus_{n \in \mathbb{Z}} A$$

equipped with the convolution product

$$f_1 * f_2(n) := \sum_{m \in \mathbb{Z}} f_1(m) \alpha^m \big(f_2(n - m) \big). \tag{5.3}$$

It is easy to check that this turns $U_{\mathrm{alg}}(A, \alpha)$ into a bornological algebra.

Exercise 5.4. Characterise $U_{\mathrm{alg}}(A, \alpha)$ by a universal property.

For example, if $A = \mathbb{C}$ and $\alpha = \mathrm{id}_{\mathbb{C}}$, then $U_{\mathrm{alg}}(A, \alpha) = \mathbb{C}[\mathbb{Z}] = \mathbb{C}[u, u^{-1}]$ is the algebra of Laurent polynomials. Thus $U_{\mathrm{alg}}(A, \alpha)$ is almost never a local Banach algebra, even if A is one. To remedy this, we consider ℓ^1- and smooth crossed products. These can only be defined under additional hypotheses on α.

First let A be a Banach algebra and let $\alpha \in \mathrm{Aut}(A)$ be an *isometric* automorphism. Then (5.3) turns $\ell^1(\mathbb{Z}, A)$ into a Banach algebra. We denote this Banach algebra crossed product by $U_{\ell^1}(\mathbb{Z}, A)$. Now let A be a local Banach algebra. We say that an automorphism $\alpha \in \mathrm{Aut}(A)$ *generates a uniformly bounded representation of* $\mathbb{Z}$ if $S^{(\alpha)} := \bigcup_{n \in \mathbb{Z}} \alpha^n(S)$ is bounded for all bounded subsets $S \subseteq A$. Notice that $S^{(\alpha)}$ is the smallest subset of A that satisfies $S \subseteq S^{(\alpha)}$ and $\alpha(S^{(\alpha)}) = S^{(\alpha)}$. Therefore, an equivalent characterisation for uniformly bounded representations is that any bounded subset is contained in one that is invariant under α and α^{-1}.

Lemma 5.5. *Let $\alpha \in \mathrm{Aut}(A)$ be an automorphism of a local Banach algebra that generates a uniformly bounded representation of $\mathbb{Z}$. Then A is an increasing union of a directed set of α-invariant Banach subalgebras $(A_S)_{S \in \mathfrak{S}}$, such that the restriction of α to each A_S is an isometric isomorphism.*

Proof. It suffices to prove that any bounded subset S of A is absorbed by a submultiplicative complete disk T with $\alpha(T) = T$ because such subsets are exactly the closed unit balls of Banach subalgebras of A on which α acts isometrically. First, we embed S in $S^{(\alpha)}$, which is again bounded. There is $r > 0$ for which $S_2 := \bigcup_{n=1}^{\infty} (r S^{(\alpha)})^n$ remains bounded. Finally, we take the smallest complete disk containing S_2. This has all the properties we need. $\qquad\square$

Lemma 5.5 makes it easy to extend the definition of $U_{\ell^1}(A, \alpha)$ to local Banach algebras with a uniformly bounded representation of $\mathbb{Z}$: we simply let $U_{\ell^1}(A, \alpha)$ be the increasing union of the Banach algebras $U_{\ell^1}(A_S, \alpha|_{A_S})$, where the system $(A_S)_{S \in \mathfrak{S}}$ is constructed as in Lemma 5.5. Notice that the underlying bornological vector space of $U_{\ell^1}(A, \alpha)$ is $\ell^1(\mathbb{Z}, A) = \ell^1(\mathbb{Z}) \mathbin{\hat{\otimes}} A$ by Lemma 2.9. We will also use the *smooth crossed product* $U_{\mathscr{S}}(A, \alpha)$, which is the dense subalgebra $\mathscr{S}(\mathbb{Z}, A) \subseteq \ell^1(\mathbb{Z}, A)$ equipped with the usual bornology (compare §2.3.4).

Proposition 5.6. *Let A be a local Banach algebra and let $\alpha \in \mathrm{Aut}(A)$ generate a uniformly bounded representation of $\mathbb{Z}$. Then $U_{\mathscr{S}}(A, \alpha)$ is an isoradial subalgebra of $U_{\ell^1}(A, \alpha)$. Therefore, it is a local Banach algebra and has the same K-theory as $U_{\ell^1}(A, \alpha)$.*

Let B be a C^*-algebra and let $\beta \in \mathrm{Aut}(B)$ be a $*$-isomorphism. Then $U_{\mathscr{S}}(B, \beta)$ is an isoradial subalgebra of $U_{C^*}(B, \beta)$ and has the same K-theory as $U_{C^*}(B, \beta)$.

Proof. Since the construction of $U_{\dots}$ is compatible with increasing unions, we may assume without loss of generality that A itself is a Banach algebra with an isometric automorphism α. The compact group $\mathbb{T} := \mathbb{R}/\mathbb{Z}$ acts on $U_{\ell^1}(A, \alpha)$ by the dual action $t \cdot f(m) := \exp(2\pi i m t) f(m)$ for all $m \in \mathbb{Z}$, which is a continuous action by algebra automorphisms. By Example 2.53, the smooth elements for this action form an isoradial subalgebra, which therefore has the same K-theory. We can also characterise smooth elements by the condition that the powers of the generator of the representation of $\mathbb{T}$, which are given by $D^k f(m) = (2\pi i m)^k \cdot f(m)$, remain bounded for all $k \in \mathbb{N}$ (see [85]). Using this, one easily identifies the subalgebra of smooth elements with $U_{\mathscr{S}}(A, \alpha)$. This finishes the proof for local Banach algebras. The assertion for C^*-algebras is proved similarly. In order to describe the smooth elements for the dual action on $U_{C^*}(B, \beta)$, we use the bounded embeddings of Banach spaces $\ell^1(\mathbb{Z}, B) \subseteq U_{C^*}(B, \beta) \subseteq C_0(\mathbb{Z}, B)$ and $\mathrm{SE}\,\ell^1(\mathbb{Z}, B) = \mathscr{S}(\mathbb{Z}, B) = \mathrm{SE}\,C_0(\mathbb{Z}, B)$. $\square$

Lemma 5.7. *Let A be a bornological algebra and let $\alpha \in A$ generate a uniformly bounded representation of $\mathbb{Z}$. Let B be a bornological algebra equipped with a bounded algebra homomorphism $j \colon A \to B$ and an invertible multiplier v such that $vj(a)v^{-1} = j(\alpha(a))$ for all $a \in A$ and such that the set of linear maps $b \mapsto \varepsilon_n b \cdot v^n$ for $n \in \mathbb{Z}$ is uniformly bounded in $\mathrm{Hom}(B, B)$ for any $(\varepsilon_n) \in \mathscr{S}(\mathbb{Z})$. Then there is a unique bounded homomorphism $f \colon U_{\mathscr{S}}(A, \alpha) \to B$ such that $f \circ j_U = j$ and $f(u) = v$; in addition, the above conditions on (B, j, v) hold for $(U_{\mathscr{S}}(A, \alpha), j_U, u)$.*

Proof. We may write $f \in U_{\mathscr{S}}(A, \alpha)$ as $\sum_{m \in \mathbb{Z}} j_U(f(m)) \cdot u^m$. Therefore, the growth condition on v suffices to get a bounded homomorphism f with the required properties. We omit the verification that u itself satisfies this growth condition. $\square$

5.1.1 Crossed Toeplitz algebras

To simplify notation, we define crossed Toeplitz algebras only for unital A. As with crossed products, there are several variants. We first introduce the C^*-algebraic one.

Let A be a unital C^*-algebra and let $\alpha \in \mathrm{Aut}(A)$ be a $*$-automorphism. We let $\mathcal{T}_{C^*}(A, \alpha)$ be the universal C^*-algebra equipped with an essential $*$-homomorphism $j_{\mathcal{T}} \colon A \to \mathcal{T}(A, \alpha)$ and an isometry $v \in \mathcal{T}(A, \alpha)$ such that

$$v^* j_{\mathcal{T}}(a) v = j_{\mathcal{T}}(\alpha(a)) \qquad \text{for all } a \in A.$$

If A were not unital, we would only have an isometric multiplier v. We call $\mathcal{T}_{C^*}(A, \alpha)$ the *crossed Toeplitz C^*-algebra* of the C^*-dynamical system (A, α). It follows from the universal property of $\mathcal{T}_{C^*}(A, \alpha)$ that we have a natural $*$-homomorphism $\pi \colon \mathcal{T}_{C^*}(A, \alpha) \to U_{C^*}(A, \alpha)$ such that $\pi \circ j_{\mathcal{T}} = j_U$ and $\pi(v) = u^*$. We warn the reader that Joachim Cuntz instead uses $\mathcal{T}_{C^*}(A, \alpha^{-1})$ in [33].

It is very convenient to describe $\mathcal{T}_{C^*}(A, \alpha)$ as a subalgebra of $\mathcal{T}_{C^*} \widehat{\otimes}_{C^*}$ $U_{C^*}(A, \alpha)$; here $\mathcal{T}_{C^*} \subseteq \mathcal{L}(\ell^2(\mathbb{N}))$ is the usual Toeplitz C^*-algebra, which is generated by the unilateral shift S, and $\widehat{\otimes}_{C_*}$ denotes the maximal C^*-tensor product. (Since $\mathcal{T}_{C^*}$ is a nuclear C^*-algebra, all C^*-tensor norms agree in this case.)

Proposition 5.8. *If A is unital, then $\mathcal{T}_{C^*}(A, \alpha)$ is naturally isomorphic to the C^*-subalgebra of $\mathcal{T}_{C^*} \widehat{\otimes}_{C_*} U_{C^*}(A, \alpha)$ generated by $1 \otimes j_U(A)$ and $S \otimes u^*$.*

Proof. The universal property yields a natural map

$$f \colon \mathcal{T}_{C^*}(A, \alpha) \to \mathcal{T}_{C^*} \widehat{\otimes}_{C^*} U_{C^*}(A, \alpha)$$

with $f \circ j_T(a) = 1 \otimes j_U(a)$ for all $a \in A$ and $f(v) = S \otimes u^*$. It remains to show that it is faithful.

For this, we first check that the projection $1 - vv^*$ in $\mathcal{T}_{C^*}(A, \alpha)$ commutes with $j_T(A)$. More precisely, we check the equivalent assertion $(1 - vv^*)j_T(a)v = 0$ for all $a \in A$. This follows from the computation

$$\big((1 - vv^*)j_T(a)v\big)^* \cdot (1 - vv^*)j_T(a)v = v^*j_T(a^*a)v - v^*j_T(a^*)vv^*j_T(a)v$$
$$= j_T\big(\alpha(a^*a)\big) - j_T\big(\alpha(a^*)\big)j_T\big(\alpha(a)\big) = 0.$$

Now we choose a faithful essential $*$-representation of $\mathcal{T}_{C^*}(A, \alpha)$ on a Hilbert space $\mathcal{H}$. This is determined by an essential $*$-representation $\pi \colon A \to \mathcal{L}(\mathcal{H})$ and an isometry $V \in \mathcal{L}(\mathcal{H})$ with $V^*\pi(a)V = \pi(\alpha(a))$ for all $a \in A$. We get another representation on $\ell^2(\mathbb{Z}, \mathcal{H})$ if we let v act by $1 \otimes V$ and $j_T(a)$ act by $\pi'(a)\varphi(n) := \pi(\alpha^n(a)) \cdot \varphi(n)$. This representation is still faithful because it contains the original one as a direct summand.

Now we represent $\mathbb{Z}$ on $\ell^2(\mathbb{Z}) \otimes \mathcal{H}$ by the left regular representation, which is generated by the unitary operator $u\varphi(n) := \varphi(n - 1)$. Using the vanishing of $(1 - vv^*)j_T(a)v = j_T(a)v - vj_T \circ \alpha(a)$ checked above, we get that the isometry $u^* \otimes V$ on $\ell^2(\mathbb{Z}, \mathcal{H})$ commutes with $\pi'(A)$. Thus (π', u^*) and $u^* \otimes V$ generate commuting representations of $U_{C^*}(A, \alpha)$ and $\mathcal{T}_{C^*}$ on $\ell^2(\mathbb{Z}, \mathcal{H})$. These combine to a representation of $\mathcal{T}_{C^*} \widehat{\otimes}_{C^*} U_{C^*}(A, \alpha)$, whose composition with f is a faithful representation of $\mathcal{T}_{C^*}(A, \alpha)$. Therefore, f is injective. $\square$

This description of $\mathcal{T}_{C^*}(A, \alpha)$ makes sense in other categories of algebras as well. We let $\mathcal{T}_{\mathrm{alg}}(A, \alpha)$ and $\mathcal{T}_{\mathcal{S}}(A, \alpha)$ be the closed subalgebras of $\mathcal{T}_{\mathrm{alg}} \widehat{\otimes} U_{\mathrm{alg}}(A, \alpha)$ and $\mathcal{T}_{\mathcal{S}} \widehat{\otimes} U_{\mathcal{S}}(A, \alpha)$ that are generated by $1 \otimes j_U(A)$ and

$$v := S \otimes u^*, \qquad w := S^* \otimes u.$$

Since $U_{\mathcal{S}}(A, \alpha)$ and $\mathcal{T}_{\mathcal{S}}$ are local Banach algebras, so is $\mathcal{T}_{\mathcal{S}}(A, \alpha)$.

To analyse the structure of these Toeplitz algebras, we first consider the purely algebraic case. We apply the quotient map

$$\mathcal{T}_{\mathrm{alg}} \widehat{\otimes} U_{\mathrm{alg}}(A, \alpha) \to \mathbb{C}[t, t^{-1}] \widehat{\otimes} U_{\mathrm{alg}}(A, \alpha).$$

The images of $S^* \otimes u$ and $S \otimes u^*$ in this quotient become inverse to each other. Together with A they generate a representation of $U_{\mathrm{alg}}(A, \alpha)$. It is easy to check that the image of $\mathcal{T}_{\mathrm{alg}}(A, \alpha)$ in $\mathbb{C}[t, t^{-1}] \widehat{\otimes} U_{\mathrm{alg}}(A, \alpha)$ is isomorphic as a bornological algebra to $U_{\mathrm{alg}}(A, \alpha)$. That is, we have constructed a quotient mapping

$$\pi_{\mathrm{alg}} \colon \mathcal{T}_{\mathrm{alg}}(A, \alpha) \to U_{\mathrm{alg}}(A, \alpha).$$

It remains to describe $\ker \pi_{\mathrm{alg}} = \mathcal{T}_{\mathrm{alg}}(A, \alpha) \cap \left(\mathbb{M}_\infty \widehat{\otimes} U_{\mathrm{alg}}(A, \alpha) \right)$.

Notice that the idempotent elements $v^n w^n$ are of the form $S^n (S^*)^n \otimes 1$ and therefore commute with $j_{\mathcal{T}}(A)$. Let

$$a E_{mn} := v^m j_{\mathcal{T}}(a)(1 - vw)w^n = E_{mn} \otimes \alpha^{-m}(a) u^{n-m} \in \mathcal{T}_{\mathrm{alg}}(A, \alpha).$$

One checks easily that $a_0 E_{m_0 n_0} \cdot a_1 E_{m_1 n_1} = \delta_{n_0 m_1} a_0 a_1 E_{m_0 n_1}$, that is, these elements generate a subalgebra isomorphic to $\mathbb{M}_\infty(A)$. Moreover,

$$j_{\mathcal{T}}(a_0) \cdot a_1 E_{mn} = \alpha^m(a_0) a_1 E_{mn}, \qquad v j_{\mathcal{T}}(a) w = j_{\mathcal{T}}\left(\alpha^{-1}(a)\right) - \alpha^{-1}(a) E_{00}.$$

A routine computation now shows that $\mathbb{M}_\infty(A) \cong \ker \pi_{\mathrm{alg}}$. Thus we get an extension of bornological algebras $\mathbb{M}_\infty(A) \rightarrowtail \mathcal{T}_{\mathrm{alg}}(A, \alpha) \twoheadrightarrow U_{\mathrm{alg}}(A, \alpha)$. Doing some additional estimates, one checks that we also have an extension of bornological algebras $\mathcal{K}_{\mathscr{S}}(A) \rightarrowtail \mathcal{T}_{\mathscr{S}}(A, \alpha) \twoheadrightarrow U_{\mathscr{S}}(A, \alpha)$.

5.2 The Pimsner–Voiculescu exact sequence

Theorem 5.9. *Let A be a local Banach algebra over $\mathbb{C}$ with a uniformly bounded representation of $\mathbb{Z}$. Then there is a cyclic six-term exact sequence*

$$\begin{array}{ccccc}
K_0(A) & \xrightarrow{\mathrm{id} - \alpha_*} & K_0(A) & \xrightarrow{j_{U*}} & K_0\big(U_{\mathscr{S}}(A, \alpha)\big) \\
\uparrow & & & & \downarrow \\
K_1\big(U_{\mathscr{S}}(A, \alpha)\big) & \xleftarrow{j_{U*}} & K_1(A) & \xleftarrow{\mathrm{id} - \alpha_*} & K_1(A).
\end{array}$$

Here α_* and j_{U*} denote the maps on K-theory that are induced by $\alpha \colon A \to A$ and $j_U \colon A \to U(A, \alpha)$. There is a similar exact sequence for $U_{\ell^1}(A, \alpha)$, and also for $U_{C^*}(A, \alpha)$ if A is a C^*-algebra and α is a $*$-automorphism.

The proof is based on the *Toeplitz extension*

$$\mathcal{K}_{\mathscr{S}}(A) \xrightarrow{\iota} \mathcal{T}_{\mathscr{S}}(A, \alpha) \xrightarrow{\pi} U_{\mathscr{S}}(A, \alpha). \tag{5.10}$$

In the situation of Theorem 5.9, all algebras in (5.10) are local Banach algebras. Recall that j_U lifts to a bounded homomorphism $j_{\mathcal{T}} \colon A \to \mathcal{T}_{\mathscr{S}}(A, \alpha)$.

Proposition 5.11. *The induced map* $j_{T,*}\colon \mathrm{K}_*(A) \to \mathrm{K}_*\bigl(\mathcal{T}_{\mathscr{S}}(A,\alpha)\bigr)$ *is an isomorphism. Moreover, if we compose the map on K-theory induced by the embedding* $A \to \mathcal{K}_{\mathscr{S}}(A) \subseteq \mathcal{T}_{\mathscr{S}}(A,\alpha)$, $a \mapsto aE_{00}$, *and the inverse of* $j_{T,*}$, *then we get the map* $\mathrm{id} - \alpha_*$ *on* $\mathrm{K}_*(A)$.

Before we prove Proposition 5.11, we show how it yields our theorem:

Proof of Theorem 5.9. In the long exact sequence for the extension (5.10), we may replace

$$\mathrm{K}_*\bigl(\mathcal{K}_{\mathscr{S}}(A)\bigr) \cong \mathrm{K}_*(A), \qquad \mathrm{K}_*\bigl(\mathcal{T}_{\mathscr{S}}(A,\alpha)\bigr) \cong \mathrm{K}_*(A)$$

by stability and Proposition 5.11. The map $\pi_*\colon \mathrm{K}_*\bigl(\mathcal{T}_{\mathscr{S}}(A,\alpha)\bigr) \to \mathrm{K}_*\bigl(\mathcal{U}_{\mathscr{S}}(A,\alpha)\bigr)$ becomes $(j_U)_*\colon \mathrm{K}_*(A) \to \mathrm{K}_*\bigl(\mathcal{U}_{\mathscr{S}}(A,\alpha)\bigr)$ because $\pi \circ j_T = j_U$. Proposition 5.11 shows that $\iota_*\colon \mathrm{K}_*\bigl(\mathcal{K}_{\mathscr{S}}(A)\bigr) \to \mathrm{K}_*\bigl(\mathcal{T}_{\mathscr{S}}(A,\alpha)\bigr)$ becomes $\mathrm{id} - \alpha_*\colon \mathrm{K}_*(A) \to \mathrm{K}_*(A)$. $\square$

Proof of Proposition 5.11. We use a quasi-homomorphism (see §3.1.1) to construct the inverse of

$$j_{T*}\colon \mathrm{K}_*(A) \to \mathrm{K}_*\bigl(\mathcal{T}_{\mathscr{S}}(A,\alpha)\bigr).$$

Let $f_+\colon \mathcal{T}_{\mathscr{S}}(A,\alpha) \to \mathcal{T}_{\mathscr{S}}(A,\alpha)$ be the identity automorphism. Although the isometry $S \otimes 1$ does not belong to the multiplier algebra of $\mathcal{T}_{\mathscr{S}}(A,\alpha) \subseteq \mathcal{T}_{\mathscr{S}} \widehat{\otimes} \mathcal{U}_{\mathscr{S}}(A,\alpha)$, conjugation by it defines an algebra endomorphism $f_-\colon \mathcal{T}_{\mathscr{S}}(A,\alpha) \to \mathcal{T}_{\mathscr{S}}(A,\alpha)$ because

$$(S \otimes 1)\bigl(1 \otimes j_U(a)\bigr)(S^* \otimes 1) = 1 \otimes j_U(a) - aE_{00},$$
$$(S \otimes 1)(S \otimes u^*)(S^* \otimes 1) = S \otimes u^* - E_{10},$$
$$(S \otimes 1)(S^* \otimes u)(S^* \otimes 1) = S^* \otimes u - E_{01}.$$

These formulas also imply that $f_+(x) - f_-(x) \in \mathcal{K}_{\mathscr{S}}(A)$ for all $x \in \mathcal{T}_{\mathscr{S}}(A,\alpha)$, that is, (f_+, f_-) is a quasi-homomorphism

$$(f_\pm)\colon \mathcal{T}_{\mathscr{S}}(A,\alpha) \rightrightarrows \mathcal{T}_{\mathscr{S}}(A,\alpha) \triangleright \mathcal{K}_{\mathscr{S}}(A).$$

Since K-theory for local Banach algebras is split-exact and stable, we get an induced map

$$(f_\pm)_*\colon \mathrm{K}_*\bigl(\mathcal{T}_{\mathscr{S}}(A,\alpha)\bigr) \to \mathrm{K}_*\bigl(\mathcal{K}_{\mathscr{S}}(A)\bigr) \cong \mathrm{K}_*(A).$$

We claim that this map is inverse to j_{T*}.

First we compute the composition of $(f_\pm)_*$ with the map

$$\mathrm{K}_*(A) \to \mathrm{K}_*\bigl(\mathcal{T}_{\mathscr{S}}(A,\alpha)\bigr)$$

induced by the embedding $a \mapsto aE_{00}$. This is the map $\mathrm{K}_*(A) \to \mathrm{K}_*\bigl(\mathcal{K}_{\mathscr{S}}(A)\bigr)$ induced by the quasi-homomorphism $a \mapsto f_\pm(aE_{00})$. Since $f_+(aE_{00}) = aE_{00}$ and $f_-(aE_{00}) = \alpha(a)E_{11}$, we are dealing with a pair of bounded homomorphisms $A \to \mathcal{K}_{\mathscr{S}}(A)$. Hence (3.7) yields $(f_\pm)_* = \mathrm{id} - \alpha_*\colon \mathrm{K}_*(A) \to \mathrm{K}_*(A)$ as desired.

Similarly, $(f_\pm)_* \circ j_{T*} \colon \mathrm{K}_*(A) \to \mathrm{K}_*\big(\mathcal{K}_{\mathscr{S}}(A)\big)$ is induced by the quasi-homomorphism $(f_+ \circ j_T, f_- \circ j_T)$ from A to $\mathcal{K}_{\mathscr{S}}(A)$. This is the orthogonal sum of the degenerate quasi-homomorphism $(f_- \circ j_T, f_- \circ j_T)$ and $(i, 0)$ where

$$i \colon A \to \mathcal{K}_{\mathscr{S}}(A), \qquad a \mapsto a E_{00},$$

is the stabilisation homomorphism. Now (3.4) and (3.7) show that $(f_\pm)_* \circ j_{T*} = i_* \colon \mathrm{K}_*(A) \to \mathrm{K}_*\big(\mathcal{K}_{\mathscr{S}}(A)\big)$. Since we invert i_* to identify $\mathrm{K}_*\big(\mathcal{K}_{\mathscr{S}}(A)\big) \cong \mathrm{K}_*(A)$, we obtain $(f_\pm \circ j_T)_* = \mathrm{id}$ as a map on $\mathrm{K}_*(A)$.

It remains to compute $j_{T*} \circ (f_\pm)_* \colon \mathrm{K}_*\big(\mathcal{T}_{\mathscr{S}}(A, \alpha)\big) \to \mathrm{K}_*\big(\mathcal{T}_{\mathscr{S}}(A, \alpha)\big)$. Before we can compose j_T and $f_\pm$, we must extend j_T to a larger domain. This requires an analogue of the algebra $\widehat{\mathcal{T}} \subseteq \mathcal{T} \widehat{\otimes} \mathcal{T}$ from our proof of Bott periodicity.

The double Toeplitz algebra $\mathcal{T}\mathcal{T}_{\mathscr{S}}(A, \alpha)$ is defined as the closed subalgebra of $\mathcal{T}_{\mathscr{S}} \widehat{\otimes} \mathcal{T}_{\mathscr{S}} \widehat{\otimes} U_{\mathscr{S}}(A, \alpha)$ that is generated by $1 \otimes 1 \otimes j_U(a)$ and

$$v_1 := S \otimes 1 \otimes u^*, \qquad v_2 := 1 \otimes S \otimes u^*,$$
$$w_1 := S^* \otimes 1 \otimes u, \qquad w_2 := 1 \otimes S^* \otimes u.$$

This algebra is isomorphic to $\mathcal{T}_{\mathscr{S}} \widehat{\otimes} \mathcal{T}_{\mathscr{S}} \widehat{\otimes} A$ if $\alpha = 1$.

Clearly, $\mathcal{T}\mathcal{T}_{\mathscr{S}}(A, \alpha)$ contains two copies of $\mathcal{T}_{\mathscr{S}}(A, \alpha)$, which are generated by $1 \otimes 1 \otimes j_U(A)$ together with v_1, w_1 and with v_2, w_2, respectively. Let

$$l_1, l_2 \colon \mathcal{T}_{\mathscr{S}}(A, \alpha) \to \mathcal{T}\mathcal{T}_{\mathscr{S}}(A, \alpha)$$

be the resulting embeddings. These are the restrictions of the natural maps

$$\bar{l}_1, \bar{l}_2 \colon \mathcal{T} \widehat{\otimes} U_{\mathscr{S}}(A, \alpha) \to \mathcal{T} \widehat{\otimes} \mathcal{T} \widehat{\otimes} U_{\mathscr{S}}(A, \alpha)$$

that send $x \otimes y$ to $x \otimes 1 \otimes y$ and $1 \otimes x \otimes y$, respectively. Using $1 - v_1 w_1 = E_{00} \otimes 1 \otimes 1$, one checks that $\mathcal{T}\mathcal{T}_{\mathscr{S}}(A, \alpha)$ contains $\mathcal{K}_{\mathscr{S}} \widehat{\otimes} \mathcal{T}_{\mathscr{S}}(A, \alpha)$ as an ideal. The intersection of this ideal with $l_1\big(\mathcal{T}_{\mathscr{S}}(A, \alpha)\big)$ is equal to $\mathcal{K}_{\mathscr{S}} \widehat{\otimes} \mathcal{K}_{\mathscr{S}}(A)$. Thus

$$\widehat{\mathcal{T}}_{\mathscr{S}} := \mathcal{K}_{\mathscr{S}} \widehat{\otimes} \mathcal{T}_{\mathscr{S}}(A, \alpha) + l_1\big(\mathcal{T}_{\mathscr{S}}(A, \alpha)\big) \subseteq \mathcal{T}\mathcal{T}_{\mathscr{S}}(A, \alpha)$$

fits into an extension of local Banach algebras

$$\mathcal{K}_{\mathscr{S}} \widehat{\otimes} \mathcal{T}_{\mathscr{S}}(A, \alpha) \rightarrowtail \widehat{\mathcal{T}}_{\mathscr{S}} \twoheadrightarrow U_{\mathscr{S}}(A, \alpha).$$

The map $\mathcal{K}_{\mathscr{S}}(j_T) \colon \mathcal{K}_{\mathscr{S}}(A) \to \mathcal{K}_{\mathscr{S}} \widehat{\otimes} \mathcal{T}_{\mathscr{S}}(A, \alpha)$ is the restriction of l_1, that is, we have a morphism of extensions

$$
\begin{array}{ccccc}
\mathcal{K}_{\mathscr{S}}(A) & \rightarrowtail & \mathcal{T}_{\mathscr{S}}(A, \alpha) & \twoheadrightarrow & U_{\mathscr{S}}(A, \alpha) \\
\downarrow{\scriptstyle \mathcal{K}_{\mathscr{S}}(j_T)} & & \downarrow{\scriptstyle l_1} & & \| \\
\mathcal{K}_{\mathscr{S}} \widehat{\otimes} \mathcal{T}_{\mathscr{S}}(A, \alpha) & \rightarrowtail & \widehat{\mathcal{T}}_{\mathscr{S}}(A, \alpha) & \twoheadrightarrow & U_{\mathscr{S}}(A, \alpha).
\end{array}
$$

By Proposition 3.3, the map $(j_T)_* \circ (f_\pm)_*$ on $K_*(\mathcal{T}_{\mathscr{S}}(A,\alpha))$ is induced by the quasi-homomorphism

$$(l_1 \circ f_+, l_1 \circ f_-) \colon \mathcal{T}_{\mathscr{S}}(A,\alpha) \rightrightarrows \widehat{\mathcal{T}}_{\mathscr{S}}(A,\alpha) \triangleright \mathcal{K}_{\mathscr{S}} \,\widehat{\otimes}\, \mathcal{T}_{\mathscr{S}}(A,\alpha).$$

We must show that this yields the identity map on K-theory. We use once again the same self-adjoint unitary operator U_0 as in the proof of Bott periodicity (see Figure 4.1 on page 68). Unlike $U_1 \otimes 1_A$, the operator $U_0 \otimes 1_A$ belongs to $\widehat{\mathcal{T}}_{\mathscr{S}}(A,\alpha)$ (this is a homogeneity property). We have already constructed a ∗-homomorphism $\varphi \colon \mathcal{T}_{\mathscr{S}} \to C^\infty([0,1], \mathcal{T}_{\mathscr{S}} \,\widehat{\otimes}\, \mathcal{T}_{\mathscr{S}})$ such that $\varphi_0(\mathrm{S}) = \mathrm{S} \otimes 1$ and $\varphi_1(\mathrm{S}) = U_0 \circ (\mathrm{S} \otimes 1)$. Moreover, $\varphi_t(\mathrm{S})$ for $t \in [0,1]$ is a linear combination of $\mathrm{S} \otimes 1$ and $U_0 \circ (\mathrm{S} \otimes 1)$. Hence

$$\varphi \,\widehat{\otimes}\, \mathrm{id}_{U_{\mathscr{S}}(A,\alpha)} \colon \mathcal{T}_{\mathscr{S}} \,\widehat{\otimes}\, U_{\mathscr{S}}(A,\alpha) \to C^\infty([0,1], \mathcal{T}_{\mathscr{S}} \,\widehat{\otimes}\, \mathcal{T}_{\mathscr{S}} \,\widehat{\otimes}\, U_{\mathscr{S}}(A,\alpha))$$

restricts to a bounded homomorphism

$$\varphi \colon \mathcal{T}_{\mathscr{S}}(A,\alpha) \to C^\infty([0,1], \widehat{\mathcal{T}}_{\mathscr{S}}(A,\alpha)).$$

One checks that $\varphi_t - \varphi_s$ maps into $\mathcal{K}_{\mathscr{S}} \,\widehat{\otimes}\, \mathcal{T}_{\mathscr{S}}(A,\alpha)$ for all $s,t \in [0,1]$. Therefore, we get a quasi-homomorphism

$$(\varphi, l_1 \circ f_-) \colon \mathcal{T}_{\mathscr{S}}(A,\alpha) \rightrightarrows C^\infty([0,1], \widehat{\mathcal{T}}_{\mathscr{S}}(A,\alpha)) \triangleright C^\infty([0,1], \mathcal{K}_{\mathscr{S}} \,\widehat{\otimes}\, \mathcal{T}_{\mathscr{S}}(A,\alpha)).$$

By homotopy invariance, the quasi-homomorphisms $(\varphi_0, l_1 \circ f_-)$ and $(\varphi_1, l_1 \circ f_-)$ induce the same map on K-theory. Since $\varphi_0 = l_1 \circ f_+$, we may replace the quasi-homomorphism $(l_1 \circ f_+, l_1 \circ f_-)$ by $(\varphi_1, l_1 \circ f_-)$.

Finally, we observe that φ_1 restricted to $\mathcal{T}_{\mathscr{S}}(A,\alpha)$ is a direct sum of $l_1 \circ f_-$ and $E_{00} \,\widehat{\otimes}\, l_2$; this follows from Figure 4.1 as in the proof of Bott periodicity. Now Proposition 3.3 yields that $(\varphi_1, l_1 \circ f_-)_*$ is equal to the map $K_*(\mathcal{T}_{\mathscr{S}}(A,\alpha)) \to K_*(\mathcal{K}_{\mathscr{S}} \,\widehat{\otimes}\, \mathcal{T}_{\mathscr{S}}(A,\alpha))$ that is induced by the stabilisation homomorphism $E_{00} \,\widehat{\otimes}\, l_2$. Thus $(j_T)_* \circ (f_\pm)_*$ is the identity map on $K_*(\mathcal{T}_{\mathscr{S}}(A,\alpha))$ as desired. $\square$

Example 5.12. Let $\mathbb{Z}$ act on $C(\mathbb{S}^1)$ by the rotation ϱ_ϑ with angle $2\pi\vartheta$ for $\vartheta \in [0,1]$. If $\vartheta = 0$, then the action is trivial, so that $U_{C^*}(C(\mathbb{S}^1), \varrho_0) \cong C(\mathbb{T}^2)$. For nontrivial ϑ, the crossed product is called a *noncommutative torus* or a *rotation algebra* and denoted A_ϑ. Since $C(\mathbb{S}^1)$ is the universal C^*-algebra generated by a single unitary, A_ϑ is the universal C^*-algebra generated by two unitaries U, V that satisfy the commutation relation $UV = \exp(2\pi i \vartheta)VU$.

Since all rotations are homotopic, the map α_* on $K_*(C(\mathbb{S}^1))$ is the identity map. Hence we get $K_0(A_\vartheta) \cong K^0(\mathbb{T}^2) \cong \mathbb{Z}^2$ and $K_1(A_\vartheta) \cong K^1(\mathbb{T}^2) \cong \mathbb{Z}^2$ by Example 4.12 and Theorem 5.9. The class of the unit element is one of the generators of $K_0(A_\vartheta)$, and the classes of U and V are generators of $K_1(A_\vartheta)$. To prove that V is a generator, one has to check that the index map for the crossed Toeplitz extension maps $[V] \mapsto [1_{C(\mathbb{S}^1)}]$. The other generator of $K_0(A_\vartheta)$ is harder to describe explicitly; this is done in [40, §VI.2].

5.2.1 Some consequences of the Pimsner–Voiculescu Theorem

The following corollary makes precise the assertion that $\mathrm{K}_*(U_{\mathscr{S}}(A, \alpha))$ only involves the induced action of the automorphism α on $\mathrm{K}_*(A)$:

Corollary 5.13. *Let A_1 and A_2 be local Banach algebras equipped with automorphisms α_1 and α_2 that generate uniformly bounded representations of $\mathbb{Z}$. Let $f\colon A_1 \to A_2$ be a bounded homomorphism that intertwines α_1 and α_2. If f induces an isomorphism on K-theory, $\mathrm{K}_*(A_1) \cong \mathrm{K}_*(A_2)$, then so does the induced homomorphism $U_{\mathscr{S}}(f)\colon U_{\mathscr{S}}(A_1, \alpha_1) \to U_{\mathscr{S}}(A_2, \alpha_2)$.*
*   Analogous statements hold for U_{ℓ^1}, and for U_{C^*} if A_1, A_2 are C^*-algebras and α_1, α_2 are $*$-automorphisms.*

Proof. The map f induces a morphism between the crossed Toeplitz extensions (5.10) for (A_1, α_1) and (A_2, α_2). This induces a natural transformation between the Pimsner–Voiculescu exact sequences. The maps $\mathrm{K}_*(A_1) \to \mathrm{K}_*(A_2)$ are isomorphisms by assumption. The induced map $\mathrm{K}_*(U_{\mathscr{S}}(A_1, \alpha_1)) \to \mathrm{K}_*(U_{\mathscr{S}}(A_2, \alpha_2))$ is an isomorphism as well by the Five Lemma. $\qquad\square$

As an application, we consider *deformations* of automorphisms. A family of automorphisms $(\alpha_t)_{t \in [0,1]}$ is called *continuous* if $\alpha f(t) := \alpha_t f(t)$ defines an automorphism of $C([0,1], A)$. We say that the continuous family $(\alpha_t)_{t \in [0,1]}$ generates a *uniformly bounded representation of $\mathbb{Z}$* if α does so. For instance, this holds if A is a Banach algebra and α_t is a continuous family of isometric automorphisms.

Corollary 5.14. *Let A be a local Banach algebra and let $(\alpha_t)_{t \in [0,1]}$ be a continuous family of automorphisms of A that generates a uniformly bounded representation of $\mathbb{Z}$. Then there is a natural isomorphism*

$$\mathrm{K}_*\big(U_{\mathscr{S}}(A, \alpha_0)\big) \cong \mathrm{K}_*\big(U_{\mathscr{S}}(A, \alpha_1)\big).$$

Analogous statements hold for U_{ℓ^1} and U_{C^*} (if defined).

Proof. Equip $C([0,1], A)$ with the automorphism α. Since K-theory is homotopy invariant, the evaluation homomorphisms $\mathrm{ev}_t\colon C([0,1], A) \to A$ for $t = 0, 1$ induce isomorphisms on K-theory. It follows from Corollary 5.13 that the induced maps $U_{\mathscr{S}}(C([0,1], A), \alpha) \to U_{\mathscr{S}}(A, \alpha_t)$ for $t = 0, 1$ are isomorphisms as well. Combining them, we get an isomorphism $\mathrm{K}_*\big(U_{\mathscr{S}}(A, \alpha_0)\big) \cong \mathrm{K}_*\big(U_{\mathscr{S}}(A, \alpha_1)\big)$. $\qquad\square$

This explains why the K-theory of the rotation algebras A_ϑ computed in Example 5.12 does not depend on ϑ.

5.3 A glimpse of the Baum–Connes conjecture

The structural properties of the K-theory of crossed products discussed in §5.2.1 are important because they generalise to more general crossed products. First, we consider crossed products for actions of $\mathbb{Z}^n$ for $n \geq 1$. This situation can be reduced to the case $n = 1$ because we can write a crossed product by $\mathbb{Z}^n$ by taking n crossed products by $\mathbb{Z}$.

Exercise 5.15. Show that the assertions in Corollary 5.13 and 5.14 extend to crossed products by $\mathbb{Z}^n$.

Hence noncommutative $2n$-tori have the same K-theory as $\mathbb{T}^{2n}$ for any $n \geq 1$ (compare Example 5.12). Together with Example 4.12, this yields the K-theory of all noncommutative tori.

The Baum–Connes conjecture deals with the K-theory of the reduced group C^*-algebra $C^*_{\mathrm{red}}(G)$ and reduced crossed products $C^*_{\mathrm{red}}(G, A)$, where G is a locally compact group acting strongly continuously on a C^*-algebra A; we briefly call A together with this action a G-C^*-algebra; we usually omit α from our notation.

We warn the reader that the Baum–Connes conjecture is no longer conjectured to hold for all reduced crossed products because there are known counterexamples. Therefore, it seems better to speak of the *Baum–Connes question* or the *Baum–Connes property*. Our treatment of this question is quite different from the traditional one, which can be found in [122]. It is more closely related to the approach of [87], but more elementary.

First we briefly recall the definitions of full and reduced crossed products (see [98] for more details); this generalises the construction in §5.1 for $G = \mathbb{Z}$. Let G be a locally compact group and let A be a G-C^*-algebra. We define a convolution product and an involution on $L^1(G, A)$ by

$$f_1 * f_2(g) := \int_G f_1(h)\alpha_h\big(f_2(h^{-1}g)\big)\,\mathrm{d}h, \qquad f^*(g) := \overline{\alpha_g\big(f(g^{-1})\big)}\Delta(g^{-1}),$$

where Δ denotes the modular function of G, which is a certain group homomorphism $G \to \mathbb{R}_{>0}$ that measures the deviation of a left-invariant Haar measure on G from being right-invariant as well. We have $\Delta = 1$ if the group G is compact or discrete.

The *full crossed product C^*-algebra* $C^*(G, A)$ is defined as the C^*-completion of $L^1(G, A)$ with respect to the largest possible C^*-norm. That is, we consider the family of all C^*-semi-norms on $L^1(G, A)$, take its supremum, which turns out to be a C^*-norm, and complete. By construction, any $*$-homomorphism $L^1(G, A) \to B$ into a C^*-algebra B extends to a $*$-homomorphism $C^*(G, A) \to B$. If $A = \mathbb{C}$ with trivial action of G, then we get the *full group C^*-algebra* $C^*(G) := C^*(G, \mathbb{C})$.

The *reduced crossed product* $C^*_{\mathrm{red}}(G, A)$ is the completion of $L^1(G, A)$ with respect to another C^*-norm, which we get from a particularly obvious $*$-representation of $L^1(G, A)$. Explicitly, let $\pi\colon A \to \mathcal{L}(\mathcal{H})$ be a faithful $*$-representation of A. The same formula that defines the convolution in $L^1(G, A)$ defines a bilinear map $L^1(G, A) \times L^2(G, \mathcal{H}) \to L^2(G, \mathcal{H})$. The computations that show that $L^1(G, A)$ is a $*$-algebra also show that this is a $*$-representation. We get $C^*_{\mathrm{red}}(G, A)$ by taking the norm completion of $L^1(G, A)$ in this $*$-representation. This does not depend on the choice of π. In particular, $C^*_{\mathrm{red}}(G)$ is the closure of $L^1(G)$ in $\mathcal{L}\big(L^2(G)\big)$, where $L^1(G)$ acts on $L^2(G)$ by convolution on the left.

If the group G is amenable, then the full and reduced crossed products coincide [98]. Since Abelian groups are amenable, the distinction between $C^*(G, A)$ and $C^*_{\mathrm{red}}(G, A)$ does not arise in the Pimsner–Voiculescu exact sequence.

In our more general situation, we no longer have an analogue of the Toeplitz C^*-algebra. Therefore, no confusion can arise if we write $C^*_{(\mathrm{red})}$ instead of $U_{C^*_{(\mathrm{red})}}$.

If A is merely a local Banach algebra with a uniformly bounded, continuous action of G, then we may still define the L^1-*crossed product* $L^1(G, A)$ in the same way as in the C^*-algebra case treated above. But there is no good general analogue of the smooth crossed product. This means that if A is a C^*-algebra, then $L^1(G, A)$, $C^*_{\mathrm{red}}(G, A)$, and $C^*(G, A)$ may all have different K-theories.

The *Baum–Connes conjecture* and the *Bost conjecture* deal with the K-theory of $C^*_{\mathrm{red}}(G, A)$ and $L^1(G, A)$, respectively. Since they predict the same answer in both cases, we may hope for $C^*_{\mathrm{red}}(G, A)$ and $L^1(G, A)$ to have the same K-theory. In contrast, there are many examples where $C^*_{\mathrm{red}}(G, A)$ and $C^*(G, A)$ have different K-theory. At the moment, we cannot compute $\mathrm{K}_*\big(C^*(G, A)\big)$ unless it agrees with $\mathrm{K}_*\big(C^*_{\mathrm{red}}(G, A)\big)$. Since the treatment of $L^1(G, A)$ and $C^*_{\mathrm{red}}(G, A)$ is very similar, we will only write down the details in the latter case.

The results of §5.2.1 lead to the following question: *Does* $\mathrm{K}_*(A) = 0$ *imply that* $\mathrm{K}_*\big(C^*_{\mathrm{red}}(G, A)\big) = 0$? We have seen that this question has a positive answer if G is $\mathbb{Z}^n$ for some $n \in \mathbb{N}_{\geq 1}$. Similarly, one can show that the answer is positive if G is $\mathbb{R}^n$ for some $n \in \mathbb{N}_{\geq 1}$. For $n = 1$, this is equivalent to Connes' Thom Isomorphism Theorem 10.12.

There are, however, counterexamples to the above question where $G = \mathbb{Z}/2$ is the 2-element group [99]. The reason is that there exists a space X and two homotopic actions α_0, α_1 of $\mathbb{Z}/2$ on X for which $\mathrm{K}^*_{\mathbb{Z}/2}(X, \alpha_t)$ are different for $t = 0, 1$. Reversing the argument in the proof of Corollary 5.14, this provides the desired counterexample. Less complicated counterexamples can be constructed where A is a UHF C^*-algebra.

The K-theory of crossed products by compact groups is hard to compute in the sense that there are very few *general* results that provide a *complete* computation; instead, general theorems like the Atiyah–Segal Completion Theorem [99] only provide partial answers. At the same time, it is often possible to compute such K-theory groups *by hand*. In contrast, such direct computations are hard for crossed products by groups like $\mathbb{Z}^n$, but here the general theory helps us out.

Our failure for compact groups forces us to amend our question:

Does vanishing of $\mathrm{K}_*\big(C^*_{\mathrm{red}}(H, A)\big)$ *for all compact subgroups* $H \subseteq G$ *imply* $\mathrm{K}_*\big(C^*_{\mathrm{red}}(G, A)\big) = 0$?

It is shown in [87] that this question is equivalent to the Baum–Connes question *with arbitrary coefficients*. That is, the Baum–Connes conjecture correctly predicts $\mathrm{K}_*\big(C^*_{\mathrm{red}}(G, A)\big)$ for all G-C^*-algebras A if and only if the above question has a positive answer. The conceptual framework in which this statement should be understood is that of localisation of triangulated categories (see Chapter 13).

For the time being, we avoid mentioning triangulated categories and follow instead
a more concrete and elementary approach (which is inspired by constructions for
general triangulated categories in [26]).

It is crucial for our approach to try to compute $\mathrm{K}_*\big(C^*_{\mathrm{red}}(G, A)\big)$ for all A,
not just $\mathrm{K}_*(C^*_{\mathrm{red}}G)$. This allows us to "decompose" A into simpler building blocks
(we even decompose $\mathbb{C}$, which is not particularly simple from our point of view).
These simple building blocks fall into two subcategories $\mathfrak{CI}$ and $\mathfrak{N}$.

Here $\mathfrak{N}$ consists of those G-C^*-algebras A with $\mathrm{K}_*\big(C^*_{\mathrm{red}}(H, A)\big) = 0$ for all
compact subgroups $H \subseteq G$. (It is better to replace $\mathfrak{N}$ by the class $\mathfrak{CC}$ used in [87];
we introduce $\mathfrak{CC}$ later in Chapter 13 because its definition requires bivariant Kasparov theory.) Our question is whether $\mathrm{K}_*\big(C^*_{\mathrm{red}}(G, A)\big) = 0$ for all $A \in \mathfrak{N}$.

Let A, B be G-C^*-algebras and let $f\colon A \to B$ be a $*$-homomorphism. We
say that f *vanishes on equivariant K-theory for compact subgroups* if the induced
map

$$f_*\colon \mathrm{K}_*\big(C^*_{\mathrm{red}}(H, A)\big) \to \mathrm{K}_*\big(C^*_{\mathrm{red}}(H, B)\big)$$

vanishes for all compact subgroups $H \subseteq G$. This notion is inspired by [26]. We may
use it to get objects of $\mathfrak{N}$; if $(A_n, \varphi_n)_{n \in \mathbb{N}}$ is an inductive system of G-C^*-algebras,
where the maps φ_n vanish on equivariant K-theory for compact subgroups, then
$\varinjlim A_n$ belongs to $\mathfrak{N}$ because reduced crossed products and K-theory commute
with inductive limits.

Definition 5.16. If $H \subseteq G$ is a compact subgroup and A is an H-C^*-algebra, then
we let $\mathrm{Ind}_H^G(A)$ be the subalgebra of $C_0(G, A)$ of all functions that are invariant
under the action of H defined by $(h \cdot f)(g) := h \cdot \big(f(gh)\big)$; the group G acts on
$\mathrm{Ind}_H^G(A)$ by left translations, $g_1 \cdot f(g_2) := f(g_1^{-1}g_2)$.

A G-C^*-algebras is called *compactly induced* if it is of this form; let $\mathfrak{CI}$ be
the class of all direct sums of compactly induced G-C^*-algebras.

More generally, one can define $\mathrm{Ind}_H^G A$ if $H \subseteq G$ is closed, but the definition
has to be modified slightly. Compactly induced coefficient algebras are particularly
nice because of the following theorem:

Theorem 5.17 (Green's Imprimitivity Theorem). *If $H \subseteq G$ is a closed subgroup,
then $C^*_{\mathrm{red}}(G, \mathrm{Ind}_H^G A)$ and $C^*_{\mathrm{red}}(H, A)$ are Morita–Rieffel equivalent, and*

$$\mathrm{K}_*\big(C^*_{\mathrm{red}}(G, \mathrm{Ind}_H^G A)\big) \cong \mathrm{K}_*\big(C^*_{\mathrm{red}}(H, A)\big) \cong \mathrm{K}_*^H(A).$$

The last isomorphism is the Green–Julg Theorem (see [10, Theorem 11.7.1]).
Actually, Green's original formulation of the imprimitivity theorem deals with
full crossed products. A proof for reduced crossed products can be found in [74,
Theorem 3.6]. Therefore, the computation of $\mathrm{K}_*\big(C^*_{\mathrm{red}}(G, A)\big)$ for $A \in \mathfrak{CI}$ reduces
to the computation of H-equivariant K-theory for compact subgroups $H \subseteq G$. As
we have observed above, we are resigned to computing such groups by hand.

The following theorem decomposes an arbitrary G-C^*-algebra into building
blocks in $\mathfrak{CI}$ and $\mathfrak{N}$. We will prove it in §5.3.2.

Theorem 5.18. *Let G be a locally compact group and let A be a separable G-C^*-algebra. Then there exists a G-C^*-algebra B together with an increasing filtration by ideals $(\mathfrak{F}_n B)_{n \in \mathbb{N}}$ such that $\bigcup \mathfrak{F}_n B$ is dense in B and such that*

(1) $\mathfrak{F}_0 B \cong A \, \widehat{\otimes}_{C^*} \, \mathcal{K}(\mathcal{H})$ *for a certain G-Hilbert space $\mathcal{H}$;*

(2) $\mathfrak{F}_{n+1} B / \mathfrak{F}_n B$ *belongs to $\mathfrak{CI}$ for all $n \in \mathbb{N}$;*

(3) *the inclusion maps $\mathfrak{F}_n B \to \mathfrak{F}_{n+1} B$ vanish on equivariant K-theory for compact subgroups for all $n \in \mathbb{N}$;*

(4) *the extensions $\mathfrak{F}_n B \rightarrowtail B \to B / \mathfrak{F}_n B$ have G-equivariant completely positive contractive sections for all $n \in \mathbb{N}$.*

Before we sketch the proof of this result, we explain how it reduces the computation of $\mathrm{K}_*\big(C^*_{\mathrm{red}}(G, A)\big)$ for general A to the special cases of coefficients in $\mathfrak{CI}$ and $\mathfrak{N}$. As we have observed above, it follows from (3) that $B \in \mathfrak{N}$. By assumption, the subquotients $\mathfrak{F}_{n+1} B / \mathfrak{F}_n B$ for $n \in \mathbb{N}$ belong to $\mathfrak{CI}$. Therefore, we consider $\mathrm{K}\big(C^*_{\mathrm{red}}(G, \_)\big)$ for these coefficient algebras as *input data* for our computation.

It follows from (4) that the extensions $\mathfrak{F}_n B \rightarrowtail \mathfrak{F}_{n+k} B \to \mathfrak{F}_{n+k} B / \mathfrak{F}_n B$ for $n, k \in \mathbb{N}$, give rise to exact sequences of C^*-algebras

$$C^*_{\mathrm{red}}(G, \mathfrak{F}_{n+k} B / \mathfrak{F}_n B) \rightarrowtail C^*_{\mathrm{red}}(G, \mathfrak{F}_{n+k+1} B / \mathfrak{F}_n B) \twoheadrightarrow C^*_{\mathrm{red}}(G, \mathfrak{F}_{n+k+1} B / \mathfrak{F}_{n+k} B)$$

for all $k \geq 1$, $n \in \mathbb{N}$. Using the resulting K-theory long exact sequences, we may try to compute the K-theory groups of $C^*_{\mathrm{red}}(G, \mathfrak{F}_{n+k} B / \mathfrak{F}_n B)$ for $k \geq 2$, $n \in \mathbb{N}$, by induction on k, starting with the case $k = 1$, which is part of our input data. Letting $k \to \infty$, we get the K-theory of $C^*(G, B / \mathfrak{F}_0 B)$. As in §4.3.1, we may organise this computation in terms of a spectral sequence (see also [112]).

Similarly, using the extension of C^*-algebras

$$C^*_{\mathrm{red}}(G, \mathfrak{F}_0 B) \rightarrowtail C^*_{\mathrm{red}}(G, B) \twoheadrightarrow C^*_{\mathrm{red}}(G, B / \mathfrak{F}_0 B),$$

we get a long exact sequence of the form

$$\cdots \to \mathrm{K}_{*+1}\big(C^*_{\mathrm{red}}(G, B)\big) \to \mathrm{K}_{*+1}\big(C^*_{\mathrm{red}}(G, B / \mathfrak{F}_0 B)\big)$$
$$\to \mathrm{K}_*\big(C^*_{\mathrm{red}}(G, A)\big) \to \mathrm{K}_*\big(C^*_{\mathrm{red}}(G, B)\big) \to \cdots .$$

Definition 5.19. The connecting map

$$\mathrm{K}_{*+1}\big(C^*_{\mathrm{red}}(G, B / \mathfrak{F}_0 B)\big) \to \mathrm{K}_*\big(C^*_{\mathrm{red}}(G, A)\big)$$

in the above long exact sequence is called the *assembly map* for (G, A). We say that A has the *Baum–Connes property* if the assembly map for (G, A) is invertible or, equivalently, $\mathrm{K}_*\big(C^*_{\mathrm{red}}(G, B)\big) = 0$.

It follows from the results of [87] that this map is equivalent to the usual Baum–Connes assembly map (see [122]). Thus the Baum–Connes property above is equivalent to the usual formulation as well. There are counterexamples where the Baum–Connes property fails. We do not yet understand these counterexamples well enough to compute $K_*\big(C^*_{\mathrm{red}}(G, B)\big)$ in such cases.

Our proof of Theorem 5.18 will be constructive, that is, we will write down a candidate for the G-C^*-algebra B and its filtration $\mathfrak{F}_n B$. But this explicit candidate is quite huge and therefore not useful for actual computations; this is not surprising because the theorem applies to all groups. As a result, the above description of the Baum–Connes assembly map is not really practical. For the time being, we can only say that the computation of $K_*\big(C^*_{\mathrm{red}}(G, A)\big)$ for general coefficient algebras can, *in principle*, be done in three steps:

- compute $K_*\big(C^*_{\mathrm{red}}(H, A)\big)$ for compact subgroups $H \subseteq G$;

- compute $K_*\big(C^*_{\mathrm{red}}(G, B)\big)$ for $B \in \mathfrak{N}$; usually we show that it vanishes;

- chase through the long exact sequences as above.

The third step is evidently *topological*. If A is commutative, say $A = \mathbb{C}$, then the first step may also be considered as purely topological. In contrast, the second step does not appear to be tractable by topological considerations. The first and third step do not depend on the choice of the crossed product (there is an analogue of the Green–Julg Theorem for L^1-crossed products as well). The second part is the only one where the choice of crossed product becomes relevant.

5.3.1   Toeplitz cones

We prepare for the proof of Theorem 5.18.

Let $f \colon A \to B$ be a $*$-homomorphism between two C^*-algebras (an analogous construction works for bounded homomorphisms between local Banach algebras). Then we define the *Toeplitz cone C^*-algebra* $\mathcal{T}_{C^*}(f)$ of f by the pull-back diagram

$$
\begin{array}{ccccc}
B \,\widehat{\otimes}_{C^*}\, \mathcal{K}(\ell^2\mathbb{N}) & \rightarrowtail & B \,\widehat{\otimes}_{C^*}\, \mathcal{T}^0_{C^*} & \longrightarrow & B \,\widehat{\otimes}_{C^*}\, C_0(\mathbb{R}) \\
\Big\| & & \bar{f} \Big\uparrow & & f \widehat{\otimes}\,\mathrm{id} \Big\uparrow \\
B \,\widehat{\otimes}_{C^*}\, \mathcal{K}(\ell^2\mathbb{N}) & \rightarrowtail & \mathcal{T}_{C^*}(f) & \longrightarrow\!\!\!\!\rightarrow & A \,\widehat{\otimes}_{C^*}\, C_0(\mathbb{R}).
\end{array}
$$

Using Bott periodicity, we get a six-term exact sequence

$$
\begin{array}{ccccc}
K_0(B) & \longrightarrow & K_0\big(\mathcal{T}_{C^*}(f)\big) & \longrightarrow & K_1(A) \\
\Big\uparrow & & & & \Big\downarrow \\
K_0(A) & \longleftarrow & K_1\big(\mathcal{T}_{C^*}(f)\big) & \longleftarrow & K_1(B),
\end{array}
\tag{5.20}
$$

which is called *dual Puppe sequence*. Using the naturality of the index map and our description of the Bott periodicity map, we may identify the vertical maps in (5.20) with $\mathrm{K}_*(f)$. The exact sequence (5.20) is similar to the Puppe exact sequence (2.34). The Toeplitz cone $\mathcal{T}_{C^*}(f)$ in (5.20) and the mapping cone in (2.34) have the same K-theory up to a dimension shift.

5.3.2 Proof of the decomposition theorem

Lemma 5.21. *Let G be a locally compact group and let A be a G-C^*-algebra. Then there is an extension of G-C^*-algebras of the form*

$$A \,\widehat{\otimes}_{C^*}\, \mathcal{K}(\mathcal{H}) \stackrel{\iota}{\rightarrowtail} E_A \stackrel{\pi}{\twoheadrightarrow} P_A$$

for a certain separable Hilbert space $\mathcal{H}$ equipped with a unitary representation of G and some $P_A \in \mathfrak{CI}$ such that the embedding ι vanishes on equivariant K-theory for compact subgroups. Moreover, this extension has a G-equivariant, completely positive contractive section.

Proof. We only give the proof in the case where G has a compact open subgroup; for instance, this covers discrete groups. For general groups, the construction of such an extension requires a certain amount of Kasparov theory.

By hypothesis, there is a discrete proper G-space X such that any compact subgroup of G fixes a point in X. Consider the embedding $C_0(X) \to \mathcal{K}(\ell^2 X)$ by pointwise multiplication operators. If $H \subseteq G$ is a (compact) subgroup, then $C^*_{\mathrm{red}}(H, \mathcal{K}(\ell^2 X)) \cong C^*_{\mathrm{red}}(H) \,\widehat{\otimes}_{C^*}\, \mathcal{K}(\ell^2 X)$ because the action of G on $\mathcal{K}(\ell^2 X)$ is inner. Hence the H-equivariant K-theory of $\mathcal{K}(\ell^2 X)$ is the same as for a one-point space with trivial action of H. Since any compact subgroup H fixes a point in X, we conclude that the embedding $C_0(X) \to \mathcal{K}(\ell^2 X)$ induces a *surjective* map on H-equivariant K-theory for all compact subgroups $H \subseteq G$. Now we form the Toeplitz cone over this map as in §5.3.1. This yields an extension

$$\mathcal{K}\big(\ell^2(X \times \mathbb{N})\big) \rightarrowtail E \twoheadrightarrow C_0(\mathbb{R} \times X).$$

It follows from the exact sequence (5.20) that the map $\mathcal{K}(\ell^2 X \times \mathbb{N}) \to E$ vanishes on equivariant K-theory for compact subgroups. The whole argument still goes through unchanged if we tensor everything with A. This finishes the proof. $\qquad\square$

For the next step, it is convenient to replace the Hilbert space $\ell^2(X \times \mathbb{N})$ by $\mathcal{H}_1 := \ell^2(X \times \mathbb{N}) \oplus \mathbb{C}$, where G acts trivially on $\mathbb{C}$. We use the additional G-fixed unit vector to embed $A \to A \widehat{\otimes}_{C^*} \mathcal{K}(\mathcal{H}_1)$. We still get an extension of G-C^*-algebras $A \,\widehat{\otimes}_{C^*}\, \mathcal{K}(\mathcal{H}_1) \rightarrowtail E_A \twoheadrightarrow P_A$ with the same properties as in Lemma 5.21. Now we apply the same construction to E_A instead of A to get an extension

$$E_A \,\widehat{\otimes}_{C^*}\, \mathcal{K}(\mathcal{H}_1) \rightarrowtail E_A^2 \twoheadrightarrow PE_A.$$

Thus E_A^2 comes with a filtration by ideals $E_A \,\widehat{\otimes}_{C^*}\, \mathcal{K}(\mathcal{H}_1)$ and $A \,\widehat{\otimes}_{C^*}\, \mathcal{K}(\mathcal{H}_1 \,\widehat{\otimes}\, \mathcal{H}_1)$, such that the subquotients belong to $\mathfrak{CI}$ and the embeddings of the ideals vanish on equivariant K-theory for compact subgroups.

Iterating this construction, we get a sequence of G-C^*-algebras E_A^n with longer and longer filtrations.

Since we have added the trivial representation to $\mathcal{H}$, we have canonical maps $E_A^n \to E_A^{n+1}$ for all $n \in \mathbb{N}$. We let B be the direct limit of this inductive system; we let $\mathfrak{F}_n B$ be an appropriate stabilisation of E_A^n, so that $\mathfrak{F}_n B$ is an ideal in B. We leave it as an exercise to check that B with this filtration has the required properties. This finishes the proof of Theorem 5.18.

The same argument still works for L^1-crossed products if G has a compact open subgroup.

Chapter 6

Towards bivariant K-theory: how to classify extensions

Many important maps between K-theory groups are constructed as index maps of certain extensions. We have seen one instance of this in our proof of Bott periodicity, where we have constructed the periodicity isomorphism as such an index map. As the notation suggests, more examples arise in index theory. Often it is important to compose index maps with homomorphisms or with other index maps. For such purposes, it is useful to have a (graded) category in which ordinary bounded algebra homomorphisms and extensions give morphisms (of degrees 0 and 1, respectively). In this chapter, we construct such a category, which is denoted ΣHo, and show that it is *triangulated*. This additional structure allows us to treat long exact sequences efficiently. Moreover, many important constructions in topology and homological algebra may be rephrased in the language of triangulated categories and then carry over to ΣHo and related categories.

We mostly follow the construction of bivariant K-theories for locally convex algebras in [36, 37, 39]. But our presentation differs in three aspects. First, we treat bornological algebras instead of locally convex algebras, which is mostly a change in notation. Secondly, we postpone the stabilisation by compact operators, which will only appear in Chapter 7; this simplifies the exposition. Thirdly, we rearrange some proofs to take advantage of the triangulated category structure.

6.1 Some tricks with smooth homotopies

Although we do not treat cyclic homology here, we want to make sure that the category we construct is compatible with periodic cyclic homology. Periodic cyclic homology is not invariant under continuous homotopies: we need homotopies that are sufficiently differentiable. Although Hölder continuity would suffice, we use smooth homotopies here.

Definition 6.1. Let A and B be bornological algebras and let $f_0, f_1 \colon A \to B$ be bounded homomorphisms. A *smooth homotopy* between f_0 and f_1 is a bounded homomorphism $f \colon A \to C^\infty([0,1], B)$ with $\mathrm{ev}_t \circ f = f_t$ for $t = 0, 1$. We call f_0 and f_1 *smoothly homotopic* if such a smooth homotopy exists.

We claim that this is an equivalence relation. Reflexivity and symmetry are evident; transitivity requires some work because we need conditions on the derivatives at the end points in order for the concatenation of two smooth homotopies to be smooth again. Fortunately, there is an easy trick to resolve this difficulty.

Let $B[0,1] \subseteq C^\infty([0,1], B)$ be the closed subalgebra of functions $[0,1] \to B$ whose nth derivatives at 0 and 1 vanish for all $n \geq 1$. Let $\varrho \colon [0,1] \to [0,1]$ be a strictly increasing smooth function with $\varrho(0) = 0$, $\varrho(1) = 1$, and $\varrho \in B[0,1]$. Such functions exist. We get a map

$$\varrho^* \colon C^\infty([0,1], B) \to B[0,1], \qquad f \mapsto f \circ \varrho$$

with $\varrho^* f(t) = f(t)$ for $t = 0, 1$ for all $f \in C^\infty([0,1], B)$. Thus two bounded homomorphisms $f_0, f_1 \colon A \to B$ are smoothly homotopic if and only if there is a bounded homomorphism $f \colon A \to B[0,1]$ such that $f(0) = f_0$ and $f(1) = f_1$.

Definition 6.2. Given $F_0, F_1 \colon A \to B[0,1]$ with $F_0(1) = F_1(0) \colon A \to B$, their *concatenation* is the bounded homomorphism $F_0 \bullet F_1 \colon A \to B[0,1]$ defined by

$$F_0 \bullet F_1(a)(t) := \begin{cases} F_0(a)(2t) & \text{for } t \leq 1/2, \\ F_1(a)(2t-1) & \text{for } t \geq 1/2. \end{cases}$$

The existence of concatenation shows that smooth homotopy is an equivalence relation on the set of bounded algebra homomorphisms $A \to B$. We let $\langle A, B \rangle$ be the associated set of equivalence classes, and we write $\langle f \rangle \in \langle A, B \rangle$ for the equivalence class of $f \colon A \to B$.

Let $B(0,1) \subseteq B[0,1]$ be the closed ideal of functions vanishing at 0 and 1. The closed ideals $B(0,1]$ and $B[0,1)$ are defined similarly.

We have $B[0,1]/B(0,1) \cong B \oplus B$ via evaluation at 0 and 1. Since these constructions are frequently used in the following, we abbreviate

$$SB := B(0,1), \qquad CB := B(0,1].$$

By construction, we have an extension of bornological algebras

$$SB \rightarrowtail CB \twoheadrightarrow B, \tag{6.3}$$

which is called the *cone extension* over B.

We define $S^n B = B(0,1)^n$ and $C^n B = B(0,1]^n$ by iterating the functors $B \mapsto SB, CB$. We identify $S^n B$ with the algebra of smooth functions from the n-dimensional cube $[0,1]^n$ to B that vanish together with all derivatives on the boundary. We have $C_c^\infty(M, B) \cong C_c^\infty(M) \,\hat{\otimes}\, B$ for any smooth manifold (Example 2.10). This yields $S^l C^k B \cong (S^l C^k \mathbb{C}) \,\hat{\otimes}\, B$ for all $k, l \in \mathbb{N}$.

Lemma 6.4. *Concatenation defines a group structure on* $\langle A, SB \rangle$.

The group structures on $\langle A, S^n B \rangle$ *that we get from concatenation in different variables agree and are Abelian if* $n \geq 2$.

We may view $\langle A, S^n B \rangle$ as the nth homotopy group of the "space" of bounded algebra homomorphisms $A \to B$ (this space does not carry a topology). The lemma then becomes a familiar assertion about homotopy groups of spaces.

Proof. It is easy to see that $F_0 \bullet F_1$ respects smooth homotopy and hence descends to a map $\bullet \colon \langle A, SB \rangle \times \langle A, SB \rangle \to \langle A, SB \rangle$. This product is associative by Figure 6.1. The class of $0 \colon A \to SB$ is an identity element for $\bullet$: appropri-

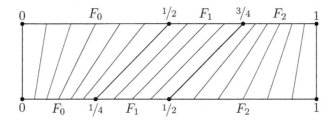

Figure 6.1: Homotopy between $F_0 \bullet (F_1 \bullet F_2)$ and $(F_0 \bullet F_1) \bullet F_2$

ate reparametrisations as in Figure 6.1 yield the necessary smooth homotopies between $F \bullet 0$, F, and $0 \bullet F$. Define f^{-1} for $f \colon A \to SB$ by

$$f^{-1}(a)(t) := f(a)(1-t).$$

We claim that $\langle f \bullet f^{-1} \rangle = \langle 0 \rangle$. First we use a smooth homotopy from $f \bullet f^{-1}$ to $\varrho^* f \bullet \varrho^* (f^{-1})$. Then we connect this to 0 via the smooth homotopy

$$F(a)(s,t) := \begin{cases} f\big(s\varrho(2t)\big) & \text{for } t \leq 1/2, \\ f\big(s\varrho(2-2t)\big) & \text{for } t \geq 1/2. \end{cases}$$

Each suspension generates a group structure on $\langle A, S^n B \rangle$. To show that these n group structures are all equal and Abelian, it suffices to treat the case $n = 2$. The necessary smooth homotopies are illustrated in Figure 6.2. $\square$

In the following, we often write $+$ instead of $\bullet$ and $-f$ instead of f^{-1}.

Definition 6.5. A bornological algebra A is *smoothly contractible* if $\langle \mathrm{id}_A \rangle = \langle 0 \rangle$.

Exercise 6.6. A bornological algebra A is smoothly contractible if and only if the cone extension over A splits by a bounded homomorphism $A \to A(0,1]$.

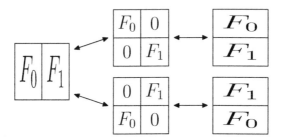

Figure 6.2: Commutativity of the concatenation

6.2   Tensor algebras and classifying maps for extensions

Definition 6.7. Let V be a bornological vector space. Let $V^{\hat{\otimes}n}$ for $n \in \mathbb{N}_{\geq 1}$ be the complete projective bornological tensor product of n copies of V (see §2.1.2). Let

$$TV := \bigoplus_{n=1}^{\infty} V^{\hat{\otimes}n} = V \oplus (V \hat{\otimes} V) \oplus (V \hat{\otimes} V \hat{\otimes} V) \oplus \cdots \qquad (6.8)$$

equipped with the direct sum bornology. That is, a subset of TV is bounded if and only if it is bounded in $\bigoplus_{n=1}^{N} V^{\hat{\otimes}n}$ for some $N \in \mathbb{N}$. We define a multiplication in TV by concatenation of tensors:

$$(v_1 \otimes \cdots \otimes v_n) \cdot (v_{n+1} \otimes \cdots \otimes v_{n+m}) := v_1 \otimes \cdots \otimes v_n \otimes v_{n+1} \otimes \cdots \otimes v_{n+m}.$$

This defines a bounded bilinear map $TV \times TV \to TV$. It is clearly associative, so that TV becomes a bornological algebra; it is called the *tensor algebra* of V. Let $\sigma_V \colon V \to TV$ be the natural bounded linear map that identifies V with the first direct summand in (6.8).

Lemma 6.9. *The map $\sigma_V \colon V \to TV$ is the universal bounded linear map from V into a bornological algebra, that is, any bounded linear map $f \colon V \to B$ from V into a bornological algebra B factors uniquely as $f = \hat{f} \circ \sigma_V$ for a bounded algebra homomorphism $\hat{f} \colon TV \to B$. This universal property determines (TV, σ_V) uniquely up to natural isomorphism.*

Proof. Any bounded algebra homomorphism $\hat{f} \colon TV \to B$ with $\hat{f} \circ \sigma_V = f$ satisfies $\hat{f}(v_1 \otimes \cdots \otimes v_n) := f(v_1) \cdots f(v_n)$. Conversely, this formula defines a bounded algebra homomorphism $TV \to B$ by the universal property of $\hat{\otimes}$. $\qquad \square$

Exercise 6.10. Use its universal property to show that TA is smoothly contractible for any A.

Definition 6.11. Let A be a bornological algebra. Let $\pi_A \colon TA \to A$ be the unique bounded algebra homomorphism lifting $\mathrm{id}_A \colon A \to A$, that is, $\pi_A \circ \sigma_A = \mathrm{id}_A$ and $\pi_A(a_1 \otimes \cdots \otimes a_n) = a_1 \cdots a_n$. Let $JA := \ker \pi_A \subseteq TA$, with the subspace bornology. The resulting extension of bornological algebras $JA \rightarrowtail TA \twoheadrightarrow A$ is called the *tensor algebra extension of A*. It has the natural bounded linear section $\sigma_A \colon A \to TA$.

Example 6.12. The tensor algebra over $\mathbb{C}$ is isomorphic to $t\,\mathbb{C}[t] \subseteq \mathbb{C}[t]$ because $\mathbb{C}^{\otimes n} \cong \mathbb{C}$ for all $n \geq 1$. We have $J\mathbb{C} \cong (1-t)t\,\mathbb{C}[t]$ because $\pi_{\mathbb{C}} \colon T\mathbb{C} \to \mathbb{C}$ is evaluation at 1. Once $\dim V \geq 2$, the tensor algebra TV becomes noncommutative.

Definition 6.13. An extension of bornological algebras is called *semi-split* if it has a bounded linear section.

Definition 6.14. A *morphism-extension* from A to I is a diagram of the form

$$
\begin{array}{c}
A \\
\downarrow f \\
I \rightarrowtail E \twoheadrightarrow B,
\end{array}
$$

where $I \rightarrowtail E \twoheadrightarrow B$ is a semi-split extension.

Any morphism-extension may be completed to a morphism of extensions

$$
\begin{array}{ccccc}
JA & \rightarrowtail & TA & \xrightarrow{\pi_A} & A \\
\downarrow \gamma & & \downarrow \tau & & \downarrow f \\
I & \rightarrowtail & E & \longrightarrow\!\!\!\!\rightarrow & B.
\end{array}
\tag{6.15}
$$

To get $\tau \colon TA \to E$, choose a bounded linear section $s \colon B \to E$ and apply the universal property of TA formulated in Lemma 6.9 to $s \circ f \colon A \to E$. Then γ is the restriction of τ.

Definition 6.16. If γ and τ make (6.15) commute, then we call γ a *classifying map* for the given morphism-extension. If $f = \mathrm{id}_A$, we call it a classifying map for the extension $I \rightarrowtail E \twoheadrightarrow A$.

Lemma 6.17. *The classifying map of a morphism-extension is unique up to smooth homotopy.*

Proof. Composition with σ_A yields a bijection between bounded algebra homomorphisms $\tau \colon TA \to E$ and bounded linear maps $A \to E$ by Lemma 6.9; the homomorphism τ makes the right square in (6.15) commute if and only if $\tau \circ \sigma_A \colon A \to E$ lifts f. Thus the possible choices for τ in (6.15) are in bijection with bounded linear maps $A \to E$ lifting f. We may join two such liftings $l_0, l_1 \colon A \to E$ by the smooth homotopy $l \colon A \to C^{\infty}([0,1], E)$, $l := (1-t)l_0 + tl_1$. This induces a bounded homomorphism $\tau \colon TA \to C^{\infty}([0,1], E)$ which provides a smooth homotopy between τ_0 and τ_1. Its restriction to JA is the desired smooth homotopy between γ_0 and γ_1. $\qquad\square$

This lemma allows us to speak of *the* classifying map of a morphism-extension as long as we only care about its smooth homotopy class.

The tensor algebra extension is functorial, that is, a bounded algebra homomorphism $f\colon A \to B$ induces a morphism of extensions

$$
\begin{array}{ccccc}
JA & \rightarrowtail & TA & \xrightarrow{\pi_A} & A \\
\downarrow{\scriptstyle Jf} & & \downarrow{\scriptstyle Tf} & & \downarrow{\scriptstyle f} \\
JB & \rightarrowtail & TB & \xrightarrow{\pi_B} & B.
\end{array}
$$

In particular, this includes the functoriality of TA and JA.

Lemma 6.18. *The classifying map $JA \to I$ of a morphism-extension*

$$
\begin{array}{ccc}
 & & A \\
 & & \downarrow{\scriptstyle f} \\
I & \rightarrowtail E \twoheadrightarrow & B
\end{array}
$$

is the composite of Jf and the classifying map γ of the extension $I \rightarrowtail E \twoheadrightarrow B$.

Proof. The commuting diagram

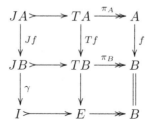

shows that $\gamma \circ Jf$ is a classifying map for the morphism-extension.   □

Lemma 6.19. *If $f_0, f_1\colon A \to B$ are smoothly homotopic, then so are the induced maps $Jf_0, Jf_1\colon JA \to JB$. Hence we get a map $\langle A, B \rangle \to \langle JA, JB \rangle$.*

Proof. Let $f\colon A \to B[0,1]$ be the smooth homotopy between f_0 and f_1. The classifying map for the morphism-extension

$$
\begin{array}{ccc}
 & & A \\
 & & \downarrow{\scriptstyle f} \\
(JB)[0,1] \rightarrowtail (TB)[0,1] \twoheadrightarrow & B[0,1]
\end{array}
$$

provides a smooth homotopy between Jf_0 and Jf_1; notice that the row is a semi-split extension.   □

We define $J^k A$ for $k \in \mathbb{N}_{\geq 1}$ by iterating the functor J. This algebra occurs when we study extensions of length k. Any such extension can be obtained by *splicing k extensions* $I_{j+1} \rightarrowtail E_j \twoheadrightarrow I_j$ for $j = 0, \ldots, k-1$ to a diagram

$$I_k \to E_{k-1} \to E_{k-2} \to \cdots \to E_1 \to E_0 \to I_0. \tag{6.20}$$

By induction on j, a bounded algebra homomorphism $f \colon A \to I_0$ gives rise to classifying maps $J^j A \to I_j$ for $j = 0, \ldots, k$. The final map $J^k A \to I_k$ is called the *classifying map of the length-k-extension* (6.20).

Exercise 6.21. Show that a bounded algebra homomorphism $\gamma \colon J^k A \to I_k$ is a classifying map for (6.20) if and only if it fits into a commuting diagram

$$
\begin{array}{ccccccccc}
J^k A & \longrightarrow & T J^{k-1} A & \longrightarrow & T J^{k-2} A & \longrightarrow & \cdots & \longrightarrow & T J^0 A & \longrightarrow & A \\
\downarrow{\scriptstyle\gamma} & & \downarrow & & \downarrow & & & & \downarrow & & \downarrow{\scriptstyle f} \\
I_k & \longrightarrow & E_{k-1} & \longrightarrow & E_{k-2} & \longrightarrow & \cdots & \longrightarrow & E_0 & \longrightarrow & I_0.
\end{array}
$$

The functor $\_ \mathbin{\widehat{\otimes}} B$ is exact for semi-split extensions, that is, if A and B are bornological algebras, then we have a semi-split bornological algebra extension $(JA) \mathbin{\widehat{\otimes}} B \rightarrowtail (TA) \mathbin{\widehat{\otimes}} B \twoheadrightarrow A \mathbin{\widehat{\otimes}} B$. We let $\kappa_{A,B} \colon J(A \mathbin{\widehat{\otimes}} B) \to (JA) \mathbin{\widehat{\otimes}} B$ be its classifying map.

Definition 6.22. Let $\kappa^k_{A,B} \colon J^k(A \mathbin{\widehat{\otimes}} B) \to (J^k A) \mathbin{\widehat{\otimes}} B$ be the classifying map of the length-k-extension

$$(J^k A) \mathbin{\widehat{\otimes}} B \to (T J^{k-1} A) \mathbin{\widehat{\otimes}} B \to (T J^{k-2} A) \mathbin{\widehat{\otimes}} B \to \cdots \to (TA) \mathbin{\widehat{\otimes}} B \to A \mathbin{\widehat{\otimes}} B.$$

Definition 6.23. The cone extension in (6.3) is semi-split: the map $a \mapsto a \otimes \varrho$ with ϱ as in §6.1 is a bounded linear section. Constructing classifying maps for the cone extension, we can lift a map $f \colon A \to B$ to a map $\Lambda(f) \colon JA \to SB$. Iterating Λ, we get maps $\Lambda^k(f) \colon J^k A \to S^k B$. We abbreviate

$$\lambda^k_A := \Lambda^k(\mathrm{id}_A) \colon J^k A \to S^k A, \qquad \lambda_A := \Lambda(\mathrm{id}_A) \colon JA \to SA.$$

We define morphism-extensions of length k and their classifying maps in the obvious fashion.

The map $\Lambda^k(f)$ is the classifying map of the morphism-extension

$$
\begin{array}{ccccccccc}
& & & & & & & & A \\
& & & & & & & & \downarrow{\scriptstyle f} \\
S^k B & \longrightarrow & CS^{k-1}B & \longrightarrow & CS^{k-2}B & \longrightarrow & \cdots & \longrightarrow & CS^0 B & \longrightarrow & B;
\end{array}
$$

we get its lower row by splicing the cone extensions $S^{j+1}B \rightarrowtail CS^j B \twoheadrightarrow S^j B$ for $j = 0, \ldots, k-1$.

Exercise 6.24. Consider the classifying maps $\gamma\colon JA \to I$ and $\gamma'\colon JI \to I'$ of two morphism-extensions

$$
\begin{array}{ccc}
& A & \\
& \downarrow f & \\
I \rightarrowtail E \twoheadrightarrow B, & & I' \rightarrowtail E' \twoheadrightarrow B'.
\end{array}
$$

$$
\begin{array}{c}
I \\
\downarrow f \\
\end{array}
$$

Pull back the extension $I' \rightarrowtail E' \twoheadrightarrow B'$ along the homomorphism $I \to B'$ to get an extension $I' \rightarrowtail E'' \twoheadrightarrow I$, which is again semi-split. Splice the latter with $I \rightarrowtail E \twoheadrightarrow B$ to get a semi-split extension from B to I' of length 2. Check that the composite map $\gamma' \circ J\gamma\colon J^2A \to I'$ of our classifying maps is the classifying map of the resulting length 2 morphism-extension

$$
\begin{array}{c}
A \\
\downarrow f \\
I' \rightarrowtail E' \longrightarrow E \twoheadrightarrow B.
\end{array}
$$

Sometimes we are interested in morphism-extensions where the extension is not semi-split or, even worse, I is only a generalised ideal in E, so that E/I is not a bornological algebra. Nevertheless, under some technical conditions we can still associate a classifying map to such a morphism-extension:

Definition 6.25. Let A, K, L be bornological algebras and assume that K is a generalised ideal in L. Let $f\colon A \to L$ be a bounded linear map and assume that $\omega_f(x,y) := f(x)f(y) - f(xy)$ defines a bounded bilinear map $A \times A \to K$ (not just $A \times A \to L$). Then we call the diagram

a *singular morphism-extension*.

Lemma 6.26. *Consider a singular morphism-extension as above. Let $E := K \oplus A$ with multiplication*

$$(k_1, a_1) \cdot (k_2, a_2) := (k_1 \cdot k_2 + f(a_1) \cdot k_2 + k_1 \cdot f(a_2) + \omega_f(a_1, a_2), a_1 \cdot a_2).$$

This bilinear map is associative and bounded, and the coordinate embedding and projection $K \rightarrowtail E \twoheadrightarrow A$ provide a semi-split bornological algebra extension, which has a classifying map $\gamma_f\colon JA \to I$ called the classifying map of the singular morphism-extension.

If $f'\colon A \to L$ differs from f by a bounded map $A \to K$, then γ_f and $\gamma_{f'}$ are smoothly homotopic.

Proof. The multiplication in E is defined so that the map

$$E \to L \oplus A, \qquad (k,a) \mapsto (k + f(a), a)$$

is an algebra homomorphism. Since this map is injective, the multiplication is associative. It is bounded by assumption, and the maps $K \rightarrowtail E \twoheadrightarrow A$ form an extension of bornological algebras. Hence we get the desired map $\gamma_f \colon JA \to I$. If $f' = f + \delta$ for a bounded linear map $\delta \colon A \to K$, then the map $E \to E'$, $(k,a) \mapsto (k - \delta(a), a)$ defines a bounded algebra isomorphism, which is compatible with the maps $K \rightarrowtail E, E' \twoheadrightarrow A$. Hence we get two isomorphic extensions, which therefore have smoothly homotopic classifying maps. $\qquad\square$

The proof shows that we get the extension $K \rightarrowtail E \twoheadrightarrow A$ by pulling back $K \rightarrowtail L \twoheadrightarrow L/K$ along $q \circ f \colon A \to L/K$. Hence the classifying map of Lemma 6.26 agrees with the usual one for a non-singular morphism-extension. Furthermore, we see that $\gamma_f \colon JA \to K$ is the restriction of the map $TA \to L$ associated to the bounded linear map f by the universal property of TA.

6.3 The suspension-stable homotopy category

Our goal is to construct a category $\Sigma\mathrm{Ho}$ in which the two suspension functors J and S become equivalences. Thus any bornological algebra should be isomorphic to a suspension of some other object. As in the construction of the stable homotopy category in topology, we achieve this by adjoining *formal* desuspensions.

Thus objects of $\Sigma\mathrm{Ho}$ are pairs (A, n), where A is a bornological algebra and $n \in \mathbb{Z}$. We think of (A, n) as the nth formal suspension of A. Thus we define the *suspension automorphism* $\Sigma \colon \Sigma\mathrm{Ho} \to \Sigma\mathrm{Ho}$ by $\Sigma(A, n) := (A, n + 1)$ on objects.

The set $\Sigma\mathrm{Ho}\big((A, m), (B, n)\big)$ of morphisms $(A, m) \to (B, n)$ is defined by

$$\Sigma\mathrm{Ho}\big((A, m), (B, n)\big) := \varinjlim_{k \to \infty} \langle J^{m+k}A, S^{n+k}B \rangle, \qquad (6.27)$$

where we only allow $k \in \mathbb{N}$ with $k + m \geq 0$ and $k + n \geq 0$ and where we form the inductive limit with respect to the operator

$$\Lambda \colon \langle J^{m+k}A, S^{n+k}B \rangle \to \langle J(J^{m+k}A), S(S^{n+k}B) \rangle = \langle J^{m+k+1}A, S^{n+k+1}B \rangle$$

constructed in Definition 6.23. The operator Λ is clearly "natural". Formalising this, we obtain the following relations:

$$\langle \Lambda(f) \rangle = \langle S(f) \circ \Lambda(\mathrm{id}_{J^{m+k}A}) \rangle = \langle \Lambda(\mathrm{id}_{S^{n+k}B}) \circ J(f) \rangle. \qquad (6.28)$$

Lemma 6.4 shows that $\langle J^{m+k}A, S^{n+k}B \rangle$ is a group for $n + k \geq 1$ and an Abelian group for $n + k \geq 2$. It follows from (6.28) that Λ is a group homomorphism whenever $n + k \geq 1$. Thus the limit $\Sigma\mathrm{Ho}\big((A, m), (B, n)\big)$ carries a canonical Abelian group structure. The suspension automorphism Σ on $\Sigma\mathrm{Ho}$ becomes an additive functor by letting it act identically on morphisms.

The composition in the category $\Sigma\mathrm{Ho}$ is defined as follows. Represent elements of $\Sigma\mathrm{Ho}\big((A_1, m_1), (A, m)\big)$ and $\Sigma\mathrm{Ho}\big((A, m), (A_3, m_3)\big)$ by bounded algebra homomorphisms

$$f_1 \colon J^{m_1+k_1} A_1 \to S^{m+k_1} A, \qquad f_2 \colon J^{m+k_2} A \to S^{m_3+k_2} A_3$$

with $k_1, k_2 \in \mathbb{N}$. Define maps

$$\kappa_A^{k,l} \colon J^k S^l A \to S^l J^k A$$

as in Definition 6.22. To simplify notation, let

$$n_1 := m + k_1, \qquad n_2 := m + k_2, \qquad A_1' := J^{m_1+k_1} A_1, \qquad A_3' := S^{m_3+k_2} A_3.$$

Let $f_1 \# f_2 \in \Sigma\mathrm{Ho}\big((A_1, m_1), (A_3, m_3)\big)$ be the composition of the homotopy classes

$$J^{n_2} A_1' \xrightarrow{\langle J^{n_2}(f_1) \rangle} J^{n_2} S^{n_1} A \xrightarrow{(-1)^{n_1 \cdot n_2} \langle \kappa_A^{n_2,n_1} \rangle} S^{n_1} J^{n_2} A \xrightarrow{\langle S^{n_1}(f_2) \rangle} S^{n_1} A_3'.$$

The sign $(-1)^{n_1 n_2}$ is necessary to cancel the signs that we get by permuting the coordinates. We often drop the brackets $\langle \_ \rangle$ from our notation to avoid clutter.

We want to show that this defines a category structure on $\Sigma\mathrm{Ho}$. It is clear that $f_1 \# f_2$ only depends on $\langle f_1 \rangle$ and $\langle f_2 \rangle$. In order for the product to be well-defined on $\Sigma\mathrm{Ho}$, we also need compatibility with the inductive limit in (6.27). This amounts to the relations $\Lambda(f_1) \# f_2 = \Lambda(f_1 \# f_2) = f_1 \# \Lambda(f_2)$. Of course, the proofs depend on some formal properties of the maps κ and Λ. Before we verify the details, we compute the product in some important special cases.

View id_{SA} and λ_A (Definition 6.23) as morphisms $(SA, m) \leftrightarrow (A, m+1)$ with $k = -m$ in (6.27). Then $\lambda_A \# \mathrm{id}_{SA}$ and $\mathrm{id}_{SA} \# \lambda_A$ are the composite maps

$$JA \xrightarrow{\lambda_A} SA \xrightarrow{\kappa_A^{0,1}} SA \xrightarrow{\mathrm{id}_{SA}} SA,$$

$$JSA \xrightarrow{J(\mathrm{id}_{SA})} JSA \xrightarrow{-\kappa_A^{1,1}} SJA \xrightarrow{S(\lambda_A)} S^2 A.$$

Clearly, the first one is $\langle \lambda_A \rangle$, the second one is $-\langle S(\lambda_A) \circ \kappa_A^{1,1} \rangle$.

Lemma 6.29. *We have* $\langle S\lambda_A \circ \kappa_A^{1,1} \rangle = -\langle \lambda_{SA} \rangle$ *for all bornological algebras* A.

Observe that $\lambda_A = \Lambda(\mathrm{id}_A)$ and $\lambda_{SA} = \Lambda(\mathrm{id}_{SA})$ represent the identity morphisms on A and SA. We will see later that id_A remains the identity morphism in $\Sigma\mathrm{Ho}$. Thus we get an isomorphism $(SA, m) \cong (A, m+1)$ in $\Sigma\mathrm{Ho}$.

Proof. The maps $\kappa_A^{1,1} \colon JSA \to SJA$ and $\lambda_A \colon JA \to SA$ are the classifying maps of the extensions $SJA \rightarrowtail STA \twoheadrightarrow SA$ and $SA \rightarrowtail CA \twoheadrightarrow A$. The commuting

diagram

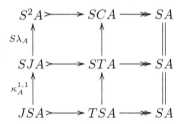

shows that $S\lambda_A \circ \kappa_A^{1,1}$ is the classifying map of the extension $S^2A \rightarrowtail SCA \twoheadrightarrow SA$ in the top row. Similarly, λ_{SA} is the classifying map of the extension $S^2A \rightarrowtail CSA \twoheadrightarrow SA$. These two extensions are isomorphic via the flip $CSA \cong SCA$. The flip operator on $S^2\mathbb{C}$ is homotopic to $\vartheta_{SC} \otimes \mathrm{id}_{SC}$, where ϑ is the orientation-reversal map, because the coordinate flip on $\mathbb{R}^2$ has determinant -1. This implies the assertion of the lemma. $\qquad\square$

Similarly, we may view id_{JA} and λ_A as morphisms $(A, m+1) \leftrightarrow (JA, m)$ with $k = -m$ in (6.27); and $\mathrm{id}_{JA}\#\lambda_A$ and $\lambda_A\#\mathrm{id}_{JA}$ are the composite maps

$$JA \xrightarrow{\mathrm{id}_{JA}} JA \xrightarrow{\kappa_A^{1,0}} JA \xrightarrow{\lambda_A} SA,$$

$$J^2A \xrightarrow{J(\lambda_A)} JSA \xrightarrow{-\kappa_A^{1,1}} SJA \xrightarrow{S(\mathrm{id}_{JA})} SJA.$$

The first one is $\langle\lambda_A\rangle$, the second one $-\langle\kappa_A^{1,1} \circ J(\lambda_A)\colon J^2A \to SJA\rangle$.

Lemma 6.30. *We have $\langle\kappa_A^{1,1} \circ J(\lambda_A)\rangle = -\langle\lambda_{JA}\rangle$ for all bornological algebras A.*

Thus $(A, m+1) \cong (JA, m)$ in $\Sigma\mathrm{Ho}$. Roughly speaking, both J and S are naturally equivalent to the suspension automorphism Σ on $\Sigma\mathrm{Ho}$.

Proof. The commutative diagram

$$
\begin{array}{ccccccccc}
SJA & \rightarrowtail & STA & \twoheadrightarrow & SA & \rightarrowtail & CA & \twoheadrightarrow & A \\
{\scriptstyle\kappa_A^{1,1}}\big\uparrow & & \big\uparrow & & \big\| & & \big\| & & \big\| \\
JSA & \rightarrowtail & TSA & \twoheadrightarrow & SA & \rightarrowtail & CA & \twoheadrightarrow & A \\
{\scriptstyle J(\lambda_A)}\big\uparrow & & \big\uparrow & & {\scriptstyle\lambda_A}\big\uparrow & & \big\uparrow & & \big\| \\
J^2A & \rightarrowtail & TJA & \twoheadrightarrow & JA & \rightarrowtail & TA & \twoheadrightarrow & A
\end{array}
$$

shows that $\kappa_A^{1,1} \circ J(\lambda_A)$ is the classifying map of the length-2-extension in the top row. Similarly, λ_{JA} is the classifying map of the length-2-extension

$$SJA \rightarrowtail CJA \twoheadrightarrow JA \rightarrowtail TA \twoheadrightarrow A.$$

We get $-\lambda_{JA}$ by composing with orientation-reversal on $(0,1)$. Since this replaces $(0,1]$ by $[0,1)$, $-\lambda_{JA}$ is the classifying map of the length-2-extension

$$SJA \rightarrowtail JA[0,1) \twoheadrightarrow JA \rightarrowtail TA \twoheadrightarrow A.$$

Now we construct a length-2-extension of A that admits morphisms to the extensions classified by $\kappa_A^{1,1} \circ J(\lambda_A)$ and $-\lambda_{JA}$. Let

$$I_1 := \{f \in CTA \mid f(1) \in JA\} = \ker(\pi_A \circ \mathrm{ev}_1 \colon CTA \twoheadrightarrow A),$$
$$E_1 := \{(f_1, f_2) \in TA(0,1] \oplus JA[1,2) \mid f_1(1) = f_2(1)\}.$$

The projection to $TA(0,1] = CTA$ provides a semi-split surjection $E_1 \twoheadrightarrow I_1$ with kernel $JA(1,2) \cong SJA$. Thus we get a length-2-extension of the form

$$SJA \rightarrowtail E_1 \twoheadrightarrow I_1 \rightarrowtail CTA \twoheadrightarrow A. \tag{6.31}$$

We map $CTA \to TA$ and $CTA \to CA$ by ev_1 and $C\pi_A \bullet 0$ (where $\bullet$ denotes concatenation). The restrictions of these maps to $I_1 \to JA$ and $I_1 \to SA$ lift to maps $E_1 \to JA[1,2) \cong JA[0,1)$ and $E_1 \to TA(0,2) \cong STA$.

Thus we get morphisms from (6.31) to the extensions that are classified by $-\lambda_{JA}$ and $\kappa_A^{1,1} \circ J(\lambda_A)$. These are smoothly homotopic to id_{SJA} on SJA. Since classifying maps of (higher length) extensions are unique up to smooth homotopy, we get $-\langle\lambda_{JA}\rangle = \langle\kappa_A^{1,1} \circ J(\lambda_A)\rangle$. □

Lemma 6.32. *We have $\Lambda(f_1)\#f_2 = \Lambda(f_1\#f_2) = f_1\#\Lambda(f_2)$. Thus we get a well-defined map*

$$\Sigma\mathrm{Ho}\big((A_1, m_1), (A, m)\big) \times \Sigma\mathrm{Ho}\big((A, m), (A_3, m_3)\big) \to \Sigma\mathrm{Ho}\big((A_1, m_1), (A_3, m_3)\big)$$

by $[f_1], [f_2] \mapsto [f_1\#f_2]$.

Proof. Using the naturality of Λ formulated in (6.28), we get

$$(-1)^{n_1 n_2}\Lambda(f_1\#f_2) = \Lambda\big(S^{n_1}(f_2) \circ \kappa_A^{n_2, n_1} \circ J^{n_2}(f_1)\big)$$
$$= \Lambda(S^{n_1}(f_2) \circ \kappa_A^{n_2, n_1}) \circ J^{n_2+1}(f_1) = S^{n_1+1}(f_2) \circ \Lambda(\kappa_A^{n_2, n_1}) \circ J^{n_2+1}(f_1),$$

$$(-1)^{n_1 n_2} f_1\#\Lambda(f_2) = (-1)^{n_1} S^{n_1}(\Lambda f_2) \circ \kappa_A^{n_2+1, n_1} \circ J^{n_2+1}(f_1)$$
$$= (-1)^{n_1} S^{n_1+1}(f_2) \circ S^{n_1}(\lambda_{J^{n_2}A}) \circ \kappa_A^{n_2+1, n_1} \circ J^{n_2+1}(f_1),$$

$$(-1)^{n_1 n_2}\Lambda(f_1)\#f_2 = (-1)^{n_2} S^{n_1+1}(f_2) \circ \kappa_A^{n_2, n_1+1} \circ J^{n_2}(\Lambda f_1)$$
$$= (-1)^{n_2} S^{n_1+1}(f_2) \circ \kappa_A^{n_2, n_1+1} \circ J^{n_2}(\lambda_{S^{n_1}A}) \circ J^{n_2+1}(f_1).$$

Thus $\Lambda(f_1\#f_2) = f_1\#\Lambda(f_2)\rangle = \Lambda(f_1)\#f_2$ follows once we have

$$\Lambda(\kappa_A^{n_2, n_1}) = (-1)^{n_1} S^{n_1}(\lambda_{J^{n_2}A}) \circ \kappa_A^{n_2+1, n_1} = (-1)^{n_2} \kappa_A^{n_2, n_1+1} \circ J^{n_2}(\lambda_{S^{n_1}A}). \tag{6.33}$$

We simplify these equations by decreasing n_1 and n_2. Since classifying maps for higher length extensions are defined iteratively, we get

$$\kappa_A^{i,j} = \kappa_{J^{i-1}A}^{1,j} \circ J(\kappa_A^{i-1,j}) = \cdots$$
$$= J^0(\kappa_{J^{i-1}A}^{1,j}) \circ J^1(\kappa_{J^{i-2}A}^{1,j}) \circ J^2(\kappa_{J^{i-3}A}^{1,j}) \circ \cdots \circ J^{i-1}(\kappa_A^{1,j}). \quad (6.34)$$

When combined with (6.28), we get

$$\Lambda(\kappa_A^{n_2,n_1}) = \lambda_{S^{n_1}J^{n_2}A} J(\kappa_A^{n_2,n_1})$$
$$= \lambda_{S^{n_1}J^{n_2}A} \circ J^1(\kappa_{J^{n_2-1}A}^{1,n_1}) \circ J^2(\kappa_{J^{n_2-2}A}^{1,n_1}) \circ J^3(\kappa_{J^{n_2-3}A}^{1,n_1}) \circ \cdots \circ J^{n_2}(\kappa_A^{1,n_1}).$$

Thus $\Lambda(\kappa_A^{n_2,n_1}) = (-1)^{n_1} S^{n_1}(\lambda_{J^{n_2}A}) \circ \kappa_A^{n_2+1,n_1}$ becomes equivalent to

$$\lambda_{S^{n_1}J^{n_2}A} = (-1)^{n_1} S^{n_1}(\lambda_{J^{n_2}A}) \circ \kappa_{J^{n_2}A}^{1,n_1}. \quad (6.35)$$

Since we can extract tensor factors one after another, we have

$$\kappa_A^{i,j} = S^{j-1}(\kappa_A^{i,1}) \circ S^{j-2}(\kappa_{SA}^{i,1}) \circ \cdots \circ S^0(\kappa_{S^{j-1}A}^{i,1}). \quad (6.36)$$

Using (6.36) and Lemma 6.29, we can prove (6.35) by induction on n_1. This finishes the proof that $\Lambda(f_1 \# f_2) = f_1 \# \Lambda(f_2)$.

Similarly, we may use (6.36) to rewrite

$$\Lambda(\kappa_A^{n_2,n_1}) = S(\kappa_A^{n_2,n_1}) \circ \lambda_{J^{n_2}S^{n_1}A}$$
$$= S^{n_1}(\kappa_A^{n_2,1}) \circ S^{n_1-1}(\kappa_{SA}^{n_2,1}) \circ \cdots \circ S^1(\kappa_{S^{n_1-1}A}^{n_2,1}) \circ \lambda_{J^{n_2}S^{n_1}A}.$$

Then the equation $\Lambda(\kappa_A^{n_2,n_1}) = (-1)^{n_2} \kappa_A^{n_2,n_1+1} \circ J^{n_2}(\lambda_{S^{n_1}A})$ becomes equivalent to

$$\lambda_{J^{n_2}S^{n_1}A} = (-1)^{n_2} \kappa_{S^{n_1}A}^{n_2,1} \circ J^{n_2}(\lambda_{S^{n_1}A}). \quad (6.37)$$

This is proved by induction on n_2 using (6.34) and Lemma 6.30. $\square$

Lemma 6.38. *The composition in ΣHo is associative, and $\mathrm{id}_A \colon A \to A$ represents the identity morphism on (A, m) in ΣHo.*

Proof. We write down $(f_1 \# f_2) \# f_3$ and $f_1 \# (f_2 \# f_3)$ and compare the results. Associativity is equivalent to the commutativity in ΣHo of the diagram

$$
\begin{array}{ccccc}
J^{l+m}S^k B & \longrightarrow & J^m S^k J^l B & \xrightarrow{\ J^m S^k f_2\ } & J^m S^{k+n} C \\
\downarrow & & & & \downarrow \\
S^k J^{l+m} B & \xrightarrow{\ S^k J^m f_2\ } & S^k J^m S^n C & \longrightarrow & S^{k+n} J^m C,
\end{array}
$$

where the unlabelled maps are induced by $\kappa_{\cdots}^{\cdots}$ and where $f_2 \colon J^l B \to S^n C$. The maps $\kappa_A^{i,j}$ are natural. This means formally that $\kappa_C^{m,k} \circ J^m S^k(f) = S^k J^m(f) \circ \kappa_B^{m,k}$. Together with (6.34), this implies the commutativity of the above diagram and hence the associativity of $\#$. It is trivial that $\mathrm{id}_A \# f = f$ and $f \# \mathrm{id}_A = f$ for all morphisms f in ΣHo; hence id_A represents the identity morphism in ΣHo. $\square$

Thus ΣHo is a category. Notice that we write the composition of morphisms in the unusual order where $f_1 f_2$ means f_1 *before* f_2. One justification for this is that the functor $X \mapsto C(X)$ from spaces to C^*-algebras is contravariant, so that an algebra homomorphism $A \to B$ may be viewed as a map from the noncommutative space underlying B to the corresponding object for A.

It is clear that the formal suspension Σ defines an automorphism of the category ΣHo. Lemmas 6.29 and 6.30 show that Σ is naturally isomorphic to the functors that send (A, m) to (SA, m) and (JA, m), respectively. Furthermore, if a class in $\Sigma\mathrm{Ho}\big((A, m), (B, n)\big)$ is represented by $f \colon J^{m+k}(A) \to S^{n+k}(B)$, then we can write it as the composite

$$(A, m) \xrightarrow{\cong} (J^{m+k}(A), -k) \xrightarrow{f} (S^{n+k}(B), -k) \xrightarrow{\cong} (B, n). \qquad (6.39)$$

The following exercise explains why cone and tensor algebra extensions play such a crucial role:

Exercise 6.40. We call an extension $I \rightarrowtail E \twoheadrightarrow A$ *smoothly contractible* if E is smoothly contractible. Such extensions are important because we expect their boundary maps to be isomorphisms.

Cone and tensor algebra extensions are smoothly contractible. Conversely, any such extension lies between the tensor algebra extension and the cone extension in the sense that there exist morphisms of extensions

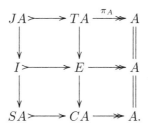

Lemma 6.41. *The category ΣHo is additive. For any objects* (A_1, m), (A_2, m), (D, n), *we have*

$$\Sigma\mathrm{Ho}\big((D, n), (A_1 \oplus A_2, m)\big) \cong \Sigma\mathrm{Ho}\big((D, n), (A_1, m)\big) \oplus \Sigma\mathrm{Ho}\big((D, n), (A_2, m)\big), \qquad (6.42)$$

$$\Sigma\mathrm{Ho}\big((A_1 \oplus A_2, m), (D, n)\big) \cong \Sigma\mathrm{Ho}\big((A_1, m), (D, n)\big) \oplus \Sigma\mathrm{Ho}\big((A_2, m), (D, n)\big). \qquad (6.43)$$

Proof. A category is *additive* if it has a zero object and finite products and if its morphism spaces carry Abelian group structures, such that the composition is additive in each variable. We have already defined the group structure on the morphism spaces in ΣHo. It is clear that $(f_1, f_2) \mapsto f_1 \# f_2$ is additive in the variable f_2. Equation (6.42) follows from the natural isomorphism $S^n B \oplus S^n B \cong S^n(B \oplus B)$. Thus $(A_1 \oplus A_2, m)$ is a direct product for (A_1, m) and (A_2, m). Using

Lemma 6.29, we can replace any pair of objects of ΣHo by an isomorphic pair with the same entry $m \in \mathbb{Z}$ in the second variable. Thus we get a direct product for any pair of objects. We also have a zero object, namely, the bornological algebra $\{0\}$. Hence the category ΣHo has finite products.

Additivity of the composition of morphisms in the variable f_1 reduces to the assertion that the composite map

$$\langle JA, SB \rangle \xrightarrow{J} \langle J^2A, JSB \rangle \xrightarrow{\kappa_*} \langle J^2A, SJB \rangle$$

is a group homomorphism for all bornological algebras A, B. This follows if we encode the group structure as a map $SB \oplus SB \to SB$. Thus ΣHo is an additive category.

Now we get (6.43) because this holds in any additive category. We recall the argument. Let $\mathfrak{C}$ be an additive category. Let x_1, x_2, y be objects of $\mathfrak{C}$ and let $x := x_1 \times x_2$. We have to prove

$$\mathfrak{C}(x, y) \cong \mathfrak{C}(x_1, y) \times \mathfrak{C}(x_2, y).$$

We identify $\mathfrak{C}(x, y)$ with the space of natural transformations between the representable functors $\mathfrak{C}(\_, x)$ and $\mathfrak{C}(\_, y)$ (by composition, any element of $\mathfrak{C}(x, y)$ yields such a natural transformation, and conversely any natural transformation is of this form). By assumption, $\mathfrak{C}(\_, x) \cong \mathfrak{C}(\_, x_1) \times \mathfrak{C}(\_, x_2)$. Moreover, this is an isomorphism of Abelian groups, and all natural transformations are additive (because the composition is bi-additive). Therefore, natural transformations from $\mathfrak{C}(\_, x)$ to $\mathfrak{C}(\_, y)$ correspond bijectively to pairs of natural transformations from $\mathfrak{C}(\_, x_1)$ and $\mathfrak{C}(\_, x_2)$ to $\mathfrak{C}(\_, y)$. The latter are equivalent to pairs of elements of $\mathfrak{C}(x_1, y)$ and $\mathfrak{C}(x_2, y)$. Thus $\mathfrak{C}(x, y) \cong \mathfrak{C}(x_1, y) \times \mathfrak{C}(x_2, y)$ as desired. $\square$

6.3.1   Behaviour for infinite direct sums

Lemma 6.41 only deals with *finite* sums. Infinite sums are more problematic. We can write an infinite direct sum as a direct limit of finite sums: if A_i, $i \in I$, is a set of bornological algebras, then

$$\bigoplus_{i \in I} A_i \cong \varinjlim_{F} \bigoplus_{i \in F} A_i,$$

where F runs through the directed set of finite subsets of I. Unfortunately, inductive limits of bornological algebras do not remain inductive limits in ΣHo. Now we explain this problem.

Lemma 6.44. *The functors $A \mapsto TA, JA$ commute with inductive limits.*

Proof. First we check that the tensor algebra functor — as a functor from bornological vector spaces to bornological algebras — commutes with arbitrary direct limits (we do not need inductive systems here). Let (A_i) be a diagram of bornological algebras. Bounded algebra homomorphisms $T(\varinjlim A_i) \to B$ correspond to bounded linear maps $\varinjlim A_i \to B$ by the universal property of the tensor algebra. The latter correspond to compatible families of bounded linear maps $A_i \to B$ by the universal property of direct limits, which again correspond to compatible families of bounded algebra homomorphisms $TA_i \to B$. Thus the tensor algebra functor commutes with direct limits.

Unlike general direct limits, inductive limits are compatible with semi-split extensions. Hence the assertion for J follows from the assertion about T and the semi-split exact sequence $JA \rightarrowtail TA \twoheadrightarrow A$. $\qquad\square$

The group $\Sigma\mathrm{Ho}(\varinjlim A_i, B)$ is generated by bounded algebra homomorphisms $J^k(\varinjlim A_i) \to S^k B$. Here we may replace $J^k(\varinjlim A_i)$ by $\varinjlim J^k(A_i)$. Nevertheless, $\langle J^k(\varinjlim_i A_i), S^k B\rangle$ may differ from $\varprojlim_i \langle J^k A_i, S^k B\rangle$. To see the problem, consider a compatible family of smooth homotopy classes $\langle \alpha_i \rangle \in \langle J^k A_i, S^k B\rangle$. Compatibility means that various diagrams commute *up to smooth homotopy*. But to get a homomorphism $\varinjlim J^k A_i \to S^k B$, these diagrams have to commute exactly. In general, it is not possible to fix this.

This problem with inductive limits is no surprise because it appears in various other contexts like homological algebra and homotopy theory. Therefore, we should not expect inductive limits of bornological algebras to give inductive limits in $\Sigma\mathrm{Ho}$ in general. But the inductive systems that we get from direct sums are more special, and we may hope that this problem does not occur for them.

Nevertheless, there is another obstacle that is less expected and that already occurs for countable direct sums. Namely, the morphism spaces in $\Sigma\mathrm{Ho}$ are defined as direct limits, and these do not commute with projective limits in general. Let (A_n) be a sequence of bornological algebras and let $\alpha_n \in \Sigma\mathrm{Ho}(A_n, B)$ be represented by a bounded algebra homomorphism

$$J^{m(n)} A_n \to S^{m(n)} B$$

with certain $m(n)$. Suppose that the minimal such $m(n)$ goes to ∞ for $n \to \infty$. Then $(\alpha_n) \in \prod \Sigma\mathrm{Ho}(A_n, B)$ cannot come from an element of $\Sigma\mathrm{Ho}(\bigoplus A_n, B)$.

As a result, we cannot say anything positive about the behaviour of $\Sigma\mathrm{Ho}$ for infinite direct sums. This is a rather serious problem with the definition of $\Sigma\mathrm{Ho}$ because, as we shall see in Chapter 13, direct sums play an important role for the Universal Coefficient Theorem and the construction of the Baum–Connes assembly map as a localisation.

By the way, similar problems occur for infinite direct products and projective limits. Since inductive limits for bornological algebras are more fundamental than projective limits, we do not discuss this issue here.

6.3.2   An alternative approach

We may also describe $\Sigma\mathrm{Ho}$ without tensor algebras, using morphism-extensions and their composition instead. We relax our definition of morphism-extension and also allow diagrams $(\varphi, \mathbf{E}, \psi)$ of the form

$$
\begin{array}{c}
A \\
\downarrow \varphi \\
K \rightarrowtail D_1 \rightarrow \cdots \rightarrow D_n \twoheadrightarrow Q \\
\downarrow \psi \\
B,
\end{array}
$$

where the middle row $\mathbf{E}$ is a semi-split bornological algebra extension of length n. We let $\operatorname{Ext}^n(A, B)$ be the set of *equivalence classes* of morphism-extensions $A \to B$ of length n with respect to the equivalence relation *generated* by the following two elementary moves:

1. two morphism-extensions

$$
\begin{array}{c}
A \\
\downarrow \varphi_1 \\
K \rightarrowtail D_1 \rightarrow \cdots \rightarrow D_n \twoheadrightarrow Q \\
\downarrow \psi \\
B
\end{array}
\qquad
\begin{array}{c}
A \\
\downarrow \varphi \\
K' \rightarrowtail D'_1 \rightarrow \cdots \rightarrow D'_n \twoheadrightarrow Q' \\
\downarrow \psi_2 \\
B
\end{array}
$$

are equivalent if there is a commuting diagram

$$
\begin{array}{c}
A \\
\downarrow \varphi_1 \\
K \rightarrowtail D_1 \rightarrow \cdots \rightarrow D_n \twoheadrightarrow Q \\
\downarrow \psi_1 \quad \downarrow \qquad \qquad \downarrow \quad \downarrow \varphi_2 \\
K' \rightarrowtail D'_1 \rightarrow \cdots \rightarrow D'_n \twoheadrightarrow Q' \\
\downarrow \psi_2 \\
B
\end{array}
$$

with $\varphi = \varphi_2 \circ \varphi_1$ and $\psi = \psi_2 \circ \psi_1$;

2. two morphism-extensions $(\varphi, \mathbf{E}, \psi)$ and $(\varphi, \mathbf{E}, \psi')$ are considered equivalent if ψ is smoothly homotopic to ψ' (we get the same equivalence relation if we allow smooth homotopies for the first homomorphism φ instead).

Thus two morphism-extensions are equivalent if they can be connected by a chain of such elementary equivalences.

Exercise 6.45. The construction of classifying maps identifies $\operatorname{Ext}^n(A, B)$ with $\langle J^n A, B \rangle$.

We compose two morphism-extensions

$$
\begin{array}{ccc}
 & A & \\
 & \downarrow{\scriptstyle\varphi} & \\
K \rightarrowtail D_1 \longrightarrow \cdots \longrightarrow D_n \twoheadrightarrow Q & \\
\downarrow{\scriptstyle\psi} & \\
B &
\end{array}
\qquad
\begin{array}{ccc}
 & B & \\
 & \downarrow{\scriptstyle\varphi'} & \\
K' \rightarrowtail D_1' \longrightarrow \cdots \longrightarrow D_m' \twoheadrightarrow Q' & \\
\downarrow{\scriptstyle\psi'} & \\
C &
\end{array}
$$

as follows. First we construct the pull-back extension

$$
\begin{array}{ccccccc}
K'' & \rightarrowtail & D_1'' & \longrightarrow \cdots \longrightarrow & D_m'' & \twoheadrightarrow & K \\
\downarrow{\scriptstyle\kappa} & & \downarrow & & \downarrow & & \downarrow{\scriptstyle\varphi'\circ\psi} \\
K' & \rightarrowtail & D_1' & \longrightarrow \cdots \longrightarrow & D_m' & \twoheadrightarrow & Q' \\
\downarrow{\scriptstyle\psi'} & & & & & & \\
C. & & & & & &
\end{array}
$$

The *composition* (*Yoneda product*) of our morphism-extensions is the length-$n+m$ morphism-extension

$$
\begin{array}{ccc}
 & & A \\
 & & \downarrow{\scriptstyle\varphi} \\
K'' \rightarrowtail D_1'' \longrightarrow \cdots \longrightarrow D_m'' \longrightarrow D_1 \longrightarrow \cdots \longrightarrow D_n \twoheadrightarrow Q \\
\downarrow{\scriptstyle\psi'\circ\kappa} & & \\
C. & &
\end{array}
$$

This defines an associative product

$$
\mathrm{Ext}^n(A_1, A_2) \times \mathrm{Ext}^m(A_2, A_3) \to \mathrm{Ext}^{n+m}(A_1, A_3).
$$

The product with the cone extension $SB \rightarrowtail CB \twoheadrightarrow B$ yields a natural map $\mathrm{Ext}^n(A, B) \to \mathrm{Ext}^{n+1}(A, SB)$. Thus we get an inductive system $\mathrm{Ext}^n(A, S^n B)$, $n \in \mathbb{N}$. Exercise 6.45 shows that its direct limit is $\Sigma\mathrm{Ho}(A, B)$.

6.4   Exact triangles in the suspension-stable homotopy category

The category $\Sigma\mathrm{Ho}$ behaves in many respects like the classical stable homotopy category in topology. There are many standard tools for doing homological computations in the classical stable homotopy category, including long exact sequences, spectral sequences, and localisation. These computations can be formalised in the framework of triangulated categories.

A *triangulated category* is an additive category with a suspension automorphism and a class of *exact triangles*, subject to certain axioms, which we will recall below. This notion is due to Jean-Louis Verdier [123], who invented it in order to clarify the properties of derived categories; it covers a wide range of other situations as well. We are going to exhibit ΣHo as a triangulated category.

There is one particularly complicated axiom, the *Octahedral Axiom*, which is mainly needed in order to localise triangulated categories. Since we will not use it, we postpone its verification until Chapter 13.

We have already seen that ΣHo is an additive category with a suspension automorphism Σ. Our description of exact triangles is based on mapping cones. We could use semi-split extensions as well: we will see that they give rise to the same class of exact triangles. But proofs become more difficult for this alternative choice. We have already met mapping cones in §2.2.3 in connection with the Puppe exact sequence. Since we only allow smooth homotopies in the category ΣHo, we slightly modify the definition of the mapping cone.

Definition 6.46. Let $f \colon A \to B$ be a bounded algebra homomorphism. Its *mapping cone* is redefined to be the bornological algebra

$$C(f) := \{(a, b) \in A \oplus C(B) \mid f(a) = b(1)\},$$

where $C(B) = B(0, 1]$. In particular, $C(\mathrm{id}_B) \cong C(B)$. We have natural maps

$$\begin{aligned} \iota_f &\colon S(B) \to C(f), & b &\mapsto (0, b), \\ \varepsilon_f &\colon C(f) \to A, & (a, b) &\mapsto a. \end{aligned}$$

Using the isomorphism $(S(B), m) \cong \Sigma(B, m)$, we get maps

$$\Sigma(B, m) \xrightarrow{(-1)^m \iota_f} (C(f), m) \xrightarrow{\varepsilon_f} (A, m) \xrightarrow{f} (B, m)$$

for any $m \in \mathbb{Z}$; such diagrams are called *mapping cone triangles*.

A *triangle* in ΣHo is a diagram in ΣHo of the form $\Sigma X \to Y \to Z \to X$; a *morphism of triangles* is a commuting diagram of the form

$$\begin{array}{ccccccc} \Sigma X & \longrightarrow & Y & \longrightarrow & Z & \longrightarrow & X \\ \downarrow{\scriptstyle \Sigma\xi} & & \downarrow{\scriptstyle \eta} & & \downarrow{\scriptstyle \zeta} & & \downarrow{\scriptstyle \xi} \\ \Sigma X' & \longrightarrow & Y' & \longrightarrow & Z' & \longrightarrow & X'. \end{array}$$

Notice that the map $\Sigma X \to \Sigma X'$ is required to be the suspension of ξ: a triangle has only three independent vertices.

Definition 6.47. A triangle in ΣHo is called *exact* if it is isomorphic (as a triangle) to a mapping cone triangle.

Theorem 6.48. *The category ΣHo with the additional structure described above is triangulated.*

We have to check the axioms (TR0)–(TR4) for a triangulated category (see [93, 123]). We will recall these axioms as we go along. Before we start, we should discuss the issue of opposite categories. The axioms of a triangulated category are tailored for categories of topological spaces. Since the functor from spaces to algebras is contravariant, many constructions in ΣHo become more transparent in the opposite category, where a bounded algebra homomorphism $A \to B$ is viewed as a morphism from B to A. For this reason, we have altered the statements of the axioms slightly, reversing arrows in several places. This does not make a difference because the notion of a triangulated category is self-dual, that is, the opposite category of a triangulated category inherits a canonical triangulated category structure.

Axiom 6.49 (TR0). *The class of exact triangles is closed under isomorphism.*
Triangles of the form $\Sigma X \to 0 \to X \xrightarrow{\mathrm{id}_X} X$ *are exact for any object* X.

Proof. The first assertion is trivial. The triangle $\Sigma X \to 0 \to X \xrightarrow{\mathrm{id}_X} X$ is exact for any object $X = (A, m)$ because $C(\mathrm{id}_A) \cong C(A)$ is smoothly contractible and hence $(C(\mathrm{id}_A), m) \cong 0$ in ΣHo for all $m \in \mathbb{Z}$. □

Axiom 6.50 (TR1). *Any morphism* $f \colon A \to B$ *in* ΣHo *is contained in some exact triangle* $\Sigma B \to C \to A \xrightarrow{f} B$.

Proof. We represent $f \colon (A, m) \to (B, n)$ by a bounded homomorphism

$$\hat{f} \colon J^{m+k}(A) \to S^{n+k}(B).$$

As in (6.39), we have $(A, m) \cong (J^{m+k}(A), -k)$ and $(B, n) \cong (S^{n+k}(B), -k)$. Hence the mapping cone triangle

$$\Sigma\big(S^{n+k}(B), -k\big) \to (C(\hat{f}), -k) \to (J^{m+k}(A), -k) \to (S^{n+k}(B), -k)$$

for $\hat{f}$ is isomorphic to a triangle that contains f. □

Definition 6.51. Given a triangle

$$\triangle = \big(\Sigma B \xrightarrow{\iota} C \xrightarrow{\varepsilon} A \xrightarrow{f} B\big),$$

we define a *rotated triangle* $R(\triangle)$ by

$$R(\triangle) := \big(\Sigma A \xrightarrow{-\Sigma f} \Sigma B \xrightarrow{-\iota} C \xrightarrow{-\varepsilon} A\big).$$

Notice that $R(\triangle)$ is isomorphic to $\Sigma A \xrightarrow{-\Sigma f} \Sigma B \xrightarrow{\iota} C \xrightarrow{\varepsilon} A$: the isomorphism is -1 on C and $+1$ on A and B. We cannot get rid of the signs completely.

Axiom 6.52 (TR2). *A triangle* $\triangle$ *is exact if and only if* $R(\triangle)$ *is.*

Proof. We claim that R maps exact triangles again to exact triangles. Hence $R^n(\triangle)$ is exact if $\triangle$ is exact and $n \in \mathbb{N}$. We have

$$R^3(\triangle) \cong (\Sigma^2 B \xrightarrow{-\Sigma\iota} \Sigma C \xrightarrow{\Sigma\varepsilon} \Sigma A \xrightarrow{\Sigma f} \Sigma B).$$

The right-hand side is a mapping cone triangle if and only if $\triangle$ is; here we use that the coordinate flip $S^2(B) \to S^2(B)$ represents -1 in $\Sigma \mathrm{Ho}(B, B)$. Therefore, if $\triangle$ is exact, then so is $R^{-3}(\triangle)$. Hence we are done if we prove the claim.

We may assume that $\triangle$ is a mapping cone triangle for some bounded algebra homomorphism $f: A \to B$; we disregard the formal suspension parameter $m \in \mathbb{Z}$ because it is irrelevant here. We want to show that $R(\triangle)$ is homotopy equivalent to the mapping cone triangle of the bounded homomorphism $\varepsilon_f: C(f) \to A$.

By definition, the mapping cone of ε_f is

$$C(\varepsilon_f) = \{(c, a) \in C(f) \oplus C(A) \mid \varepsilon_f(c) = a(1)\}$$
$$\cong \{(a_1, b, a_2) \in A \oplus C(B) \oplus C(A) \mid f(a_1) = b(1),\ a_1 = a_2(1)\}$$
$$\cong \{(b, a) \in C(B) \oplus C(A) \mid f(a(1)) = b(1)\}.$$

The natural maps $S(A) \to C(\varepsilon_f) \to C(f)$ send $a \mapsto (0, a)$ and $(b, a) \mapsto (a(1), b)$, respectively. We are going to construct a natural homotopy equivalence between $S(B)$ and $C(\varepsilon_f)$. We map $S(B) \to C(\varepsilon_f)$ by $b \mapsto (b, 0)$ and $C(\varepsilon_f) \to S(B)$ by $(b, a) \mapsto b \bullet f(a)^{-1}$. The composite map $S(B) \to C(\varepsilon_f) \to S(B)$ maps $b \mapsto b \bullet 0$; this is smoothly homotopic to the identity map on $S(B)$ (see §6.1). The other composition maps $(b, a) \mapsto (b \bullet f(a)^{-1}, 0)$. To connect it to the identity map, we reparametrise the second summand $C(A)$ in $C(\varepsilon_f)$ as $C(A) \cong C[1, 2)$, so that elements of $C(\varepsilon_f)$ become functions $\varphi: (0, 2) \to A \cup B$ with $\varphi(t) \in B$ for $t < 1$ and $\varphi(t) \in A$ for $t \geq 1$.

The idea of the proof is to reparametrise functions on $[0, 2]$, using f to transport function values from $[0, 1]$ to $[1, 2]$. This simple idea is complicated by the need to make derivatives vanish at 1. Let $\varrho: [0, 1] \to [0, 1]$ be a smooth function as in §6.1 and define $\varrho(1 + t) := 1 + \varrho(t)$ for $t \in [1, 2]$. We may define a bounded algebra homomorphism $H: C(\varepsilon_f) \to C^\infty([0, 1], C(\varepsilon_f))$ by

$$(H_s\varphi)(t) := \begin{cases} \varphi(\varrho(t) + \varrho(t)s) & 0 < \varrho(t) < 1/1+s, \\ f(\varphi(\varrho(t) + \varrho(t)s)) & 1/1+s \leq \varrho(t) < 1, \\ \varphi(\varrho(t) + \varrho(t)s) & 1 \leq \varrho(t) < 2/1+s, \\ 0 & 2/1+s \leq \varrho(t) < 2, \end{cases}$$

for $s \in [0, 1]$, $t \in [0, 2]$; this provides a smooth homotopy between $H_0 = \varrho^*$ and $H_1(b, a) = (\varrho^*(b \bullet f(a)), 0)$. Joining ϱ and $\mathrm{id}_{[0,1]}$ by $(1 - t)\varrho + t\,\mathrm{id}$, we get a smooth homotopy between ϱ^* and the identity map. Hence our maps between $C(\varepsilon_f)$ and $S(B)$ are inverse to each other up to smooth homotopy.

Under this smooth homotopy equivalence, the natural projection $C(\varepsilon_f) \to C(f)$ corresponds to the embedding $S(B) \to C(f)$, whereas the natural map

$S(A) \to C(\varepsilon_f)$ corresponds to the map $-S(f)\colon S(A) \to S(B)$; the sign appears because of the orientation reversal. Thus the mapping cone triangle for ε_f is smoothly homotopy equivalent to the rotated mapping cone triangle for f, as desired.   □

Axiom 6.53 (TR3). *Suppose given the solid arrows in the following diagram, and suppose that the rows are exact triangles and the right square commutes:*

$$
\begin{array}{ccccccc}
\Sigma B & \xrightarrow{\iota} & C & \xrightarrow{\varepsilon} & A & \xrightarrow{f} & B \\
\Big\downarrow{\scriptstyle\Sigma\beta} & & \Big\downarrow{\scriptstyle\gamma} & & \Big\downarrow{\scriptstyle\alpha} & & \Big\downarrow{\scriptstyle\beta} \\
\Sigma B' & \xrightarrow{\iota'} & C' & \xrightarrow{\varepsilon'} & A' & \xrightarrow{f'} & B'.
\end{array}
\tag{6.54}
$$

In such a situation, we can find a morphism $\gamma\colon C \to C'$ so that the whole diagram commutes; thus we get a morphism of triangles.

The map γ in Axiom (TR3) is usually not unique, and not even canonical.

Proof. To check this axiom in $\Sigma\mathrm{Ho}$ we may assume without loss of generality that the two rows are mapping cone triangles because any exact triangle is isomorphic to such a triangle. We represent the vertical maps α, β by bounded algebra homomorphisms $\alpha\colon J^{k+m}(A) \to S^{k+m'}(A')$ and $\beta\colon J^{k+m}(B) \to S^{k+m'}(B')$; we can choose the same k for both maps. Increasing k if necessary, we can achieve $\langle f' \circ \alpha \rangle = \langle \beta \circ f \rangle$ because $[f' \circ \alpha] = [\beta \circ f]$.

Recall that $(S^{m'+k}(B'), -k) \cong (B', m')$. Moreover, the mapping cone construction commutes with S, that is, $C(S^{m'+k}(f')) \cong S^{m'+k}(C(f'))$. Hence the mapping cone triangle $\Sigma(B', m') \to (C(f'), m') \to (A', m') \to (B', m')$ for f' is isomorphic to the mapping cone triangle $\Sigma(S^{m'+k}(B'), -k) \to (C(S^{m'+k}f'), -k) \to (S^{m'+k}(A'), -k) \to (S^{m'+k}(B'), -k)$ for $S^{m+k'}(f')$; the signs also work out. Therefore, we may assume without loss of generality that $m' + k = 0$.

The functor J does not commute with the mapping cone construction. Nevertheless, the natural projections $C(f) \to A$ and $C(f) \to C(B)$ induce natural maps $J^{k+m}(C(f)) \to J^{k+m}(A)$ and $J^{k+m}(C(f)) \to J^{k+m}(C(B)) \to CJ^{k+m}(B)$. These combine to a natural bounded homomorphism $J^{k+m}(C(f)) \to C(J^{k+m}(f))$. Even more, we get a commuting diagram

$$
\begin{array}{ccccccc}
J^{k+m}(S(B)) & \xrightarrow{J^{k+m}(\iota_f)} & J^{k+m}(C(f)) & \xrightarrow{J^{k+m}(\varepsilon_f)} & J^{k+m}(A) & \xrightarrow{J^{k+m}(f)} & J^{k+m}(B) \\
\Big\downarrow{\scriptstyle\kappa_B^{k+m,1}} & & \Big\downarrow & & \Big\| & & \Big\| \\
S(J^{k+m}B) & \xrightarrow{\iota_{J^{k+m}(f)}} & C(J^{k+m}(f)) & \xrightarrow{\varepsilon_{J^{k+m}(f)}} & J^{k+m}(A) & \xrightarrow{J^{k+m}(f)} & J^{k+m}(B).
\end{array}
$$

Hence it suffices to extend (α, β) to a morphism from the mapping cone triangle for $J^{m+k}(f)$ to the mapping cone triangle for f'. Again we may use the isomorphism $(A, m) \cong (J^{m+k}A, -k)$ to reduce to the case $m + k = 0$.

Thus the conclusion of Axiom (TR3) holds in general once it holds in the special case where the rows in the diagram (6.54) are mapping cone triangles and the vertical maps α and β are bounded algebra homomorphisms such that $\beta \circ f$ and $f' \circ \alpha$ are smoothly homotopic. That is, we also have a bounded algebra homomorphism $H \colon A \to B'[0, 1]$ with $H_0 = \beta \circ f$ and $H_1 = f' \circ \alpha$. Now $\gamma(a, b) := (\alpha(a), \beta(b) \bullet H(a))$ for $(a, b) \in C(f)$ defines a bounded algebra homomorphism $\gamma \colon C(f) \to C(f')$. By construction, $\varepsilon_{f'} \circ \gamma = \alpha \circ \varepsilon_f$ and $\gamma \circ \iota_f(b) = \iota_{f'} \circ \beta(b \bullet 0)$; the latter is smoothly homotopic to $\iota_{f'} \circ \beta$. Hence the map γ has the required properties. $\qquad \square$

Finally, there is the Octahedral Axiom (TR4), which plays an important role in connection with the localisation of triangulated categories. This axiom is easy enough to prove but complicated to state. Therefore we postpone its verification to §13.2.

6.5 Long exact sequences in triangulated categories

Let $\mathfrak{T}$ be a triangulated category with suspension automorphism Σ. Let Ab be the category of Abelian groups. For a covariant functor $F \colon \mathfrak{T} \to$ Ab, we put $F_n(A) := F(\Sigma^n A)$ for $n \in \mathbb{Z}$. For a contravariant functor $G \colon \mathfrak{T}^{\mathrm{op}} \to$ Ab, we put $G^n(A) := G(\Sigma^n A)$ for $n \in \mathbb{Z}$.

Definition 6.55. A covariant functor $F \colon \mathfrak{T} \to$ Ab is called *homological* if $F(C) \to F(A) \to F(B)$ is exact for each exact triangle $\Sigma B \to C \to A \to B$. A contravariant functor $G \colon \mathfrak{T}^{\mathrm{op}} \to$ Ab is called *cohomological* if $G(C) \leftarrow G(A) \leftarrow G(B)$ is exact for each exact triangle $\Sigma B \to C \to A \to B$.

Lemma 6.56. *Let* $\Sigma B \to C \to A \to B$ *be an exact triangle in* $\mathfrak{T}$. *For a homological functor* $F \colon \mathfrak{T} \to$ Ab, *there is a long exact sequence*

$$\cdots \to F_{n+1}(B) \to F_n(C) \to F_n(A) \to F_n(B) \to F_{n-1}(C) \to \cdots$$

that extends indefinitely in both directions.

For a cohomological functor $G \colon \mathfrak{T}^{\mathrm{op}} \to$ Ab, *there is a long exact sequence*

$$\cdots \leftarrow G^{n+1}(B) \leftarrow G^n(C) \leftarrow G^n(A) \leftarrow G^n(B) \leftarrow G^{n-1}(C) \leftarrow \cdots$$

that extends indefinitely in both directions. The maps in these two long exact sequences are induced by the given maps $\Sigma B \to C \to A \to B$.

Proof. By Axiom 6.52 (TR2), our exact triangle gives rise to a whole sequence of rotated exact triangle. When we apply the definition of a (co)homological functor to these triangles, we get the exactness of our sequence at all places. $\qquad \square$

More generally, we may consider (co)homological functors with values in another Abelian category than the category of Abelian groups. Lemma 6.56 remains valid in this generality.

The connecting maps $F_{n+1}(B) \to F_n(C)$ in Lemma 6.56 are already contained in the exact triangle. This is convenient for algebraic arguments. In contrast, the connecting map in the K-theory long exact sequence for an extension of bornological algebras (Theorem 2.33) is *constructed* from the given extension $I \to E \to Q$. But this uses the maps $I \rightarrowtail E \twoheadrightarrow Q$ themselves. We are forced to add the map $\Sigma Q \to I$ to our initial data because of the following:

Exercise 6.57. Find semi-split algebra extensions $I \rightarrowtail E_1 \twoheadrightarrow Q$ and $I \rightarrowtail E_2 \twoheadrightarrow Q$ with different classifying maps in $\Sigma \mathrm{Ho}_{-1}(Q, I)$ and a diagram

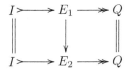

that commutes *up to smooth homotopy* and where the vertical map $E_1 \to E_2$ is an algebra isomorphism. Hence the classifying map of an extension is not yet determined by the image of the extension in the suspension-stable homotopy category. *Hint*: You may assume that E_1 and E_2 are smoothly contractible.

Now we provide some examples of (co)homological functors. For an object D of a triangulated category $\mathfrak{T}$, we define a covariant functor $F^D \colon \mathfrak{T} \to \mathrm{Ab}$ and a contravariant functor $G_D \colon \mathfrak{T}^{\mathrm{op}} \to \mathrm{Ab}$ by

$$F^D(A) := \mathfrak{T}(D, A), \qquad G_D(A) := \mathfrak{T}(A, D).$$

Proposition 6.58. *The functors F^D and G_D are (co)homological.*

Proof. Let $\Sigma B \xrightarrow{\iota} C \xrightarrow{\varepsilon} A \xrightarrow{f} B$ be an exact triangle in $\mathfrak{T}$. The Axioms 6.50 (TR1) and 6.53 (TR3) imply $f \circ \varepsilon = 0$ (see [93]). We leave the proof as an amusing exercise. Hence the composite maps $\mathfrak{T}(D, C) \to \mathfrak{T}(D, A) \to \mathfrak{T}(D, B)$ and $\mathfrak{T}(C, D) \leftarrow \mathfrak{T}(A, D) \leftarrow \mathfrak{T}(B, D)$ vanish.

Now let $x \in \ker\big(\mathfrak{T}(D, f)\big)$, that is, $f \circ x = 0$. By Axioms 6.49 (TR0) and 6.52 (TR2), we have an exact triangle $0 \to D = D \to 0$. Now we apply Axiom 6.53 (TR3) to the diagram

$$
\begin{array}{ccccccc}
\Sigma B & \xrightarrow{\iota} & C & \xrightarrow{\varepsilon} & A & \xrightarrow{f} & B \\
\big\uparrow{\scriptstyle 0} & & \big\uparrow{\scriptstyle y} & & \big\uparrow{\scriptstyle x} & & \big\uparrow{\scriptstyle 0} \\
0 & \longrightarrow & D & \xlongequal{\ \mathrm{id}_D\ } & D & \longrightarrow & 0.
\end{array}
$$

The dashed map y is the required pre-image of x.

Similarly, if $x \in \ker\big(\mathfrak{T}(\varepsilon, D)\big)$, that is, $x \circ \varepsilon = 0$, then we consider the diagram

$$
\begin{array}{ccccccc}
D & \xlongequal{\ \mathrm{id}_D\ } & D & \longrightarrow & 0 & \longrightarrow & \Sigma^{-1} D \\
\big\uparrow{\scriptstyle x} & & \big\uparrow{\scriptstyle y} & & \big\uparrow{\scriptstyle 0} & & \big\uparrow{\scriptstyle \Sigma^{-1} x} \\
A & \xrightarrow{f} & B & \xrightarrow{\Sigma^{-1}\iota} & \Sigma^{-1} C & \xrightarrow{\Sigma^{-1}\varepsilon} & \Sigma^{-1} A.
\end{array}
$$

Its rows are exact triangles by Axioms 6.49 (TR0) and 6.52 (TR2). Axiom 6.53 (TR3) yields a map $y \in \mathfrak{T}(B, D)$ with $x = y \circ f$. □

Lemma 6.59. *Let*

$$
\begin{array}{ccccccc}
\Sigma B & \xrightarrow{\iota} & C & \xrightarrow{\varepsilon} & A & \xrightarrow{f} & B \\
\downarrow{\Sigma\beta} & & \downarrow{\gamma} & & \downarrow{\alpha} & & \downarrow{\beta} \\
\Sigma B' & \xrightarrow{\iota'} & C' & \xrightarrow{\varepsilon'} & A' & \xrightarrow{f'} & B'
\end{array}
$$

be a morphism of triangles. If two of α, β, γ are isomorphisms, so is the third.

Proof. By Proposition 6.58 and Lemma 6.56, the rows of our diagram yield long exact sequences for the homological functor $\mathfrak{T}(D, \_)$ for any object D; the vertical maps yield a natural transformation between these two long exact sequences. Suppose, say, that α and β are isomorphisms. Then $\mathfrak{T}(D, \alpha)$ and $\mathfrak{T}(D, \beta)$ are invertible. The Five Lemma shows that $\mathfrak{T}(D, \gamma)$ is invertible as well, for any D. Finally, the Yoneda Lemma yields that γ is invertible. □

Lemma 6.60. *Let $f \colon A \to B$ be a morphism. Then there exists an exact triangle $\Sigma B \to C \to A \xrightarrow{f} B$, and any two such triangles are isomorphic (though not canonically).*

Proof. Existence is required in Axiom 6.50 (TR1). Suppose that we have two different such exact triangles, $\Sigma B \to C \to A \xrightarrow{f} B$ and $\Sigma B \to C' \to A \xrightarrow{f} B$. Now apply Axiom 6.53 (TR3) to get a morphism of triangles

$$
\begin{array}{ccccccc}
\Sigma B & \longrightarrow & C & \longrightarrow & A & \xrightarrow{f} & B \\
\| & & \downarrow{\gamma} & & \| & & \| \\
\Sigma B & \longrightarrow & C' & \longrightarrow & A & \xrightarrow{f} & B.
\end{array}
$$

Lemma 6.59 shows that γ is an isomorphism. □

Lemma 6.61. *Let $\Sigma B \xrightarrow{\iota} C \xrightarrow{\varepsilon} A \xrightarrow{f} B$ be an exact triangle. Then $C \cong 0$ if and only if f is invertible, and $f = 0$ if and only if ε is an epimorphism, if and only if ε is a split epimorphism — that is, there is $\sigma \colon A \to C$ with $\varepsilon \circ \sigma = \mathrm{id}_A$.*

 In the latter case, our triangle is isomorphic to the direct sum of the triangles $\Sigma B = \!\!= \Sigma B \to 0 \to B$ and $0 \to A = \!\!= A \to 0$. Any such triangle is exact.

Proof. By Proposition 6.58, we have long exact sequences

$$
\cdots \to \mathfrak{T}_{n+1}(D, B) \to \mathfrak{T}_n(D, C) \to \mathfrak{T}_n(D, A) \to \mathfrak{T}_n(D, B) \to \cdots
$$

for all objects D. We have $C \cong 0$ if and only if $\mathfrak{T}_n(D, C) = 0$ for all D, n, if and only if $f_* \colon \mathfrak{T}_n(D, A) \to \mathfrak{T}_n(D, B)$ is an isomorphism for all D, n, if and only if f is invertible. The last step uses once again the Yoneda Lemma. If ε is an epimorphism, then $f \circ \varepsilon = 0$ implies $f = 0$. Conversely, if $f = 0$, then we apply the long exact sequence above to lift the identity map in $\mathfrak{T}_0(A, A)$ to $\sigma \in \mathfrak{T}_0(A, C)$. This yields a section for ε, so that ε is a split epimorphism.

It is shown in [93] that direct sums of exact triangles are again exact. We omit the argument because this property is evident in $\Sigma\mathrm{Ho}$ anyway: direct sums of mapping cone triangles are again mapping cone triangles. □

6.6   Long exact sequences in the suspension-stable homotopy category

Now we consider long exact sequences in the category $\Sigma\mathrm{Ho}$, which is triangulated by Theorem 6.48. Since mapping cone triangles are exact by definition, the long exact sequences in Proposition 6.58 are analogues of the Puppe sequence of Theorem 2.38. Thus $\Sigma\mathrm{Ho}$ has Puppe exact sequences in both variables. Next we construct long exact sequences for semi-split extensions.

Definition 6.62. Recall that any *semi-split* extension $\mathbf{E} := (I \overset{\iota}{\rightarrowtail} E \overset{\pi}{\twoheadrightarrow} B)$ determines a classifying map $\gamma\colon JB \to I$, which yields a class $\Sigma\mathrm{Ho}(\mathbf{E}) := [\gamma]$ in $\Sigma\mathrm{Ho}\big((B, 1), (I, 0)\big)$. Diagrams in $\Sigma\mathrm{Ho}$ of the form

$$\Sigma(B, m) \xrightarrow{(-1)^m \Sigma^m [\gamma]} (I, m) \xrightarrow{[\iota]} (E, m) \xrightarrow{[\pi]} (B, m)$$

for some $m \in \mathbb{Z}$ are called *extension triangles*.

Theorem 6.63. *Extension triangles in $\Sigma\mathrm{Ho}$ are exact. Hence we have long exact sequences for semi-split extensions of the form*

$$\cdots \to \Sigma\mathrm{Ho}_{j+1}(X, I) \to \Sigma\mathrm{Ho}_{j+1}(X, E) \to \Sigma\mathrm{Ho}_{j+1}(X, B)$$
$$\to \Sigma\mathrm{Ho}_j(X, I) \to \Sigma\mathrm{Ho}_j(X, E) \to \Sigma\mathrm{Ho}_j(X, B) \to \cdots,$$

$$\cdots \leftarrow \Sigma\mathrm{Ho}^{j+1}(I, X) \leftarrow \Sigma\mathrm{Ho}^{j+1}(E, X) \leftarrow \Sigma\mathrm{Ho}^{j+1}(B, X)$$
$$\leftarrow \Sigma\mathrm{Ho}^j(I, X) \leftarrow \Sigma\mathrm{Ho}^j(E, X) \leftarrow \Sigma\mathrm{Ho}^j(B, X) \leftarrow \cdots.$$

Their connecting maps are, up to signs, composition with $\Sigma\mathrm{Ho}(\mathbf{E})$.

Here we abbreviate

$$\Sigma\mathrm{Ho}_j(A, B) := \Sigma\mathrm{Ho}(A, \Sigma^j B) = \Sigma\mathrm{Ho}\big(A, (B, j)\big)$$
$$\cong \Sigma\mathrm{Ho}\big((A, -j), (B, 0)\big) = \Sigma\mathrm{Ho}(\Sigma^{-j} A, B) =: \Sigma\mathrm{Ho}^{-j}(A, B)$$

for two bornological algebras A, B and $j \in \mathbb{Z}$.

Proof. We have a morphism of triangles

$$
\begin{array}{ccccccc}
(B, j+1) & \xrightarrow{(-1)^j[\gamma]} & (I, j) & \xrightarrow{[\iota]} & (E, j) & \xrightarrow{[\pi]} & (B, j) \\
\| & & \downarrow{[e]} & & \| & & \| \\
(B, j+1) & \xrightarrow{(-1)^j[\iota_\pi]} & (C(\pi), j) & \xrightarrow{[\varepsilon_\pi]} & (E, j) & \xrightarrow{[\pi]} & (B, j)
\end{array}
$$

in $\Sigma\mathrm{Ho}$, where the bounded homomorphism $e\colon I \to C(\pi)$ is defined by $e(x) :=$ $(x, 0)$ for all $x \in I$. We claim that e is invertible in $\Sigma\mathrm{Ho}$. This implies that the extension triangle is exact because it is isomorphic to a mapping cone triangle. Conversely, by Lemma 6.59, e has to be invertible if the extension triangle is exact.

Our proof of the claim follows the proof of [36, Satz 5.3].

The cokernel of the natural embedding $S(I) \to C(E)$ is naturally isomorphic to $C(\pi)$ via the map $q\colon C(E) \to C(\pi)$, $x \mapsto \big(x(1), C\pi(x)\big)$; it is easy to see that the kernel of q is $S(I)$ and that q has the bounded linear section

$$
C(\pi) \to C(E), \qquad (e, b) \mapsto \sigma(b) + \varrho \cdot \big(e - \sigma b(1)\big),
$$

where $\sigma\colon B \to E$ is a bounded linear section for π and $\varrho\colon [0, 1] \to [0, 1]$ is chosen as in §6.1. The semi-split extension $S(I) \rightarrowtail C(E) \twoheadrightarrow C(\pi)$ yields a classifying map $f\colon J\big(C(\pi)\big) \to S(I)$, which defines $[f] \in \Sigma\mathrm{Ho}(C(\pi), I)$.

The obvious embedding $C(I) \to C(E)$ yields a morphism of extensions

$$
\begin{array}{ccccc}
S(I) & \rightarrowtail & C(E) & \twoheadrightarrow & C(\pi) \\
\| & & \uparrow & & \uparrow e \\
S(I) & \rightarrowtail & C(I) & \twoheadrightarrow & I.
\end{array}
$$

Hence $f \circ J(e)$ is a classifying map for the cone extension in the second row, that is, $[f] \circ e = [\lambda_I] = [\mathrm{id}_I]$ as desired.

The morphism $[e] \circ [f]$ is represented by $S(e) \circ f\colon J\big(C(\pi)\big) \to S\big(C(\pi)\big)$. We define a bounded algebra homomorphism

$$
\Delta\colon C(E) \to C^2(E), \qquad \Delta f(s, t) := \begin{cases} f(s + t - 1) & \text{for } s + t \geq 1, \\ 0 & \text{otherwise,} \end{cases}
$$

where $C^2(E)$ denotes the cone over $C(E)$. Let

$$
e' := C(q) \circ \Delta\colon C(E) \to C\big(C(\pi)\big).
$$

Easy computations show that the following diagram commutes:

$$
\begin{array}{ccccc}
S\big(C(\pi)\big) & \rightarrowtail & C\big(C(\pi)\big) & \twoheadrightarrow & C(\pi) \\
\uparrow{S(e)} & & \uparrow{e'} & & \| \\
S(I) & \rightarrowtail & C(E) & \xrightarrow{q} & C(\pi).
\end{array}
$$

Hence $S(e) \circ f$ is a classifying map for the cone extension in the first row.

As above, this shows that $[e] \circ [f] = [\mathrm{id}_{C(\pi)}]$. Therefore, $[f]$ and $[e]$ are inverse to each other. This establishes the exactness of the extension triangle. Now the long exact sequences follow from Proposition 6.58.   □

Theorem 6.63 asserts that extension triangles are exact. Conversely, we claim that any exact triangle is isomorphic to an extension triangle. Therefore, we can also define the class of exact triangles in $\Sigma\mathrm{Ho}$ to consist of triangles isomorphic to extension triangles.

Let $f\colon A \to B$ be a bounded homomorphism. To construct a semi-split extension whose extension triangle is isomorphic to the mapping cone triangle for f, we use the *mapping cylinder*

$$Z(f) := \{(a,b) \in A \oplus B[0,1] \mid b(1) = f(a)\}. \tag{6.64}$$

The difference between $C(f)$ and $Z(f)$ is that we do not require b to vanish at 0. Given $b \in B$, let $\mathrm{const}\, b \in B[0,1]$ be the constant function with value b. Define natural bounded homomorphisms

$$\begin{aligned}
p_A &\colon Z(f) \to A, & (a,b) &\mapsto a,\\
j_A &\colon A \to Z(f), & a &\mapsto (a, \mathrm{const}\, f(a)),\\
\tilde{f} &\colon Z(f) \to B, & (a,b) &\mapsto b(0).
\end{aligned}$$

Then $p_A j_A = \mathrm{id}_A$ and $\tilde{f} j_A = f$. Check that $j_A p_A$ is smoothly homotopic to the identity map. Thus $Z(f)$ is smoothly homotopy equivalent to A, and this homotopy equivalence intertwines $\tilde{f}$ and f. We have a semi-split extension

$$\mathbf{E} := \left(C(f) \to Z(f) \xrightarrow{\tilde{f}} B \right)$$

and a commuting diagram

$$
\begin{array}{ccccc}
C(f) & \rightarrowtail & Z(f) & \xrightarrow{\tilde{f}} & B\\
\uparrow{\scriptstyle \iota_f} & & \uparrow & & \Big\|\\
S(B) & \rightarrowtail & C(B) & \twoheadrightarrow & B.
\end{array}
$$

Hence the classifying map of $\mathbf{E}$ is equal to $\iota_f \circ \lambda_B$. Therefore, the extension triangle for $\mathbf{E}$ is isomorphic to the mapping cone triangle of f. Thus any exact triangle is isomorphic to an extension triangle.

Corollary 6.65. *Let* $\mathbf{E} := (I \rightarrowtail E \twoheadrightarrow B)$ *be a semi-split extension with* $(E,0) \cong 0$ *in* $\Sigma\mathrm{Ho}$. *Then* $\Sigma\mathrm{Ho}(\mathbf{E}) \in \Sigma\mathrm{Ho}\big((B,1),(I,0)\big)$ *is invertible.*

Proof. This follows from Theorem 6.63, Lemma 6.61, and Axiom 6.52 (TR2).   □

Theorem 6.66. *Let* $\mathbf{E} := (I \xrightarrow{\iota} E \xrightarrow{\pi} B)$ *be a split extension, let* $\sigma\colon B \to E$ *be a section. Then* $(\iota, \sigma)\colon (I, 0) \oplus (B, 0) \to (E, 0)$ *is an isomorphism in* $\Sigma\mathrm{Ho}$.

Thus a quasi-homomorphism $f_{\pm}\colon A \rightrightarrows D \rhd B$ *induces an element*

$$\Sigma\mathrm{Ho}(f_{\pm}) \in \Sigma\mathrm{Ho}\big((A, 0), (B, 0)\big).$$

Proof. The first assertion follows from Theorem 6.63 and long exact sequences (compare Lemma 6.61). Any quasi-homomorphism $f_{\pm}$ yields an extension of bornological algebras $B \rightarrowtail D' \twoheadrightarrow A$ with two bounded sections $f'_{\pm}$. Identifying $D' \cong B \oplus A$ in $\Sigma\mathrm{Ho}$, we view $f'_{+} - f'_{-}$ as a morphism $(A, 0) \to (B, 0)$ in $\Sigma\mathrm{Ho}$. □

Corollary 6.67. *As in Theorem 2.41, we pull back a semi-split extension of bornological algebras* $I \rightarrowtail E \twoheadrightarrow Q$ *along a bounded homomorphism* $f\colon Q' \to Q$ *to an extension* $I \rightarrowtail E' \twoheadrightarrow Q'$. *Then there are associated Mayer–Vietoris sequences in both variables of the form*

$$\cdots \to \Sigma\mathrm{Ho}_{j+1}(X, E') \to \Sigma\mathrm{Ho}_{j+1}(X, E) \oplus \Sigma\mathrm{Ho}_{j+1}(X, Q') \to \Sigma\mathrm{Ho}_{j+1}(X, Q)$$
$$\to \Sigma\mathrm{Ho}_j(X, E') \to \Sigma\mathrm{Ho}_j(X, E) \oplus \Sigma\mathrm{Ho}_j(X, Q') \to \Sigma\mathrm{Ho}_j(X, Q) \to \cdots$$

$$\cdots \leftarrow \Sigma\mathrm{Ho}^{j+1}(E', X) \leftarrow \Sigma\mathrm{Ho}^{j+1}(E, X) \oplus \Sigma\mathrm{Ho}^{j+1}(Q', X) \leftarrow \Sigma\mathrm{Ho}^{j+1}(Q, X)$$
$$\leftarrow \Sigma\mathrm{Ho}^j(E', X) \leftarrow \Sigma\mathrm{Ho}^j(E, X) \oplus \Sigma\mathrm{Ho}^j(Q', X) \leftarrow \Sigma\mathrm{Ho}^j(Q, X) \leftarrow \cdots$$

Proof. The pulled back extension $I \rightarrowtail E' \twoheadrightarrow Q'$ is again semi-split. Hence we get long exact sequences for the two extensions $I \rightarrowtail E \twoheadrightarrow Q$ and $I \rightarrowtail E' \twoheadrightarrow Q'$. Now copy the proof of Theorem 2.41. □

6.7 The universal property of the suspension-stable homotopy category

We want to characterise the obvious functor from the category of bornological algebras (with bounded algebra homomorphisms as morphisms) to $\Sigma\mathrm{Ho}$ by a universal property.

Definition 6.68. A *homology theory for bornological algebras* is a sequence of covariant functors $(F_n)_{n \in \mathbb{Z}}$ from the category of bornological algebras to an Abelian category together with natural isomorphisms $F_n\big(S(A)\big) \cong F_{n+1}(A)$ for all $n \in \mathbb{Z}$, such that

(1) the functors F_n are *smoothly homotopy invariant*, that is, $F_n(f_0) = F_n(f_1)$ if $\langle f_0 \rangle = \langle f_1 \rangle$;

(2) the functors F_n are *half-exact* for semi-split extensions.

A *cohomology theory for bornological algebras* is defined dually as a sequence of contravariant functors satisfying analogous axioms.

We are particularly interested in the canonical functor from the category of bornological algebras to ΣHo. In order to formalise its properties, we have to consider functors with values in triangulated categories. More generally, let F be a covariant or contravariant functor from the category of bornological algebras to some additive category $\mathfrak{C}$. (Recall that Abelian categories and triangulated categories are additive.) The functor F is called *half-exact (for semi-split extensions)* if the functors $A \mapsto \mathfrak{C}(D, F(A))$ from the category of bornological algebras to the category of Abelian groups are half-exact in the usual sense of Definition 1.41. Similarly, we define what it means for F to be *split-exact*.

Definition 6.69. Let $\mathfrak{T}$ and $\mathfrak{T}'$ be triangulated categories with suspension automorphisms Σ and Σ', respectively. An *exact functor* $F \colon \mathfrak{T} \to \mathfrak{T}'$ is a functor $F \colon \mathfrak{T} \to \mathfrak{T}'$ together with natural isomorphisms $F(\Sigma A) \cong \Sigma' F(A)$ for all objects A of $\mathfrak{T}$, such that F maps exact triangles again to exact triangles.

Definition 6.70. A *triangulated homology theory for bornological algebras* is a functor F from the category of bornological algebras to a triangulated category $\mathfrak{T}$ together with natural isomorphisms $F(S(A)) \cong \Sigma(F(A))$, where Σ denotes the suspension automorphism on $\mathfrak{T}$, such that

(1) the functor F is *smoothly homotopy invariant*;

(2) the functors $A \mapsto \mathfrak{T}(X, F(A))$ are *half-exact* for semi-split extensions for all objects X of $\mathfrak{T}$;

(3) the functor F maps mapping cone triangles to exact triangles in $\mathfrak{T}$.

Proposition 6.71. *Let F be a functor on* BAlg *that is half-exact for semi-split extensions and invariant under smooth homotopies. Define $F_n(A) := F(S^n A)$ for $n \geq 0$. Then the functor F has long exact sequences of the form*

$$\cdots \to F_1(I) \to F_1(E) \to F_1(Q) \to F_0(I) \to F_0(E) \to F_0(Q)$$

for any semi-split extension $I \rightarrowtail E \twoheadrightarrow Q$. The functor F is split-exact. Hence it is functorial for quasi-homomorphisms.

Proof. First, we claim that we have Puppe sequences

$$\cdots \to F_{n+1}(C(f)) \to F_{n+1}(A) \xrightarrow{F_{n+1}(f)} F_{n+1}(B)$$

$$\to F_n(C(f)) \to F_n(A) \xrightarrow{F_n(f)} F_n(B) \to \cdots \to F_0(C(f)) \to F_0(A) \to F_0(B)$$

for any bounded algebra homomorphism $f \colon A \to B$. Exactness at $F_n(C(f))$ and $F_n(A)$ for $n \in \mathbb{N}$ follows from the semi-split exact sequences $S(B) \rightarrowtail C(f) \twoheadrightarrow A$ and $C(f) \rightarrowtail Z(f) \twoheadrightarrow B$, the smooth homotopy equivalence $Z(f) \sim A$, and half-exactness of F_n. Exactness at $F_{n+1}(B)$ is proved similarly.

Now we consider a semi-split algebra extension $I \xrightarrow{\iota} E \xrightarrow{\pi} Q$. We claim that the canonical embedding $e \colon I \to C(\pi)$ induces an isomorphism on F_n for all $n \in \mathbb{N}$.

Then the Puppe sequence yields the desired long exact sequence as in the proof of Theorem 6.63.

The proof of the claim follows [10, §21.4] and uses half-exactness of F_n for the following two semi-split extensions. The first one is $I \rightarrowtail C(\pi) \twoheadrightarrow CQ$; since CQ is smoothly contractible, $F_n(CQ) = 0$, so that $F_n(e) \colon F_n(I) \to F(C(\pi))$ must be surjective. The second one is of the form $CI \rightarrowtail Z(\iota) \twoheadrightarrow C(\pi)$, where $Z(\iota)$ denotes the mapping cylinder of ι. We compute

$$Z(\iota) \cong \{f \in E[0,1] \mid f(1) \in I\},$$

and the map $Z(\iota) \to C(\pi)$ simply maps $f \mapsto (f(0), \pi \circ f)$. As above, we conclude that this map induces an injective map $F_n(Z(\iota)) \to F_n(C(\pi))$. Finally, we recall that $Z(\iota)$ is smoothly homotopy equivalent to I. Hence $F_n(e)$ is both surjective and injective, as desired. This yields the desired long exact sequence.

If the extension splits, then the maps $F_n(E) \to F_n(Q)$ are split-surjective. Hence the maps $F_{n+1}(Q) \to F_n(I)$ vanish, so that the maps $F_n(I) \to F_n(E)$ are injective. Thus $F = F_0$ is split-exact. This yields functoriality for quasi-homomorphisms by §3.1.1. $\qquad\square$

Proposition 6.72. *If $(F_n)_{n \in \mathbb{Z}}$ is a homology theory for bornological algebras, then $\bar{F}(A, n) := F_n(A)$ defines a homological functor $\bar{F} \colon \Sigma\mathrm{Ho} \to \mathrm{Ab}$. Conversely, any such homological functor $\bar{F}$ arises from a unique homology theory for bornological algebras in this fashion.*

Similarly, there are natural bijections between cohomology theories for bornological algebras and cohomological functors $\Sigma\mathrm{Ho}^{\mathrm{op}} \to \mathrm{Ab}$, and between triangulated homology theories for bornological algebras and exact functors $\Sigma\mathrm{Ho} \to \mathfrak{T}$.

Proof. A homological functor $\bar{F} \colon \Sigma\mathrm{Ho} \to \mathrm{Ab}$ yields a homology theory for bornological algebras by $F_n(A) := \bar{F}(A, n)$; use Lemma 6.29 and Theorem 6.63 to check the first two conditions in Definition 6.68. Conversely, let (F_n) be a homology theory for bornological algebras. Proposition 6.71 applied to F_n for $n \ll 0$ yields a long exact sequence for a semi-split extension that extends indefinitely in both directions.

Since $F_*(TA) = 0$ and $F_*(CA) = 0$ for all A by smooth homotopy invariance, the long exact sequences for tensor algebra and cone extensions provide natural isomorphisms

$$F_m(A) \cong F_{-k}(J^{m+k}A), \qquad F_n(B) \cong F_{-k}(S^{n+k}B)$$

for all bornological algebras A, B and all $m, n, k \in \mathbb{Z}$ with $m + k, n + k \geq 0$. We use them to associate a map $F_m(A) \to F_n(B)$ to a bounded homomorphism $J^{m+k}(A) \to S^{n+k}(B)$. Using the naturality of the index maps for (F_n), we find that this construction is compatible with the inductive system in (6.27) that defines $\Sigma\mathrm{Ho}((A,m),(B,n))$. Moreover, it is compatible with the product $\#$. Thus we have turned $\bar{F}$ into a functor $\bar{F} \colon \Sigma\mathrm{Ho} \to \mathrm{Ab}$. This functor is homological because F is half-exact for semi-split extensions. Exactly the same arguments yield the assertions for cohomological and exact functors. $\qquad\square$

Thus $\Sigma\mathrm{Ho}$ is the *universal triangulated homology theory for bornological algebras*.

We can use the universal property of $\Sigma\mathrm{Ho}$ to construct functors $\Sigma\mathrm{Ho} \to \Sigma\mathrm{Ho}$. Consider, for example, the tensor product functor $\sigma_B(A) := A \mathbin{\widehat{\otimes}} B$ for a bornological algebra B; it maps semi-split extensions of bornological algebras again to such extensions, preserving their extension triangles. It also commutes with the functor S and satisfies $\sigma_B(A[0,1]) \cong \sigma_B(A)[0,1]$. Therefore, the functor $A \mapsto (A \mathbin{\widehat{\otimes}} B, 0)$ to $\Sigma\mathrm{Ho}$ is a triangulated homology theory. By Proposition 6.72, it induces a functor $\sigma_B \colon \Sigma\mathrm{Ho} \to \Sigma\mathrm{Ho}$, which acts on objects by $\sigma_B(A, m) := (A \mathbin{\widehat{\otimes}} B, m)$.

Exercise 6.73. Describe how σ_B acts on morphisms in $\Sigma\mathrm{Ho}$, using the canonical maps in Definition 6.22.

Chapter 7

Bivariant K-theory for bornological algebras

The category $\Sigma\mathrm{Ho}$ still lacks many important properties of K-theory like Bott periodicity. Therefore, it does not yet behave like a bivariant version of K-theory. Even more importantly, $A \mapsto \Sigma\mathrm{Ho}(\mathbb{C}, A)$ has nothing to do with K-theory.

We are going to improve upon $\Sigma\mathrm{Ho}$ and construct bivariant K-theories with these desirable properties. The remarkably simple recipe is to define

$$\mathrm{kk}_*^?(A, B) := \Sigma\mathrm{Ho}_*\big(\mathcal{K}_?(A), \mathcal{K}_?(B)\big)$$

for a suitable stabilisation functor $\mathcal{K}_?$. Now things get a bit technical because there are various possible choices for $\mathcal{K}_?$.

The smooth stabilisation is the smallest one that yields Bott periodicity and Pimsner–Voiculescu exact sequences. But we cannot compute the resulting group $\mathrm{kk}_*^?(\mathbb{C}, \mathbb{C})$. Since we want our bivariant theory to specialise to a reasonable topological K-theory for bornological algebras, we use larger stabilisations.

A good choice are the stabilisations $\mathcal{CK}^r(A)$ introduced in Chapter 3. The resulting bivariant K-theories do not depend on r because $\mathcal{CK}^r(A) \cong \mathcal{CK}^s(A)$ in $\Sigma\mathrm{Ho}$ for all s, r. We will see that

$$\mathrm{kk}^{\mathcal{CK}}(A, B) := \Sigma\mathrm{Ho}\big(\mathcal{CK}^r(A), \mathcal{CK}^r(B)\big) \tag{7.1}$$

has all the features we want, including computability of $\mathrm{kk}_*^{\mathcal{CK}}(\mathbb{C}, B)$.

Another good choice is to stabilise by Schatten ideals. The resulting bivariant K-theory $\mathrm{kk}^{\mathscr{L}}$ seems very similar to $\mathrm{kk}^{\mathcal{CK}}$.

In §7.1, we first provide a basic tool for comparing different stabilisations and then explain the ubiquity of the smooth stabilisation $\mathcal{K}_{\mathscr{S}}$. Then we define several bivariant K-theories in §7.2; we study their formal properties in §7.3 and relate them to algebraic K-theory in §7.4, using the homotopy invariance of stabilised algebraic K-theory proved in Chapter 3.

7.1 Some tricks with stabilisations

7.1.1 Comparing stabilisations

Our main purpose here is to show that the bivariant K-theory defined in (7.1) is independent of r.

Theorem 7.2. *Let A and I be bornological algebras, let $\iota\colon I \to A$ be an injective bounded algebra homomorphism, and suppose that the multiplication map on A defines a bounded bilinear map $A \times A \to I$. Then $[\iota]$ is invertible in $\Sigma\mathrm{Ho}(I,A)$.*

Proof. Since I is an ideal in A, we get a canonical map $A \to \mathcal{M}(I)$. Let $\varrho\colon [0,1] \to [0,1]$ be a function as in §6.1; thus $\varrho^2 - \varrho \in \mathbb{C}(0,1)$. We define a multiplication on $S(I) \oplus A$ by viewing (f,a) as the function $[0,1] \to A$, $t \mapsto f(t) + \varrho(t) \cdot a$; you should check that pointwise products of such functions are again of the same form, so that $S(I) + \varrho A$ becomes a bornological algebra. We get a semi-split extension

$$S(I) \rightarrowtail S(I) + \varrho A \twoheadrightarrow A.$$

Its classifying map $\gamma\colon J(A) \to S(I)$ defines a class in $\Sigma\mathrm{Ho}(A, I)$. We claim that $[\gamma] = [\iota]^{-1}$.

The claim follows from a commuting diagram of semi-split extensions

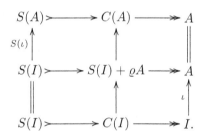

We get the vertical maps in the middle by observing that $S(I) + \varrho \cdot I = C(I)$ and $S(A) + \varrho \cdot A = C(A)$. Since classifying maps of extensions are unique up to smooth homotopy, $S(\iota) \circ \gamma\colon J(A) \to S(A)$ and $\gamma \circ J(\iota)\colon J(I) \to I$ are the classifying maps λ_A and λ_I of the cone extensions over A and I, respectively. These represent identity maps in $\Sigma\mathrm{Ho}$. □

Corollary 7.3. *The embedding $\mathcal{CK}^r(A) \to \mathcal{CK}^s(A)$ for $r \geq s \geq 0$ is invertible in $\Sigma\mathrm{Ho}$ for any bornological algebra A.*

Proof. Theorem 7.2 applies if $2s \geq r \geq s$ because of 3.23. By induction, we get the assertion for $2^n s \geq r \geq s$ and hence for all $r \geq s$. □

The stabilisations by Schatten ideals $\mathcal{K}_{\mathscr{L}^p}(A) := A \mathbin{\widehat{\otimes}} \mathscr{L}^p$ can be treated similarly:

Corollary 7.4. *For any $1 \leq p \leq q < \infty$ and any bornological algebra A, the class of the natural map $\mathcal{K}_{\mathscr{L}^p}(A) \to \mathcal{K}_{\mathscr{L}^q}(A)$ in $\Sigma\mathrm{Ho}$ is invertible.*

Now we consider the following variant of $\mathcal{CK}^r(A)$: let $\mathcal{CK}_1^r(A)$ be the set of all matrices $(T_{ij})_{i,j\in\mathbb{N}_{\geq 1}}$ for which there exists a bounded disk $S \subseteq A$ with

$$\sum_{i,j=1}^{\infty} \|v_{ij}\|_S i^a j^{r-a} (1 + \ln i)^k (1 + \ln j)^k < \infty$$

for all $a \in \mathbb{R}$, $k \in \mathbb{N}$. The only significant difference to (3.21)—which describes $\mathcal{CK}^r(A)$—is that we replace a supremum over i by a sum over i; then the symmetrisation $i \leftrightarrow j$ becomes redundant. It is straightforward to see that $\mathcal{CK}_1^r(A)$ is a bornological algebra and invariant under transposition.

Lemma 2.9 implies

$$\mathcal{CK}_1^r(A) \cong \mathcal{CK}_1^r \widehat{\otimes} A.$$

Hence we can usually drop the coefficient algebra A in arguments about $\mathcal{CK}_1^r$.

We have embeddings

$$\mathcal{CK}^{r+1+\varepsilon}(A) \subseteq \mathcal{CK}_1^r(A) \subseteq \mathcal{CK}^r(A) \tag{7.5}$$

for all $\varepsilon > 0$ because $\sum_{i=1}^{\infty} i^{-1-\varepsilon}$ converges absolutely.

Corollary 7.6. *The class of the embedding $\mathcal{CK}_1^r(A) \subseteq \mathcal{CK}^r(A)$ is an invertible element in $\Sigma\mathrm{Ho}\big(\mathcal{CK}_1^r(A), \mathcal{CK}^r(A)\big)$.*

Proof. If $r > 2$, then (7.5) shows that the multiplication on $\mathcal{CK}^r(A)$ defines a map $\mathcal{CK}^r(A) \times \mathcal{CK}^r(A) \to \mathcal{CK}_1^r(A)$, so that the assertion follows from Theorem 7.2. The same argument as above shows that the embeddings $\mathcal{CK}_1^r(A) \to \mathcal{CK}_1^s(A)$ become invertible in $\Sigma\mathrm{Ho}$, so that the assertion holds for all $r \geq 0$. $\qquad\square$

7.1.2 A general class of stabilisations

There is a general recipe for constructing stabilisations with particularly nice properties. Let $V^>$ and $V^<$ be bornological vector spaces and let

$$b: V^< \times V^> \to \mathbb{C}, \qquad (v^<, v^>) \mapsto \langle v^< \mid v^> \rangle$$

be a non-zero bounded bilinear map. Then the rule

$$(v_1^> \otimes w_1^<) \cdot (v_2^> \otimes w_2^<) := v_1^> \otimes \langle w_1^< \mid v_2^> \rangle w_2^<, \qquad v_1^>, v_2^> \in V^>, w_1^<, w_2^< \in V^<,$$

defines an associative product on $V^> \widehat{\otimes} V^<$, so that we get a bornological algebra.

Exercise 7.7. Let $A = V^> \widehat{\otimes} V^<$. Use $v^< \in V^<$, $v^> \in V^>$ with $\langle v^< \mid v^> \rangle = 1$ to construct a bounded linear A-bimodule map $A \to A \widehat{\otimes} A$ that is a section for the multiplication map. Thus the multiplication map $A \widehat{\otimes} A \to A$ is always surjective. Conclude that $\mathcal{CK}_1^r(A)$ is not of the form $V^> \widehat{\otimes} V^<$.

In most examples, $V^< = V^> = V$ and b is a non-degenerate, bounded, symmetric bilinear form on V. We write $\mathcal{K}_V$ for the bornological algebra $V \widehat{\otimes} V$ with product defined by b, and we let $\mathcal{K}_V(A) := \mathcal{K}_V \widehat{\otimes} A$.

Exercise 7.8. Consider the pairing

$$b(w, v) := \sum_{j=1}^{\infty} w_n v_n$$

for $V = \bigoplus_{n \in \mathbb{N}} \mathbb{C}$. Then $\mathcal{K}_V(A) \cong \mathbb{M}_\infty(A)$ for all A. If we use the same bilinear map b with $V = \mathbb{C}^n$ for some $n \in \mathbb{N}$, then we get $\mathcal{K}_V(A) \cong \mathbb{M}_n(A)$.

Example 7.9. If we take the same bilinear map b as in Exercise 7.8 on $V = \mathscr{S}(\mathbb{N})$ and $V = \ell^2(\mathbb{N})$, then we get the smooth compact operators $\mathcal{K}_{\mathscr{S}} = \mathscr{S}(\mathbb{N}^2)$ and the algebra $\mathscr{L}^1$ of trace class operators on the separable Hilbert space $\ell^2(\mathbb{N})$, respectively.

Example 7.10. Let V be a Banach space with Grothendieck's approximation property. Let V' be its dual space and let $b \colon V' \widehat{\otimes} V \to \mathbb{C}$ be the canonical pairing. Then $V \widehat{\otimes} V'$ may be identified with the algebra of *nuclear operators* on V (see [54]). In particular, if V is a Hilbert space, then $V = V'$ and we get the algebra $\mathscr{L}^1(\mathcal{H})$ of *trace class* operator on $\mathcal{H}$ as in Example 7.9.

Example 7.11. There is a natural isomorphism $\mathcal{K}_{V_1} \widehat{\otimes} \mathcal{K}_{V_2} \cong \mathcal{K}_{V_1 \widehat{\otimes} V_2}$, where we equip $V_1 \otimes V_2$ with the induced symmetric bilinear form. In particular, we get

$$\mathscr{L}^1 \widehat{\otimes} \mathscr{L}^1 \cong \mathcal{K}_{\ell^2(\mathbb{N}) \widehat{\otimes} \ell^2(\mathbb{N})}.$$

The projective tensor product $\ell^2(\mathbb{N}) \widehat{\otimes} \ell^2(\mathbb{N})$ is no longer a Hilbert space.

Definition 7.12. Let V_1 and V_2 be bornological vector spaces with bilinear pairings. A bounded linear map $T \colon V_1 \to V_2$ is called *isometric* if

$$\langle T(v_1) \mid T(v_2) \rangle = \langle v_1 \mid v_2 \rangle \qquad \text{for all } v_1, v_2 \in V_1.$$

Then $T \widehat{\otimes} T \widehat{\otimes} \mathrm{id}_A \colon \mathcal{K}_{V_1}(A) \to \mathcal{K}_{V_2}(A)$ is an algebra homomorphism. Such homomorphisms are called *standard*. Any $v_0 \in V$ with $\langle v_0 \mid v_0 \rangle = 1$ yields a standard homomorphism $\iota_{A,V} \colon A \to \mathcal{K}_V(A)$, which is called a *stabilisation homomorphism*.

Another important example comes from the isometries

$$V \to V \widehat{\otimes} V, \qquad v \mapsto v \widehat{\otimes} v_0, \quad v \mapsto v_0 \widehat{\otimes} v,$$

which induce standard homomorphisms $\mathcal{K}_V(A) \to \mathcal{K}_V \mathcal{K}_V(A) \cong \mathcal{K}_{V \widehat{\otimes} V}(A)$.

Definition 7.13. A functor F on the category of bornological algebras is called $\mathcal{K}_V$-*stable* if $F(\iota_{A,V}) \colon F(A) \to F(\mathcal{K}_V(A))$ is invertible for all A and all stabilisation homomorphisms (compare Definition 3.29).

Lemma 7.14. *Let* $T_0, T_1 \colon \mathcal{K}_{V_1} \to \mathcal{K}_{V_2}$ *be standard homomorphisms and let*

$$\iota \colon \mathcal{K}_{V_2} \to \mathbb{M}_2(\mathcal{K}_{V_2})$$

be the stabilisation homomorphism. Then $\iota \circ T_0$ *and* $\iota \circ T_1$ *are smoothly homotopic. Therefore,* $F(T_0) = F(T_1)$ *if* F *is* $\mathbb{M}_2$-*stable and smoothly homotopy invariant.*

Proof. Notice that $\mathbb{M}_2(\mathcal{K}_{V_2}) \cong \mathcal{K}_{V_2 \widehat{\otimes} \mathbb{C}^2}$. Let e_1, e_2 be an orthonormal basis of $\mathbb{C}^2$. We consider the smooth homotopy

$$v_1 \mapsto \sqrt{1 - t^2}\, T_0(v_1) \,\widehat{\otimes}\, e_1 + t\, T_1(v_1) \,\widehat{\otimes}\, e_2$$

between $v_1 \mapsto T_0(v_1) \widehat{\otimes} e_1$ and $v_1 \mapsto T_1(v_1) \widehat{\otimes} e_2$. This smooth homotopy consists of isometries because e_1 and e_2 are orthonormal. Concatenating this with a similar smooth homotopy from $T_1 \widehat{\otimes} e_1$ to $T_1 \widehat{\otimes} e_2$, we get a smooth homotopy of isometries $V_1 \to V_2 \widehat{\otimes} \mathbb{C}^2$ between $T_0 \widehat{\otimes} e_1$ and $T_1 \widehat{\otimes} e_1$; this yields a smooth homotopy between the associated algebra homomorphisms. $\square$

We now describe the multiplier algebra of $\mathcal{K}_V$ because we want to find inner endomorphisms of stabilisations.

A bounded linear map $T\colon V \to V$ is called *adjointable* if there is $T^*\colon V \to V$ such that $\langle Tv_1 \mid v_2 \rangle = \langle v_1 \mid T^*v_2 \rangle$ for all $v_1, v_2 \in V$; since b is non-degenerate, the operator T^* is determined uniquely by T if it exists; since b is symmetric, we have $(T^*)^* = T$. Adjointable operators on T yield multipliers of $\mathcal{K}_V$ via

$$T \cdot (v_1 \widehat{\otimes} v_2) := Tv_1 \widehat{\otimes} v_2, \qquad (v_1 \widehat{\otimes} v_2) \cdot T := v_1 \widehat{\otimes} T^*v_2.$$

Exercise 7.15. Check that any multiplier of $\mathcal{K}_V$ is of this form for some T. Thus $\mathcal{M}(\mathcal{K}_V)$ is the algebra of adjointable operators $V \to V$.

An isometric bounded linear map $T\colon V \to V$ is adjointable if and only if $T\colon V \to T(V)$ is a bornological isomorphism and $V \cong T(V) \oplus T(V)^\perp$, where $T(V)^\perp$ denotes the orthogonal complement of $T(V)$; the adjoint vanishes on $T(V)^\perp$ and is equal to T^{-1} on $T(V)$. Thus an adjointable isometry satisfies $T^*T = \mathrm{id}_V$, so that we get an inner endomorphism $\mathrm{Ad}_{T,T^*}\colon x \mapsto TxT^*$. Recall that inner endomorphisms act identically on $\mathbb{M}_2$-stable functors (Proposition 3.16).

Proposition 7.16. *Let F be a functor on* BAlg *that is $\mathbb{M}_2$-stable, and let $p \geq 1$. Then the functor $A \mapsto F\big(\mathcal{K}_{\mathscr{L}^p}(A)\big)$ is $\mathcal{K}_{\mathscr{L}^1}$- and $\mathcal{K}_{\mathscr{S}}$-stable, and $\mathbb{M}_n$-stable for all $n \in \mathbb{N} \cup \{\infty\}$.*

If $p = 1$ and if F is both $\mathbb{M}_2$-stable and smoothly homotopy invariant, this is already contained in Lemma 7.14 above.

Proof. We only write down the proof for $\mathcal{K}_{\mathscr{L}^1}$-stability. The other assertions are similar. Moreover, we may assume $A = \mathbb{C}$ for simplicity if we replace F by the functor $A' \mapsto F(A' \widehat{\otimes} A)$. We have to check that the stabilisation homomorphism $\iota\colon \mathscr{L}^p \to \mathscr{L}^1 \widehat{\otimes} \mathscr{L}^p$ induces an isomorphism on F. To get a candidate for the inverse, we use the map

$$\mu\colon \mathscr{L}^1(\mathcal{H}) \,\widehat{\otimes}\, \mathscr{L}^p(\mathcal{H}) \to \mathscr{L}^p(\mathcal{H} \,\overline{\otimes}\, \mathcal{H}) \cong \mathscr{L}^p(\mathcal{H})$$

for a separable Hilbert space $\mathcal{H}$; here $\overline{\otimes}$ denotes the Hilbert space tensor product.

We claim that $F(\mu)$ and $F(\iota)$ are inverse to each other. The endomorphism $\mu \circ \iota$ on $\mathscr{L}^p$ is the homomorphism associated to the isometry $\mathcal{H} \to \mathcal{H} \overline{\otimes} \mathcal{H} \cong \mathcal{H}$, where the first map is given by $v \mapsto v_0 \mathbin{\widehat{\otimes}} v$. Since $\mathscr{L}^p$ is an operator ideal, this isometry is a multiplier of $\mathscr{L}^p$. Thus $\mu \circ \iota$ is an inner endomorphism of $\mathscr{L}^p$. Now Proposition 3.16 yields that $F(\mu \circ \iota)$ is the identity.

The endomorphism $\iota \circ \mu$ on $\mathscr{L}^1 \mathbin{\widehat{\otimes}} \mathscr{L}^p$ is equal to the composite homomorphism

$$\mathscr{L}^1 \mathbin{\widehat{\otimes}} \mathscr{L}^p \xrightarrow{\iota' \widehat{\otimes} \mathrm{id}_{\mathscr{L}^p}} \mathscr{L}^1 \mathbin{\widehat{\otimes}} \mathscr{L}^1 \mathbin{\widehat{\otimes}} \mathscr{L}^p \xrightarrow{\mathrm{id}_{\mathscr{L}^1} \widehat{\otimes} \mu} \mathscr{L}^1 \mathbin{\widehat{\otimes}} \mathscr{L}^p,$$

where $\iota' \colon \mathscr{L}^1 \to \mathscr{L}^1 \mathbin{\widehat{\otimes}} \mathscr{L}^1$ is the stabilisation homomorphism induced by the isometry $v \mapsto v_0 \mathbin{\widehat{\otimes}} v$. The flip isomorphism Θ on $\mathcal{H} \mathbin{\widehat{\otimes}} \mathcal{H}$, $v_1 \mathbin{\widehat{\otimes}} v_2 \mapsto v_2 \mathbin{\widehat{\otimes}} v_1$, is adjointable and hence gives rise to an inner automorphism on $\mathscr{L}^1 \mathbin{\widehat{\otimes}} \mathscr{L}^1$. Therefore, $F\big((\Theta \circ \iota') \mathbin{\widehat{\otimes}} \mathrm{id}_{\mathscr{L}^p}\big) = F(\iota' \mathbin{\widehat{\otimes}} \mathrm{id}_{\mathscr{L}^p})$. Now we have

$$(\mathrm{id}_{\mathscr{L}^1} \mathbin{\widehat{\otimes}} \mu) \circ \big((\Theta \circ \iota') \mathbin{\widehat{\otimes}} \mathrm{id}_{\mathscr{L}^p}\big) = \mathrm{id}_{\mathscr{L}^1} \mathbin{\widehat{\otimes}} (\mu \circ \iota).$$

We have already seen that $\mu \circ \iota$ is an inner endomorphism on $\mathscr{L}^p$. Hence so is $\mathrm{id}_{\mathscr{L}^1} \mathbin{\widehat{\otimes}} (\mu \circ \iota)$, and $F(\iota \circ \mu)$ is the identity. $\qquad\square$

Exercise 7.17. Use Lemma 3.38 to prove that any standard homomorphism $\mathcal{K}_{\mathscr{S}} \to \mathcal{K}_{\mathscr{S}}$ is smoothly homotopic to the identity map on $\mathcal{K}_{\mathscr{S}}$.

Compare this with Lemma 7.14, where we also have to stabilise by $\mathbb{M}_2$.

7.1.3 Smooth stabilisations everywhere

Here we discuss some alternative realisations of $\mathcal{K}_{\mathscr{S}}$.

The enumeration $0, 1, -1, 2, -2, \dots$, of $\mathbb{Z}$ yields a bijection $\mathbb{N} \to \mathbb{Z}$, which induces an isometric bornological isomorphism $\mathscr{S}(\mathbb{N}) \cong \mathscr{S}(\mathbb{Z})$; here we equip $\mathscr{S}(\mathbb{Z})$ with the obvious bilinear form $\langle f_1 \mid f_2 \rangle := \sum_{n \in \mathbb{Z}} f_1(n) f_2(n)$. There are similar isometric isomorphisms $\mathscr{S}(\mathbb{N}) \cong \mathscr{S}(\mathbb{N}^n) \cong \mathscr{S}(\mathbb{Z}^n)$ for $n \geq 1$. These induce bornological algebra isomorphism

$$\mathcal{K}_{\mathscr{S}(\mathbb{N})} \cong \mathcal{K}_{\mathscr{S}(\mathbb{N}^n)} \cong \mathcal{K}_{\mathscr{S}(\mathbb{Z}^n)} \tag{7.18}$$

for all $n \in \mathbb{N}_{\geq 1}$. In particular, we have

$$\mathcal{K}_{\mathscr{S}} = \mathcal{K}_{\mathscr{S}(\mathbb{N})} \cong \mathcal{K}_{\mathscr{S}(\mathbb{N}^2)} \cong \mathcal{K}_{\mathscr{S}} \mathbin{\widehat{\otimes}} \mathcal{K}_{\mathscr{S}}.$$

The Fourier transform is an isometric bornological isomorphism $\mathscr{S}(\mathbb{Z}^n) \cong C^\infty(\mathbb{T}^n)$, where we equip $C^\infty(\mathbb{T}^n)$ with the bilinear form $b(f_1, f_2) := \int_{\mathbb{T}^n} f_1(x) f_2(x) \, \mathrm{d}x$ for the normalised Haar measure (Lebesgue measure) $\mathrm{d}x$ on $\mathbb{T}^n$.

More generally, let M be a compact Riemannian manifold and define

$$b_M \colon C^\infty(M) \times C^\infty(M) \to \mathbb{C}, \qquad b_M(f_1, f_2) := \int_M f_1(x) f_2(x) \, \mathrm{d}x,$$

where $\mathrm{d}x$ is the measure associated to the Riemannian metric. It is known that the Laplace operator on $L^2(M)$ is essentially self-adjoint and has compact resolvent. Let (φ_n) be its orthonormal eigenbasis, ordered so that the corresponding eigenvalue sequence is increasing. It is known that the map

$$\mathscr{S}(\mathbb{N}) \to L^2(M), \qquad (a_n) \mapsto \sum_{n \in \mathbb{N}} a_n \varphi_n,$$

is a bornological isomorphism onto $C^\infty(M) \subseteq L^2(M)$. It is an isometry by construction. Thus $(\mathscr{S}(\mathbb{N}), b)$ is isometrically isomorphic to $(C^\infty(M), b_M)$. In this case, we have $C^\infty(M) \mathbin{\hat{\otimes}} C^\infty(M) \cong C^\infty(M^2)$, and the product is given by the usual convolution product for integral kernels:

$$f_1 * f_2(x, y) = \int_M f_1(x, z) f_2(z, y) \, \mathrm{d}z$$

for all $f_1, f_2 \in C^\infty(M)$. Thus we get the algebra of *smoothing operators* on M.

Similarly, we may consider the Schwartz space $\mathscr{S}(\mathbb{R})$ with the bilinear form $b_{\mathbb{R}}(f_1, f_2) := \int_{\mathbb{R}} f_1(x) f_2(x) \, \mathrm{d}x$. The resulting bornological algebra $\mathcal{K}_{\mathscr{S}(\mathbb{R})}$ is isomorphic to $\mathscr{S}(\mathbb{R}^2)$ with the convolution of integral kernels as product. There is an isometric isomorphism $(\mathscr{S}(\mathbb{R}), b_{\mathbb{R}}) \cong (\mathscr{S}(\mathbb{N}), b)$. To construct it, consider the operator

$$H \colon \mathscr{S}(\mathbb{R}) \to \mathscr{S}(\mathbb{R}), \qquad H(f)(x) := x^2 f(x) - \frac{\partial^2 f}{\partial x^2}(x).$$

In physical terms, this differential operator describes the *harmonic oscillator*. This operator is usually viewed as an unbounded operator on $L^2(\mathbb{R})$. As such, it is essentially self-adjoint and has compact resolvent; its spectrum is the set $1 + 2\mathbb{N}$, and all its eigenvalues are simple. Its orthonormal eigenbasis $(\varphi_n)_{n \in \mathbb{N}}$ with $H\varphi_n = (1 + 2n) \cdot \varphi_n$ is of the form $\varphi_n(x) = P_n(x) \exp(-x^2/2)$ for suitable polynomials P_n of degree n (see [106, §10.C]). These eigenfunctions all belong to $\mathscr{S}(\mathbb{R})$; it is shown in see [106, §10.C] that the map

$$\mathscr{S}(\mathbb{N}) \to \mathscr{S}(\mathbb{R}), \qquad (a_n) \mapsto \sum_{n \in \mathbb{N}} a_n \varphi_n,$$

is an isometric bornological isomorphism.

7.2 Definition and basic properties

We define three categories $\mathrm{kk}^{\mathscr{S}}$, $\mathrm{kk}^{\mathcal{CK}}$, and $\mathrm{kk}^{\mathscr{L}}$.

The smooth stabilisation functor $A \mapsto \mathcal{K}_{\mathscr{S}}(A)$ maps semi-split extensions again to such extensions and commutes with the mapping cone construction. By the universal property of $\Sigma\mathrm{Ho}$ (Proposition 6.72), it follows that $(A, m) \mapsto (\mathcal{K}_{\mathscr{S}}(A), m)$ defines a functor $\mathcal{K}_{\mathscr{S}} \colon \Sigma\mathrm{Ho} \to \Sigma\mathrm{Ho}$. Moreover, the stabilisation homomorphisms

$$\iota_A \colon (A, m) \to (\mathcal{K}_{\mathscr{S}}(A), m)$$

yield a natural transformation from the identical functor to $\mathcal{K}_{\mathscr{S}}$. Similar remarks apply to the stabilisation functors $\mathcal{CK}^r$, $\mathcal{CK}_1^r$, and $\mathcal{K}_{\mathscr{L}^p}$.

Definition 7.19. Let A, B be objects of $\Sigma\mathrm{Ho}$ and let $r > 0$. We define

$$\mathrm{kk}_*^{\mathscr{S}}(A, B) := \Sigma\mathrm{Ho}_*\big(\mathcal{K}_{\mathscr{S}}(A), \mathcal{K}_{\mathscr{S}}(B)\big),$$

$$\mathrm{kk}_*^{\mathcal{CK}}(A, B) := \Sigma\mathrm{Ho}_*\big(\mathcal{CK}_1^r(A), \mathcal{CK}_1^r(B)\big),$$

$$\mathrm{kk}_*^{\mathscr{L}}(A, B) := \mathrm{kk}_*^{\mathscr{S}}\big(\mathcal{K}_{\mathscr{L}^1}(A), \mathcal{K}_{\mathscr{L}^1}(B)\big) = \Sigma\mathrm{Ho}_*\big(\mathcal{K}_{\mathscr{S}}\mathcal{K}_{\mathscr{L}^1}(A), \mathcal{K}_{\mathscr{S}}\mathcal{K}_{\mathscr{L}^1}(B)\big).$$

We write $\mathrm{kk}^?$ for one of these three theories if it is irrelevant which stabilisation we choose.

The usual composition product in $\Sigma\mathrm{Ho}$ turns $\mathrm{kk}_*^?$ into a $\mathbb{Z}$-graded category. The discussion above shows that there is a functor

$$\mathbf{kk}^? : \mathrm{BAlg} \to \Sigma\mathrm{Ho} \to \mathrm{kk}^?$$

that acts identically on objects and by the appropriate stabilisation functor on morphisms.

The corollaries of Theorem 7.2 show that

$$\mathcal{K}_{\mathscr{L}^p}(B) \cong \mathcal{K}_{\mathscr{L}^1}(B), \qquad \mathcal{CK}^r(A) \cong \mathcal{CK}_1^r(A) \cong \mathcal{CK}^s(A)$$

in $\Sigma\mathrm{Ho}$ for all $p \in \mathbb{R}_{\geq 1}$, $r, s \in \mathbb{R}_{\geq 1}$. Therefore, in Definition 7.19 we may use $\mathcal{K}_{\mathscr{L}^p}$ instead of $\mathcal{K}_{\mathscr{L}^1}$, and $\mathcal{CK}_1^r(A)$ instead of $\mathcal{CK}^r(A)$, and the choice of r does not matter.

Lemma 7.20. *The functor* $\mathbf{kk}^?$ *is* $\mathcal{K}_{\mathscr{S}}$-*stable and* $\mathbb{M}_n$-*stable for all* $n \in \mathbb{N} \cup \{\infty\}$. *The functor* $\mathbf{kk}^{\mathscr{L}}$ *is* $\mathcal{K}_{\mathscr{L}^1}$-*stable, whereas* $\mathbf{kk}^{\mathcal{CK}}$ *is* $\mathcal{CK}^r$- *and* $\mathcal{CK}_1^r$-*stable for all* $r \geq 0$.

Proof. The assertions about $\mathrm{kk}^{\mathscr{S}}$ and $\mathrm{kk}^{\mathcal{CK}}$ follow from Lemmas 3.36 and 3.38, those about $\mathrm{kk}^{\mathscr{L}}$ from Proposition 7.16 and the $\mathbb{M}_2$-stability of $\mathrm{kk}^{\mathscr{S}}$. □

What we denote by $\mathrm{kk}^{\mathscr{S}}$ is called $\mathrm{kk}^{\mathrm{alg}}$ in [37]. The notation $\mathrm{kk}^{\mathscr{S}}$ is more consistent with our previous notation regarding crossed products and Toeplitz algebras, which would suggest to define $\mathrm{kk}^{\mathrm{alg}}(A, B) := \Sigma\mathrm{Ho}\big(A, \mathbb{M}_\infty(B)\big)$. The definition in [37] is slightly different from ours but equivalent:

Lemma 7.21. *For each of the stabilisations* $\mathcal{K}_{\mathscr{S}}$, $\mathcal{CK}_1^r$, *and* $\mathcal{K}_{\mathscr{S}} \circ \mathcal{K}_{\mathscr{L}^1}$ *in Definition 7.19, composition with the stabilisation homomorphism* $A \to \mathcal{K}_?(A)$ *induces natural isomorphisms*

$$\Sigma\mathrm{Ho}\big(\mathcal{K}_?(A), \mathcal{K}_?(B)\big) \cong \Sigma\mathrm{Ho}\big(A, \mathcal{K}_?(B)\big).$$

Proof. In each case, we know that a functor of the form $F \circ \mathcal{K}_?$ is automatically $\mathcal{K}_?$-stable. Hence we get an isomorphism

$$\Sigma\mathrm{Ho}\big(\mathcal{K}_?(A), \mathcal{K}_?(B)\big) \cong \Sigma\mathrm{Ho}\big(\mathcal{K}_?(A), \mathcal{K}_?\mathcal{K}_?(B)\big).$$

Composing this with the functor $\mathcal{K}_?$, we get a natural map

$$\Sigma\mathrm{Ho}\big(A, \mathcal{K}_?(B)\big) \xrightarrow{\mathcal{K}_?} \Sigma\mathrm{Ho}\big(\mathcal{K}_?(A), \mathcal{K}_?\mathcal{K}_?(B)\big) \xrightarrow{\cong} \Sigma\mathrm{Ho}\big(\mathcal{K}_?(A), \mathcal{K}_?(B)\big).$$

We postpone the verification that this map is inverse to the map in the converse direction induced by the stabilisation homomorphism $A \to \mathcal{K}_?(A)$ because this will become clearer when we describe our new categories as localisations, see the proof of Theorem 13.7. $\qquad\square$

Next we turn $\mathrm{kk}^?$ into a triangulated category. The constructions are the same for all stabilisations. The suspension automorphism and the class of exact triangles in $\mathrm{kk}^?$ are defined exactly as for $\Sigma\mathrm{Ho}$: a triangle is called exact if it is isomorphic to the $\mathbf{kk}^?$-image of a mapping cone triangle.

Proposition 7.22. *This additional structure turns* $\mathrm{kk}^?$ *into a triangulated category. The functor* $\mathbf{kk}^? \colon \Sigma\mathrm{Ho} \to \mathrm{kk}^?$ *is exact and has the following universal property. Let* $F \colon \Sigma\mathrm{Ho} \to \mathfrak{T}$ *be an exact functor into a triangulated category* $\mathfrak{T}$*. Then:*

- *F factors through* $\mathbf{kk}^{\mathscr{S}}$ *if and only if F is* $\mathcal{K}_{\mathscr{S}}$*-stable;*
- *F factors through* $\mathbf{kk}^{C\mathcal{K}}$ *if and only if F is* $C\mathcal{K}^r$*-stable;*
- *F factors through* $\mathbf{kk}^{\mathscr{L}}$ *if and only if F is* $\mathbb{M}_2$*- and* $\mathcal{K}_{\mathscr{L}^1}$*-stable.*

These factorisations are unique if they exist, and similar factorisations exist for homological and cohomological functors.

Proof. Consider $\mathrm{kk}^{\mathscr{S}}$ first. The verification of Axioms TR0–3 is almost literally the same as for $\Sigma\mathrm{Ho}$. It is crucial for Axiom TR1 that the stabilisation homomorphism $A \to \mathcal{K}_{\mathscr{S}}(A)$ is invertible in $\mathrm{kk}^{\mathscr{S}}$ (Lemma 7.20). This means that the range and source of a homomorphism $(J^{m+k}\mathcal{K}_{\mathscr{S}}(A), -k) \to (S^{n+k}\mathcal{K}_{\mathscr{S}}(B), -k)$ are isomorphic to (A, m) and (B, n), respectively. As for $\Sigma\mathrm{Ho}$, we will check the Octahedral Axiom in §13.2. The exactness of the functor $\mathbf{kk}^{\mathscr{S}}$ is trivial.

Since $\mathbf{kk}^{\mathscr{S}}$ is $\mathcal{K}_{\mathscr{S}}$-stable, only $\mathcal{K}_{\mathscr{S}}$-stable functors can factor through it. Conversely, let F be $\mathcal{K}_{\mathscr{S}}$-stable. Then a morphism $\mathcal{K}_{\mathscr{S}}(A) \to \mathcal{K}_{\mathscr{S}}(B)$ induces a map $F(A) \cong F\big(\mathcal{K}_{\mathscr{S}}(A)\big) \to F\big(\mathcal{K}_{\mathscr{S}}(B)\big) \cong F(B)$. It is easy to check that this defines a functor on $\mathrm{kk}^{\mathscr{S}}$ that extends F. This is the only possible extension of F to $\mathrm{kk}^{\mathscr{S}}$. Since exact triangles in $\Sigma\mathrm{Ho}$ and $\mathrm{kk}^{\mathscr{S}}$ are defined in the same way, this extension inherits the property of being exact, homological, or cohomological, respectively.

Similar arguments work for $\mathrm{kk}^{C\mathcal{K}}$ and $\mathrm{kk}^{\mathscr{L}}$. To get the universal property of $\mathbf{kk}^{\mathscr{L}}$, we proceed in two steps and first factor F through $\mathbf{kk}^{\mathscr{S}}$ using Proposition 7.16, which shows that F is $\mathcal{K}_{\mathscr{S}}$-stable. $\qquad\square$

Theorem 7.23. *Extension triangles for semi-split extensions of bornological algebras are exact in* $\mathrm{kk}^?$*. Hence we have long exact sequences for the morphism spaces in these categories as in Theorem 6.63.*

We have Puppe exact sequences and Mayer–Vietoris sequences as in Corollary 6.67, and quasi-homomorphisms induce morphisms in $\mathrm{kk}^?$ *as in Theorem 6.66.*

Proof. This follows from the corresponding properties of $\Sigma \mathrm{Ho}$ because the functor $\mathbf{kk}^?$ is exact. $\square$

7.3 Bott periodicity and related results

In this section, we establish Bott periodicity and Pimsner–Voiculescu exact sequences for our new categories. This can be done most easily by establishing these properties for any functor with some formal properties.

Theorem 7.24. *Let F be a functor on BAlg that is smoothly homotopy invariant, half-exact, and $\mathcal{K}_{\mathscr{S}}$-stable. Then F satisfies Bott periodicity, that is, there is a natural isomorphism $F(S^2 A) \cong F(A)$ for all A.*

Proof. This is shown by going through the proof of Bott periodicity for topological K-theory of local Banach algebras in Chapter 4 and checking that all the steps still work. For any bornological algebra A, we have the smooth Toeplitz extension

$$\mathcal{K}_{\mathscr{S}}(A) \rightarrowtail \mathcal{T}_{\mathscr{S}}^0(A) \to C_0^\infty(\mathbb{S}^1 \smallsetminus \{1\}, A).$$

The reparametrisation trick from §6.1 shows that $C_0^\infty(\mathbb{S}^1 \smallsetminus \{1\}, A)$ is smoothly homotopy equivalent to SA. The long exact sequence from Proposition 6.71 involves a natural map

$$\beta \colon F_2(A) = F_1(SA) \to F(\mathcal{K}_{\mathscr{S}} A) \cong F(A)$$

because F is $\mathcal{K}_{\mathscr{S}}$-stable. It is an isomorphism if $F_*\big(\mathcal{T}_{\mathscr{S}}^0(A)\big) = 0$ for $* = 0, 1$. The proof of the corresponding assertion for K-theory in Chapter 4 only uses that K-theory is $\mathcal{K}_{\mathscr{S}}$-stable, smoothly homotopy invariant, and split-exact. Therefore, it still applies in our more general situation. $\square$

Corollary 7.25. *The category $\mathbf{kk}^?$ satisfies Bott periodicity: there are natural isomorphisms $\Sigma^2(A) \cong A$ for all objects A.*

Proof. Theorem 7.24 applies to the functor $\mathbf{kk}^?$. We get $(S^2 A, 0) \cong (A, 0)$ in $\mathbf{kk}^?$ for all bornological algebras A. This implies $\Sigma^2(A, m) \cong \Sigma^m(S^2 A, 0) \cong \Sigma^m(A, 0) = (A, m)$ for all $m \in \mathbb{Z}$, as desired. $\square$

Corollary 7.25 shows that $(A, n) \cong (S^m A, 0)$ in $\mathbf{kk}^?$, where $m \in \mathbb{N}$ is arbitrary with $n \equiv m \bmod 2$. That is, any object is isomorphic to one of the form $(A, 0)$. Therefore, the additional parameter n becomes irrelevant.

The above proof is based on a close relationship between $\mathbb{C}(0, 1)$ and $C^\infty(\mathbb{S}^1)$. If we wanted to work purely algebraically, we could try the dense subalgebras

$$t(t-1)\mathbb{C}[t] = \{f \in \mathbb{C}[t] \mid f(0) = f(1) = 0\},$$
$$(t-1)\mathbb{C}[t, t^{-1}] = \{f \in \mathbb{C}[t, t^{-1}] \mid f(1) = 0\}.$$

But these two algebras are quite different. This is why Bott periodicity fails if we merely require $\mathbb{M}_\infty$-stability, or if we replace smooth homotopy invariance by invariance under polynomial homotopies. In such a situation, the relevant Toeplitz extension involves $(t-1)\mathbb{C}[t, t^{-1}]$ instead of $t(t-1)\mathbb{C}[t]$.

Theorem 7.26. *Let $\mathfrak{C}$ be an Abelian category and let F be a covariant (or contravariant) functor from the category of bornological algebras to $\mathfrak{C}$ with the following properties:*

(1) *$F(f_0) = F(f_1)$ if f_0 and f_1 are smoothly homotopic;*

(2) *F is half-exact for semi-split extensions of bornological algebras;*

(3) *F is $\mathbb{M}_2$-stable and $\mathcal{K}_{\mathscr{L}^1}$-stable.*

Then $F = \bar{F} \circ \mathbf{kk}^{\mathscr{L}}$ for a unique (co)homological functor $\bar{F} \colon \mathbf{kk}^{\mathscr{L}} \to \mathfrak{C}$.

A natural transformation $\Phi \colon F_1(A) \to F_2(A)$ between two such functors remains natural on $\mathrm{kk}^{\mathscr{L}}$, that is, we have commuting diagrams

$$
\begin{array}{ccc}
F_1(A) & \xrightarrow{\bar{F}(f)} & F_1(B) \\
\Big\downarrow{\scriptstyle \Phi_A} & & \Big\downarrow{\scriptstyle \Phi_B} \\
F_2(A) & \xrightarrow{\bar{F}(f)} & F_2(B)
\end{array}
$$

for $f \in \mathrm{kk}_0^{\mathscr{L}}(A, B)$ and not just for bounded algebra homomorphisms.

We get analogous assertions for $\mathrm{kk}^{\mathscr{S}}$ and $\mathrm{kk}^{\mathcal{CK}}$, where we require $\mathcal{K}_{\mathscr{S}}$-stability and $\mathcal{CK}$-stability, respectively, instead of (3).

Proof. By Proposition 7.16, the functor F is $\mathcal{K}_{\mathscr{S}}$-stable as well. Hence Theorem 7.24 yields that F satisfies Bott periodicity. Therefore, we may define $F_n(A)$ also for $n < 0$ by periodicity. Now the universal property of $\Sigma\mathrm{Ho}$ (Proposition 6.72) shows that F factors through $\Sigma\mathrm{Ho}$. By Proposition 7.22, we can further factor F through $\mathrm{kk}^{\mathscr{L}}$. Similar constructions work for $\mathrm{kk}^{\mathscr{S}}$ and $\mathrm{kk}^{\mathcal{CK}}$. To get the unique extension of natural transformations, we observe that any morphism in $\mathrm{kk}^{\mathscr{L}}$ is a product of bounded algebra homomorphisms and inverses of such. Hence naturality for bounded algebra homomorphisms implies naturality for all morphisms in $\mathrm{kk}^{\mathscr{L}}$. $\qquad\square$

A similar criterion exists for exact functors $\mathrm{kk}^? \to \mathfrak{T}$, where $\mathfrak{T}$ is a triangulated category. The requirements are similar to the definition of a triangulated homology theory (Definition 6.70): in order to descend to $\mathrm{kk}^?$, we need F to be smoothly homotopy invariant, half-exact for semi-split extensions, exact on mapping cone triangles, and appropriately stable. We do not need a sequence of functors $(F_n)_{n \in \mathbb{Z}}$ as in Definition 6.70 because we get this for free from Bott periodicity.

The proof of the Pimsner–Voiculescu exact sequence can also be generalised; we use the notation introduced in §5.1.

Theorem 7.27. *Let F be a functor from BAlg to an Abelian category. Suppose that F is smoothly homotopy invariant, half-exact on semi-split extensions, and $\mathcal{K}_{\mathscr{S}}$-stable. Let A be a bornological algebra equipped with an automorphism α.*

Then the canonical map $j_T \colon A \to T_{\mathrm{alg}}(A, \alpha)$ induces an isomorphism $F(A) \cong F\big(T_{\mathrm{alg}}(A, \alpha)\big)$, and there is a natural exact sequence of the form

$$F_0(A) \xrightarrow{\ \mathrm{id} - \alpha_* \ } F_0(A) \xrightarrow{\ j_U * \ } F_0\big(U_{\mathrm{alg}}(A, \alpha)\big)$$

$$F_1\big(U_{\mathrm{alg}}(A, \alpha)\big) \xleftarrow{\ j_U * \ } F_1(A) \xleftarrow{\ \mathrm{id} - \alpha_* \ } F_1(A).$$

If α generates a uniformly bounded representation of $\mathbb{Z}$, then the embeddings $T_{\mathrm{alg}}(A, \alpha) \to T_{\mathscr{S}}(A, \alpha)$ and $U_{\mathrm{alg}}(A, \alpha) \to U_{\mathscr{S}}(A, \alpha)$ induce isomorphisms on F.

Proof. Theorem 7.26 shows that F factors through a cohomological functor on $\mathrm{kk}^{\mathscr{S}}$. Hence F is functorial for quasi-homomorphisms. Following the argument in §5.2, we show that $j_T \colon A \to T_{\mathrm{alg}}(A, \alpha)$ induces an isomorphism on F for all A. Here we need $\mathbb{M}_\infty$-stability, which follows from $\mathcal{K}_{\mathscr{S}}$-stability by Lemma 3.38. The same argument works for the embedding $j_T \colon A \to T_{\mathscr{S}}(A, \alpha)$ whenever the latter Toeplitz algebra is defined. Hence we have $F\big(T_{\mathrm{alg}}(A, \alpha)\big) \cong F(A) \cong F\big(T_{\mathscr{S}}(A, \alpha)\big)$.

Using the algebraic crossed Toeplitz extension

$$\mathbb{M}_\infty(A) \rightarrowtail T_{\mathrm{alg}}(A, \alpha) \twoheadrightarrow U_{\mathrm{alg}}(A, \alpha),$$

we then get the long exact sequence that computes $F_*\big(U_{\mathrm{alg}}(A, \alpha)\big)$. If $U_{\mathscr{S}}(A, \alpha)$ is defined, then we also get a corresponding exact sequence computing $F_*\big(U_{\mathscr{S}}(A, \alpha)\big)$, and there is a natural transformation between these two exact sequences. By the Five Lemma, we conclude that the embedding $U_{\mathrm{alg}}(A, \alpha) \to U_{\mathscr{S}}(A, \alpha)$ induces an isomorphism on F. $\qquad\square$

Theorem 7.28. *Let A be a bornological algebra and let α be an automorphism of A. The canonical map $A \to T_{\mathrm{alg}}(A, \alpha)$ is invertible in $\mathrm{kk}^?$, and there is a natural exact triangle*

$$\Sigma U_{\mathrm{alg}}(A, \alpha) \to A \xrightarrow{\ \mathrm{id} - \alpha \ } A \xrightarrow{\ j_U \ } U_{\mathrm{alg}}(A, \alpha).$$

If α generates a uniformly bounded representation of $\mathbb{Z}$, then the canonical embeddings $T_{\mathrm{alg}}(A, \alpha) \to T_{\mathscr{S}}(A, \alpha)$ and $U_{\mathrm{alg}}(A, \alpha) \to U_{\mathscr{S}}(A, \alpha)$ are invertible as well.

Proof. Apply Theorem 7.27 to the functor $\mathrm{kk}^?$. $\qquad\square$

Conversely, Theorem 7.28 yields back Theorem 7.27 because of the universal property of $\mathrm{kk}^{\mathscr{S}}$ (Theorem 7.26). Thus the two versions of the Pimsner–Voiculescu sequence are equivalent, even though Theorem 7.27 may seem more general.

Corollary 7.29. *Let (A_1, α_1) and (A_2, α_2) be bornological algebras with automorphisms. Let $f \colon A_1 \to A_2$ be a bounded algebra homomorphism that intertwines α_1 and α_2, and let $\hat{f} \colon U_{\mathrm{alg}}(A_1, \alpha_1) \to U_{\mathrm{alg}}(A_2, \alpha_2)$ be the induced homomorphism. Let F be a functor as in Theorem 7.27. If $F(f) \colon F(A_1) \to F(A_2)$ is invertible, then so is $F(\hat{f})$.*

Proof. Copy the proof of Corollary 5.13. ☐

There is an analogue of Corollary 5.14 as well; the only difference is that we now need a smooth deformation of the automorphism.

Example 7.30. For the trivial automorphism on $\mathbb{C}$, we have

$$U_{\mathrm{alg}}(\mathbb{C},\mathrm{id}) = \mathbb{C}[t,t^{-1}], \qquad U_{\mathscr{S}}(\mathbb{C},\mathrm{id}) = (\mathscr{S}(\mathbb{Z}),*) \cong C^{\infty}(\mathbb{S}^1).$$

Theorem 7.28 yields that the canonical embedding $\mathbb{C}[t,t^{-1}] \to C^{\infty}(\mathbb{S}^1)$ is invertible in $\mathrm{kk}^?$. Moreover, split-exactness easily implies $C^{\infty}(\mathbb{S}^1) \cong \mathbb{C} \oplus \Sigma\mathbb{C}$.

Example 7.31. We consider rotation algebras once again. The crossed product

$$\mathbb{C}[\mathbb{T}_{\vartheta}] := U_{\mathrm{alg}}(\mathbb{C}[t,t^{-1}], \varrho_{\vartheta})$$

consists of Laurent polynomials in the two non-commuting variables U, V, subject to the relation $UV = \exp(2\pi i\vartheta)VU$. Let $\mathscr{S}(\mathbb{T}_{\vartheta})$ be the smooth crossed product of the corresponding action on $C^{\infty}(\mathbb{S}^1)$. That is, $\mathscr{S}(\mathbb{T}_{\vartheta})$ consists of series $\sum a_{mn}U^m V^n$ with $(a_{mn}) \in \mathscr{S}(\mathbb{Z}^2)$. We claim that the natural embedding $\mathbb{C}[\mathbb{T}_{\vartheta}] \to C^{\infty}(\mathbb{T}_{\vartheta})$ becomes an invertible morphism in $\mathrm{kk}^?$.

To see this, we go via the intermediate step $U_{\mathrm{alg}}(C^{\infty}(\mathbb{S}^1), \varrho_{\vartheta})$; this is equivalent to the subalgebra $\mathbb{C}[\mathbb{T}_{\vartheta}] = U_{\mathrm{alg}}(\mathbb{C}[t,t^{-1}], \varrho_{\vartheta})$ by Corollary 7.29, and equivalent to $U_{\mathscr{S}}(C^{\infty}(\mathbb{S}^1), \varrho_{\vartheta}) = C^{\infty}(\mathbb{T}_{\vartheta})$ by the second part of Theorem 7.28.

The analogue of Corollary 5.14 yields that the algebraic and smooth rotation algebras for different ϑ become isomorphic in $\mathrm{kk}^?$. For $\vartheta = 0$, we compute

$$C^{\infty}(\mathbb{T}^2) \cong (\mathbb{C} \oplus \Sigma\mathbb{C})^{\hat{\otimes}2} \cong \mathbb{C} \oplus \mathbb{C} \oplus \Sigma\mathbb{C} \oplus \Sigma\mathbb{C} \qquad \text{in } \mathrm{kk}^?.$$

Therefore, no $\mathcal{K}_{\mathscr{S}}$-stable homology theory for bornological algebras can distinguish between $C^{\infty}(\mathbb{T}_{\vartheta})$ for some ϑ and the rather trivial object $\mathbb{C}\oplus\mathbb{C}\oplus\Sigma\mathbb{C}\oplus\Sigma\mathbb{C}$, that is, we have

$$F_*\big(C^{\infty}(\mathbb{T}_{\vartheta})\big) \cong F_*(\mathbb{C}) \oplus F_*(\mathbb{C}) \oplus F_{*+1}(\mathbb{C}) \oplus F_{*+1}(\mathbb{C}).$$

This is an instance of a *universal coefficient theorem*. We will examine universal coefficient theorems in greater detail in §13.1.1.

Since the embedding $\mathscr{S}(\mathbb{T}_{\vartheta}) \to C^*(\mathbb{T}_{\vartheta})$ is isoradial, K_0 does not distinguish between these two bornological algebras. But $\mathscr{S}(\mathbb{T}_{\vartheta})$ and $C^*(\mathbb{T}_{\vartheta})$ are not isomorphic in, say, $\mathrm{kk}^{\mathscr{L}}$. This can be seen using periodic cyclic cohomology, which is an $\mathscr{L}^1$-stable cohomology theory for bornological algebras and hence factors through $\mathrm{kk}^{\mathscr{L}}$. It is known that the above universal coefficient theorem *fails* for the periodic cyclic cohomology of $C^*(\mathbb{T}_0) \cong C(\mathbb{T}^2)$.

7.4 K-theory versus bivariant K-theory

Now we relate our bivariant theories $\mathrm{kk}^{\mathcal{CK}}$ and $\mathrm{kk}^{\mathscr{L}}$ back to K-theory. The main results are that there are natural isomorphisms

$$\mathrm{kk}_0^{\mathscr{L}}(\mathbb{C}, A) \cong \mathrm{K}_0\big(\mathcal{K}_{\mathscr{L}^p}(B)\big), \qquad \mathrm{kk}_0^{\mathcal{CK}}(\mathbb{C}, A) \cong \mathrm{K}_0\big(\mathcal{CK}^r(B)\big)$$

for all bornological algebras A and all $p > 1$, $0 < r < 1$. Thus the bivariant K-theory groups $\mathrm{kk}^?_*(A, B)$ specialise to a kind of topological K-theory of B if $A = \mathbb{C}$. This follows from the homotopy invariance results. We do not know what happens for $\mathrm{kk}^{\mathscr{S}}$.

Definition 7.32. Let A be a bornological algebra and let $1 > r > 0$. We define $\mathrm{K}_0^{\mathrm{top}}(A) := \mathrm{K}_0\big(\mathcal{CK}^r(A)\big)$.

It is also possible to use $\mathcal{K}_{\mathscr{L}^p}(A)$ instead; this leads to a similar theory with $\mathrm{kk}^{\mathscr{L}}$ playing the role of $\mathrm{kk}^{\mathcal{CK}}$. We know no examples where the two theories differ.

Theorem 7.33. *Let $p > 1$. There is a natural isomorphism*

$$\mathrm{K}_0^{\mathrm{top}}(A) \cong \mathrm{kk}_0^{\mathcal{CK}}(\mathbb{C}, A)$$

for all bornological algebras A. It maps the class of an idempotent $e \in \mathrm{Idem}\, A$ to the $\mathrm{kk}^{\mathcal{CK}}$-class of the homomorphism $\mathbb{C} \to A$, $\lambda \mapsto \lambda\, e$.

Thus $\mathrm{K}_0^{\mathrm{top}}(A)$ does not depend on the choice of $r \in (0, 1)$.

Proof. Let A be a unital algebra. Then any class in $\mathrm{K}_0(A)$ comes from an idempotent $e \in \mathbb{M}_\infty(A) \subseteq \mathcal{K}_{\mathscr{S}}(A)$ and hence gives rise to a bounded homomorphism $\mathbb{C} \to \mathcal{K}_{\mathscr{S}}(A)$. Since $\mathrm{kk}^{\mathcal{CK}}$ is $\mathbb{M}_2$-stable, conjugation by inner automorphisms operates trivially on $\mathrm{kk}_0^{\mathcal{CK}}(\mathbb{C}, A)$. Therefore, similar idempotents give rise to the same class in $\mathrm{kk}_0^{\mathcal{CK}}(\mathbb{C}, A)$. Thus we get a well-defined natural map $\mathrm{K}_0(A) \to \mathrm{kk}_0^{\mathcal{CK}}(\mathbb{C}, A)$ for unital A. Using split-exactness of K_0 and $\mathrm{kk}^{\mathcal{CK}}$ for the extension $A \rightarrowtail A_{\mathbb{C}}^+ \twoheadrightarrow \mathbb{C}$, we extend this natural transformation to non-unital algebras.

Lemma 7.20 yields a natural transformation

$$\alpha \colon \mathrm{K}_0^{\mathrm{top}}(A) := \mathrm{K}_0\big(\mathcal{CK}^r(A)\big) \to \mathrm{kk}_0^{\mathcal{CK}}\big(\mathbb{C}, \mathcal{CK}^r(A)\big) \cong \mathrm{kk}_0^{\mathcal{CK}}(\mathbb{C}, A).$$

It is easy to see that α sends the class of an idempotent $e \in \mathrm{Idem}\, A$ to the class of the associated bounded homomorphism $\varphi \colon \mathbb{C} \to A$.

We want to construct a map in the converse direction. Let $e_{\mathbb{C}} \in \mathrm{K}_0^{\mathrm{top}}(\mathbb{C})$ be the class of a rank-1 idempotent. The functor $\mathrm{K}_0^{\mathrm{top}}$ is half-exact (Theorem 1.44), $\mathbb{M}_2$-stable (easy fact), and $\mathcal{CK}^r$-stable (Corollary 3.37), and invariant under Hölder continuous homotopies and therefore under smooth homotopies (Corollary 3.33). By the universal property of $\mathbf{kk}^{\mathcal{CK}}$ (Theorem 7.26), we can factor $\mathrm{K}_0^{\mathrm{top}}$ through $\mathbf{kk}^{\mathcal{CK}}$. Thus we can define a map

$$\alpha^{-1} \colon \mathrm{kk}_0^{\mathcal{CK}}(\mathbb{C}, A) \to \mathrm{K}_0^{\mathrm{top}}(A), \qquad f \mapsto e_{\mathbb{C}} \# f.$$

The map $\alpha \colon \mathrm{K}_0^{\mathrm{top}}(A) \to \mathrm{kk}_0^{\mathcal{CK}}(\mathbb{C}, A)$ above is clearly natural for bounded algebra homomorphisms. By the universal property of $\mathrm{kk}^{\mathcal{CK}}$ (Theorem 7.26), this transformation is natural with respect to morphisms in $\mathrm{kk}^{\mathcal{CK}}$ as well. Since $\alpha(e_{\mathbb{C}})$ is the identity morphism on $\mathbb{C}$, we get $\alpha(e_{\mathbb{C}} \# f) = \mathrm{id}_{\mathbb{C}} \# f = f$ for all $f \in \mathrm{kk}_0^{\mathcal{CK}}(\mathbb{C}, A)$, that is, $\alpha \circ \alpha^{-1}$ is the identity map on $\mathrm{kk}_0^{\mathcal{CK}}(\mathbb{C}, A)$.

To finish the proof, it remains to show that the map α^{-1} above is surjective. If $e \in \mathrm{Idem}\, A$, then e generates a bounded homomorphism $\varphi\colon \mathbb{C} \to A$, which has a class in $\mathrm{kk}_0^{\mathcal{CK}}(\mathbb{C}, A)$. It is not hard to see that $\alpha^{-1}[\varphi] = [e]$, so that $[e]$ belongs to the range of α^{-1}. More generally, if $e \in \mathrm{Idem}\,\mathcal{CK}^r(A)$, then e generates a bounded homomorphism $\varphi\colon \mathbb{C} \to \mathcal{CK}^r(A)$, which determines a class $[\varphi]$ in $\mathrm{kk}_0^{\mathcal{CK}}(\mathbb{C}, A)$ because $\mathcal{CK}^r(A) \cong A$ in $\mathrm{kk}^{\mathcal{CK}}$. One can show that $\alpha^{-1}[\varphi] = [e]$, so that $[e]$ belongs to the range of α^{-1} as well. Finally, any element of $\mathrm{K}_0^{\mathrm{top}}(A)$ is represented by a formal difference of idempotents $e_+, e_- \in \mathrm{Idem}\,\mathcal{CK}^r(A)^+$ with $e_+ - e_- \in \mathcal{CK}(A)$. Using split-exactness, one shows that the class of (e_+, e_-) belongs to the range of α^{-1}. Hence α and α^{-1} are inverse to each other. $\qquad\square$

Corollary 7.34. *We have* $\mathrm{kk}_1^{\mathcal{CK}}(\mathbb{C}, \mathbb{C}) = 0$ *and* $\mathrm{kk}_0^{\mathcal{CK}}(\mathbb{C}, \mathbb{C}) \cong \mathbb{Z}$ *with generator* $[\mathrm{id}_{\mathbb{C}}]$.

By Bott periodicity, this determines the groups $\mathrm{kk}_n^{\mathcal{CK}}(\mathbb{C}, \mathbb{C})$ for all $n \in \mathbb{Z}$. Even if we only want to compute $\mathrm{kk}_*^{\mathcal{CK}}(\mathbb{C}, \mathbb{C})$, there seems no way to avoid computing $\mathrm{kk}_*^{\mathcal{CK}}(\mathbb{C}, A)$ for all A at the same time.

We can replace $\mathcal{CK}^r$ by $\mathcal{K}_{\mathscr{L}^p}$ for $p > 1$ in the above arguments and get a corresponding isomorphism

$$\mathrm{kk}^{\mathscr{L}}(\mathbb{C}, A) \cong \mathrm{K}_0\big(\mathcal{K}_{\mathscr{L}^p}(A)\big)$$

for all $p > 1$. Hence there is an analogue of Corollary 7.34 for $\mathrm{kk}^{\mathscr{L}}$. In contrast, it is unclear what happens for $\mathrm{kk}^{\mathscr{S}}$.

The isomorphism $\mathrm{K}_0^{\mathrm{top}}(A) \cong \mathrm{kk}_0^{\mathcal{CK}}(\mathbb{C}, A)$ can often be used to compute $\mathrm{K}_0^{\mathrm{top}}(A)$. But since this approach is indirect, it may be hard to find explicit generators. We illustrate this by an example:

Example 7.35. Theorem 7.33 and Example 7.31 yield $\mathrm{K}_*^{\mathrm{top}}(\mathbb{C}[\mathbb{T}_\vartheta]) = \mathbb{Z}^2$ for $* = 0, 1$, for all $\vartheta \in [0, 1]$. The two generators of K_1 and one of the generators of K_0 are easy to describe, see Example 5.12. The remaining generator for K_0 can be represented by an explicit idempotent in $\mathbb{C}[\mathbb{T}_\vartheta]$ if ϑ is *irrational*. In contrast, for $\vartheta = 0$, where we get $\mathbb{C}[\mathbb{T}_0] = \mathbb{C}[U, V, U^{-1}, V^{-1}]$, it is known that $\mathrm{K}_0\big(\mathbb{C}[\mathbb{T}_0]\big) \cong \mathbb{Z}$ (see [109, Corollary 3.2.13]). Therefore, the additional generator cannot be represented by an idempotent in a matrix algebra over $\mathbb{C}[\mathbb{T}_0]$.

7.4.1 Comparison with other topological K-theories

We want to compare $\mathrm{K}_0^{\mathrm{top}}$ with other existing topological K-theories for Banach algebras and Fréchet algebras. We consider the case of (local) Banach algebras first:

Proposition 7.36. *The stabilisation homomorphism* $A \to \mathcal{CK}^r(A)$ *induces an isomorphism* $\mathrm{K}_0(A) \xrightarrow{\cong} \mathrm{K}_0\big(\mathcal{CK}^r(A)\big) = \mathrm{K}_0^{\mathrm{top}}(A)$ *if A is a local Banach algebra.*

Proof. Both $\mathbb{M}_\infty(A)$ and $\mathcal{CK}^r(A)$ are local Banach algebras, and the embedding $\mathbb{M}_\infty(A) \to \mathcal{CK}^r(A)$ is isoradial. Hence Theorem 2.60 yields the assertion. $\qquad\square$

Next we turn our attention to locally multiplicatively convex Fréchet algebras. By definition, these are complete topological algebras whose topology can be defined by an increasing sequence of submultiplicative semi-norms. If A is such an algebra and $(\nu_n)_{n\in\mathbb{N}}$ is such a sequence of semi-norms, then the completion of A with respect to ν_n is a Banach algebra A_n. These Banach algebras form a projective system, and the maps $A_n \to A_m$ have dense range for all $n \geq m$. Its projective limit is naturally isomorphic to A. Thus any locally multiplicatively convex Fréchet algebra is a projective limit of a countable projective system of Banach algebras. Conversely, any such projective limit is a locally multiplicatively convex Fréchet algebra.

Example 7.37. Let $X = \varinjlim K_n$ be the union of an increasing sequence (K_n) of compact spaces, equipped with the direct limit topology. Then the algebra of continuous functions $X \to \mathbb{C}$ without growth restriction is $C(X) := \varprojlim C(K_n)$. This is a projective limit of Banach algebras and thus a locally multiplicatively convex Fréchet algebra.

N. Christopher Phillips [100] defines a topological K-theory for locally multiplicatively convex Fréchet algebras, which he calls *representable K-theory* because it agrees with the representable K-theory of X in the situation of Example 7.37. Its main feature is that it can be computed by a Milnor $\varprojlim^1$ sequence. Namely, if we write a locally multiplicatively convex Fréchet algebra A as a projective limit $A = \varprojlim A_n$ as above, then there is a natural exact sequence

$$\varprojlim{}^1 \mathrm{K}_{*+1}(A_n) \rightarrowtail \mathrm{K}_*(A) \twoheadrightarrow \varprojlim \mathrm{K}_*(A_n).$$

This sequence is very useful for computations.

Theorem 7.38. *For locally multiplicatively convex Fréchet algebras, $\mathrm{K}_0^{\mathrm{top}}$ is naturally isomorphic to Phillips' representable K-theory.*

Proof. We only have to show that several small modifications to Phillips' definitions do not change the resulting K-theory groups. If we did this thoroughly, we would have to repeat many of the arguments in [100], which we do not want to do. Therefore, the following argument is rather sketchy: we describe what has to be done and argue why it can be done, without actually doing it.

Let $\mathrm{K}_0^P(A)$ be the K-theory defined by Phillips. First we have to recall its definition. Let A be a locally multiplicatively convex Fréchet algebra. Let $A' := \mathbb{M}_2\big((\mathcal{K}_{\mathscr{S}}(A))^+\big)$. That is, we first stabilise A by $\mathcal{K}_{\mathscr{S}}$, then adjoin a unit to the stabilisation, and finally take 2×2-matrices. Let $I(A)$ be the set of all $e \in \mathrm{Idem}\, A'$ with $e - \left(\begin{smallmatrix}1 & 0\\ 0 & 0\end{smallmatrix}\right) \in \mathbb{M}_2\big(\mathcal{K}_{\mathscr{S}}(A)\big)$. Then $\mathrm{K}_0^P(A)$ is the set of homotopy classes of elements in $I(A)$.

The first important step is to show that $\mathrm{K}_0^P(A)$ is naturally isomorphic to the usual algebraic K-theory of $B := \mathcal{K}_{\mathscr{S}}(A)$. Since B is already matrix-stable, any element of $\mathrm{K}_0(B)$ may be represented by a pair (e_1, e_0) of idempotents in B^+ with $e_1 - e_0 \in B$. We may stabilise this to $(e_1 \oplus (1 - e_0), e_0 \oplus (1 - e_0))$. Since $e_0 \oplus (1 - e_0)$ is similar to $1 \oplus 0$, any element of $\mathrm{K}_0(B)$ is represented by a pair

$(e, 1 \oplus 0)$ where $e \in I(A)$. Thus we obtain a surjective map $I(A) \to K_0(B)$. Then one has to show that two elements of $I(A)$ are homotopic if and only if they are stably equivalent. This is the step that fails for general algebras. Even for Fréchet algebras, this is difficult because the subset of invertible elements in B need not be open.

Next one has to show that the stabilisation homomorphism $A \to \mathcal{C}\mathcal{K}^r(A)$ induces an isomorphism $K_0^P(A) \to K_0^P(\mathcal{C}\mathcal{K}^r(A))$. This is well-known for Banach algebras and follows for all locally multiplicative Fréchet algebras using the Milnor $\varprojlim^1$ sequence for K_0^P mentioned above.

Finally, we recall that the functor $A \mapsto K_0(\mathcal{C}\mathcal{K}^r(A))$ is already $\mathcal{C}\mathcal{K}^r$-stable and $\mathcal{K}_{\mathscr{S}}$-stable by Corollary 3.33. Thus

$$K_0^{\mathrm{top}}(A) = K_0(\mathcal{C}\mathcal{K}^r(A)) \cong K_0(\mathcal{K}_{\mathscr{S}}\mathcal{C}\mathcal{K}^r(A)) \cong K_0^P(\mathcal{C}\mathcal{K}^r(A)) \cong K_0^P(A). \qquad \square$$

7.5 The Weyl algebra

Definition 7.39. The *Weyl algebra* W is the universal unital algebra generated by two elements p, q that satisfy the relation $[p, q] = 1$, that is, $pq - qp = 1$. We equip it with the fine bornology.

The relation $[p, q] = 1$ is the *Heisenberg commutation relation*, and is the starting point of quantum mechanics. Thus W is a very natural example from the point of view of noncommutative algebraic geometry. We refer to [43] for its basic properties. The elements $p^m q^n$ for $m, n \in \mathbb{N}$ form a basis for W, so that $W \cong \mathbb{C}[p, q]$ as (bornological) vector spaces.

The algebra W carries no submultiplicative semi-norms because the relation $pq - qp = 1$ cannot be solved in a unital Banach algebra. The problem is that the elements pq and qp must have the same spectrum, which is a non-empty compact subset of $\mathbb{C}$. But the relation $pq - qp = 1$ implies that the spectrum of pq is invariant under translation by 1, which is impossible. Hence W is very far away from locally multiplicatively convex topological algebras and local Banach algebras.

Theorem 7.40. *The unit map* $\mathbb{C} \to W$, $\lambda \mapsto \lambda 1_W$ *is invertible in* $\mathrm{kk}^?$. *Thus* $K_*^{\mathrm{top}}(W) \cong K_*^{\mathrm{top}}(\mathbb{C})$ *is isomorphic to* $\mathbb{Z}$ *for even* $*$ *and vanishes for odd* $*$.

This result is proved in [37]. The idea is to find an algebra extension

$$I \rightarrowtail W' \twoheadrightarrow W$$

such that $W' \cong 0$ and $I \cong \Sigma\mathbb{C}$ in $\mathrm{kk}^?$. This extension is obtained by relaxing the defining relations of the Weyl algebra: the algebra W' is the universal algebra with two generators p' and q' satisfying the relations

$$(p'q' - q'p')q' = q', \qquad p'(p'q' - q'p') = p'.$$

The proofs of $W' \cong 0$ and $I \cong \Sigma\mathbb{C}$ still require some work, which we omit here.

Instead, we mention a conjecture that would contain Theorem 7.40 as a special case. Recall that a *derivation* of an algebra A is a bounded linear map $D\colon A \to A$ that satisfies the Leibniz rule

$$D(a_1 \cdot a_2) = D(a_1) \cdot a_2 + a_1 \cdot D(a_2)$$

for all $a_1, a_2 \in A$. If $\alpha\colon \mathbb{R} \to \mathrm{Aut}(A)$ is a smooth action of $\mathbb{R}$ by automorphisms, then the generator $a \mapsto \partial_t \alpha_t(a)|_{t=0}$ is a derivation.

A derivation is called *inner* if it is of the form $a \mapsto [x, a]$ for some $x \in \mathcal{M}(A)$. If $\alpha\colon \mathbb{R} \to \mathrm{Gl}_1\left(\mathcal{M}(A)\right)$ is a smooth group homomorphism, then the generator of the corresponding representation of $\mathbb{R}$ by inner automorphisms is an inner derivation.

The *crossed product* $D \ltimes A$ is equal to $A \mathbin{\widehat{\otimes}} \mathbb{C}[t]$ as a bornological vector space; the multiplication is defined so that the map $a\, t^n \mapsto \lambda(a) \circ D^n$ from $D \ltimes A$ to $\mathcal{L}(A_{\mathbb{C}}^+)$ is an algebra homomorphism; here λ denotes the left regular representation, $\lambda(a_1)(a_2) := a_1 \cdot a_2$. Equivalently, $D \ltimes A$ is the universal algebra generated by A together with an element D such that $[D, a] = D(a)$ for all $a \in A$.

The Weyl algebra is isomorphic to such a crossed product for the derivation $f \mapsto f'$ on the algebra $\mathbb{C}[t]$.

The analogue of the Baum–Connes conjecture for crossed products by derivations asserts that the canonical embedding $A \to D \ltimes A$ is invertible in $\mathrm{kk}^?$ for all derivations D. Equivalently, $D \ltimes A \cong 0$ in $\mathrm{kk}^?$ once $A \cong 0$. The general case follows from this special case as in the proof of Corollary 5.14 because any derivation is smoothly homotopic to 0 by the linear homotopy sD, $s \in [0, 1]$. Theorem 7.40 would be a special case of this conjecture because the unit map $\mathbb{C} \to \mathbb{C}[t]$ is an isomorphism in $\Sigma\mathrm{Ho}$. Unfortunately, we do not know how to prove this conjecture.

Chapter 8

A survey of bivariant K-theories

In this chapter, we briefly survey a number of alternative bivariant K-theories. Each one has its own advantages and disadvantages. While we will not give complete details (especially when it comes to Kasparov's KK-theory, which deserves, and has gotten, whole books by itself: [10, 61, 62, 67, 119]), it is helpful to know what each theory is good for and how the various theories differ from each other and from the bivariant theory developed elsewhere in these notes. The theories are:

1. Gennadi Kasparov's KK — constructed from "generalised elliptic operators." This was the first bivariant K-theory to be developed and works for C^*-algebras [71]. Kasparov's theory has been adapted to take into account symmetries such as group actions [73] and groupoid actions [77].

2. BDF-Kasparov Ext — constructed from extensions of C^*-algebras by a stable C^*-algebra, modulo split extensions. The original BDF (Brown–Douglas–Fillmore) one-variable version of [23] is constructed from C^*-algebra extensions by $\mathcal{K}$.

3. Algebraic Dual K-Theory — an algebraic analogue of one-variable Ext. This is the easiest of these theories to define.

4. Homotopy-Theoretic KK — an analogue of KK constructed using homotopy theory, with a "built-in UCT."

5. Connes–Higson E-Theory — A simpler replacement for KK, devised by Alain Connes and Nigel Higson [30], designed to eliminate certain technical difficulties that arise when working with non-nuclear C^*-algebras. This often agrees with Kasparov's theory and is somewhat easier to define; this theory also admits equivariant versions for groups and groupoids.

Of these, numbers (1), (2), and (5) make sense only for C^*-algebras, and depend on special features of C^*-algebras in order to construct the composition product. Vincent Lafforgue [76] found a way to extend the definition of Kasparov theory to Banach algebras, and this is sometimes useful (see [76]), but then we no longer have a product. In contrast, our bivariant K-theories $\mathrm{kk}^?$ have good formal properties for general bornological algebras, but they yield poor results for C^*-algebras; this is briefly discussed in Example 7.31.

(3) and (4) make sense for arbitrary Banach (and even for many Fréchet) algebras. But Kasparov's KK is by far the most important, because of the way it "fits" both with classical index theory and with "exotic" index theory like Mishchenko–Fomenko theory.

We will start with (3) and (4) because they can be defined out of one-variable K-theory. But before we define algebraic dual K-theory, we need to introduce K-theory with coefficients, which is also useful in many other contexts.

When it comes to KK and E, it is important to note that we can modify the definitions of $\mathrm{kk}^?$ to get bivariant K-theories that agree with Kasparov theory and E-theory for separable C^*-algebras. Furthermore, there are equivariant versions of our theories with respect to a group action.

Since the constructions are quite similar to that of $\mathrm{kk}^?$, we only outline the necessary changes and leave it to the reader to check that everything works as expected. The basic ingredients are:

(1) a category of algebras (with additional structure like a bornology);

(2) a notion of homotopy; this dictates what should be the suspension and cone functors SA, CA;

(3) a class of algebra extensions and a tensor algebra adapted to it;

(4) a stabilisation functor.

These ingredients must satisfy various conditions, which we do not formalise here. Another useful but optional ingredient is an exterior product operation that plays the role of the tensor product $A \mathbin{\widehat{\otimes}} B$ for bornological algebras.

In our construction of $\mathrm{kk}^?$, the category of algebras is the category of bornological algebras with bounded algebra homomorphisms as morphisms; the notion of homotopy is smooth homotopy; the class of algebra extensions is the class of semi-split extensions; the stabilisation functor is $\mathcal{K}_\mathscr{S}$, $\mathcal{CK}^r$, or $\mathcal{K}_{\mathscr{L}^1} \circ \mathcal{K}_\mathscr{S}$.

We may modify this setup and consider locally convex topological algebras instead of bornological algebras; this is the setting used in [36, 37, 39]. This case is almost literally the same. Since both theories are so similar, we do not discuss this modification here.

To get KK and E, we work in the category of separable C^*-algebras and use continuous homotopy and the C^*-stabilisation. Depending on whether we want to construct KK or E, we either use extensions with a completely positive contractive section or all extensions of C^*-algebras.

In the equivariant case, we consider the category of C^*-algebras with a strongly continuous action of a group G instead, with G-equivariant $*$-homomorphisms as morphisms; we use the same notion of homotopy; the stabilisation is modified to allow representations of G on Hilbert space, and the class of extensions consists of all extensions with a G-equivariant completely positive contractive section for KK^G, or of all extensions for E^G.

8.1 K-Theory with coefficients

In algebraic topology, we need homology with coefficients in $\mathbb{Z}/m$ or $\mathbb{Q}$, not just with coefficients in $\mathbb{Z}$. Similarly, it is useful to introduce K-theory with coefficients. Instead of doing this in complete generality, we only define the cases we need (which suffice for all applications we are aware of), namely, K-theory with coefficients in $\mathbb{Q}$, $\mathbb{Z}/m$ for $m \in \mathbb{N}_{\geq 1}$, and $\mathbb{Q}/\mathbb{Z} \cong \varinjlim \mathbb{Z}/m$, where the limit is taken over the set of positive integers m partially ordered by divisibility.

K-theory with coefficients in $\mathbb{Q}$ is simplest.

Definition 8.1. Let A be a local Banach algebra. Since $\mathbb{Q}$ is torsion-free, thus flat as a $\mathbb{Z}$-module, we can simply define $\mathrm{K}_*(A; \mathbb{Q}) := \mathrm{K}_*(A) \otimes_{\mathbb{Z}} \mathbb{Q}$. Since tensoring over $\mathbb{Z}$ with $\mathbb{Q}$ is an exact functor, it preserves long exact sequences. Thus we get a theory with the same properties as (integral) K-theory, except that we have $\mathrm{K}_0(\mathbb{C}; \mathbb{Q}) = \mathbb{Q}$ instead of $\mathbb{Z}$, and all K-groups in the theory become rational vector spaces.

There is an alternative way to define $\mathrm{K}_*(A; \mathbb{Q})$. Recall that $\mathbb{Q}$ can be realised as an inductive limit of copies of $\mathbb{Z}$, either abstractly as $\varinjlim \mathbb{Z}$, where the limit is taken over the set of all injective homomorphisms from $\mathbb{Z}$ to itself, or more concretely, as the limit of the sequence

$$\mathbb{Z} \xrightarrow{2} \mathbb{Z} \xrightarrow{2\cdot 3} \mathbb{Z} \xrightarrow{2\cdot 3\cdot 5} \mathbb{Z} \xrightarrow{2\cdot 3\cdot 5\cdot 7} \mathbb{Z} \to \cdots , \tag{8.2}$$

where multiplication by each prime eventually occurs infinitely many times in the sequence. Thus $\mathrm{K}_*(A; \mathbb{Q}) = \mathrm{K}_*(A) \otimes_{\mathbb{Z}} \mathbb{Q}$ is the inductive limit

$$\mathrm{K}_*(A) \xrightarrow{2} \mathrm{K}_*(A) \xrightarrow{2\cdot 3} \mathrm{K}_*(A) \xrightarrow{2\cdot 3\cdot 5} \mathrm{K}_*(A) \to \cdots .$$

There is still another way to think about this, motivated by the construction of UHF (uniformly hyperfinite) algebras by Glimm [50]. Namely, form the inductive limit U of the sequence of algebras

$$\mathbb{C} \xrightarrow{\mathrm{id} \otimes 1} \mathrm{M}_2(\mathbb{C}) \xrightarrow{\mathrm{id} \otimes 1} \mathrm{M}_2(\mathbb{C}) \otimes \mathrm{M}_{2\cdot 3}(\mathbb{C})$$

$$\xrightarrow{\mathrm{id} \otimes 1} \mathrm{M}_2(\mathbb{C}) \otimes \mathrm{M}_{2\cdot 3}(\mathbb{C}) \otimes \mathrm{M}_{2\cdot 3\cdot 5}(\mathbb{C}) \to \cdots , \tag{8.3}$$

where each homomorphism in the sequence corresponds to the map

$$A \mapsto \begin{pmatrix} A & 0 & \cdots & 0 \\ 0 & A & \cdots & 0 \\ \vdots & \vdots & \ddots & \vdots \\ 0 & 0 & \cdots & A \end{pmatrix} . \tag{8.4}$$

The inductive limit should be taken in the appropriate category: in BAlg, we merely give the algebraic inductive limit the fine bornology, and in the category of C^*-algebras, we complete in the obvious C^*-norm as in [50]. Each embedding of matrix algebras multiplies the rank of each idempotent by the number of diagonal blocks, and thus induces multiplication by this number on K_0. Thus the sequence of algebras (8.3) realises the original sequence (8.2) on passage to K_0. Since K-theory commutes with direct limits, we get $K_0(U) = \mathbb{Q}$ and $K_1(U) = 0$. Furthermore, we can tensor the sequence (8.3) with A, taking bigger and bigger matrix algebras over A; passage to the limit in our category is the same as taking the completed tensor product with U. We get a natural isomorphism $K_*(A; \mathbb{Q}) \cong K_*(A \mathbin{\widehat{\otimes}}_{C^*} U)$. This provides a better realisation of K-theory with rational coefficients for some purposes.

Next we consider K-theory with finite coefficients. For this, it is useful to consider the mapping cone C_m of the map $\mathbb{C} \to \mathbb{M}_m(\mathbb{C})$ in (8.4). The mapping cone should be taken in whatever category we are working in. In a C^*-algebra context, this is

$$C_m = \left\{ (a, f) \ \middle| \ a \in \mathbb{C},\ f \in C_0\big((0,1], \mathbb{M}_m\big),\ f(1) = \begin{pmatrix} a & & \\ & \ddots & \\ & & a \end{pmatrix} \right\}$$

$$\cong \{ f \in C_0((0,1], \mathbb{M}_m) \mid f(1) \text{ diagonal} \}, \quad (8.5)$$

which sits in a C^*-algebra extension $C_0((0,1), \mathbb{M}_m) \rightarrowtail C_m \twoheadrightarrow \mathbb{C}$; in the context of bornological algebras, we use smooth functions instead, as in Definition 2.35.

Definition 8.6. Let C_m be defined as in (8.5), and let A be a local Banach algebra. Define $K_*(A; \mathbb{Z}/m) := K_*(SA \mathbin{\widehat{\otimes}} C_m)$. This fits into a natural exact sequence

$$
\begin{array}{ccccc}
K_0(A) & \xrightarrow{\ m\ } & K_0(A) & \longrightarrow & K_0(A; \mathbb{Z}/m) \\[2pt]
{\scriptstyle \partial}\Big\uparrow & & & & \Big\downarrow{\scriptstyle \partial} \\[2pt]
K_1(A; \mathbb{Z}/m) & \longleftarrow & K_1(A) & \xleftarrow[\ m\ ]{} & K_1(A)
\end{array}
\qquad (8.7)
$$

where the maps denoted m are multiplication by m (see [113] for more details). It is called the *Bockstein exact sequence* after the Russian mathematician Meer Feliksovich Bokshtein.

The exact sequence comes from the mapping cone sequence of the map that we get by tensoring (8.4) with A. This definition agrees with a more classical definition using Moore spaces (see Exercise 8.12 below).

Remark 8.8. Besides the obvious functoriality in A, K-theory mod m has an additional functoriality in m: if m_1 divides m_2, then the embedding $\mathbb{Z}/m_1 \hookrightarrow \mathbb{Z}/m_2$ corresponds to a natural map $K_*(A; \mathbb{Z}/m_1) \to K_*(A; \mathbb{Z}/m_2)$ for any local Banach algebra A. We get this from an algebra homomorphism $C_{m_1} \to C_{m_2}$ that comes from the factorisation of the unital inclusion $\mathbb{C} \hookrightarrow \mathbb{M}_{m_2}$ through the unital inclusion $\mathbb{C} \hookrightarrow \mathbb{M}_{m_1}$.

There are two possible ways to define K-theory with coefficients in $\mathbb{Q}/\mathbb{Z}$.

Definition 8.9. Let A be a local Banach algebra. Let C_∞ be the mapping cone of the unital inclusion $\mathbb{C} \hookrightarrow U$, where U is defined as before to be the inductive limit of the sequence (8.3). (The notation is justified by the fact that C_∞ is the inductive limit of the C_m's via the maps of Remark 8.8.) Define

$$K_*(A; \mathbb{Q}/\mathbb{Z}) := K_*\big(S(A \,\widehat{\otimes}\, C_\infty)\big).$$

The mapping cone extension $S(A \,\widehat{\otimes}\, U) \rightarrowtail A \,\widehat{\otimes}\, C_\infty \twoheadrightarrow A$ yields a Bockstein long exact sequence

$$
\begin{array}{ccccc}
K_0(A) & \longrightarrow & K_0(A; \mathbb{Q}) & \xrightarrow{\varrho_*} & K_0(A; \mathbb{Q}/\mathbb{Z}) \\
\partial \uparrow & & & & \downarrow \partial \\
K_1(A; \mathbb{Q}/\mathbb{Z}) & \xleftarrow{\;\varrho_*\;} & K_1(A; \mathbb{Q}) & \longleftarrow & K_1(A).
\end{array}
\tag{8.10}
$$

A second definition is based on Remark 8.8. Namely, we have functorial maps $K_*(A; \mathbb{Z}/m_1) \to K_*(A; \mathbb{Z}/m_2)$ whenever m_1 divides m_2, so that we can define

$$K_*(A; \mathbb{Q}/\mathbb{Z}) := \varinjlim K_*(A; \mathbb{Z}/m),$$

where the inductive system is indexed by $\mathbb{N}^{\geq 1}$ partially ordered by divisibility. Equivalently, we can use the inductive system

$$K_*(A; \mathbb{Z}/2) \to K_*\big(A; \mathbb{Z}/(2^2 \cdot 3)\big) \to K_*\big(A; \mathbb{Z}/(2^3 \cdot 3^2 \cdot 5)\big) \to \cdots,$$

where each prime number occurs infinitely often. (Compare the sequence (8.2).) Since inductive limits — unlike projective limits — yield an exact functor, this gives a homology theory. The description of C_∞ as an inductive limit shows that the two definitions of $K_*(A; \mathbb{Q}/\mathbb{Z})$ coincide.

Exercise 8.11. Verify that $K_*(\_; \mathbb{Q})$ and $K_*(\_; \mathbb{Q}/\mathbb{Z})$, as we defined them for local Banach algebras as $K_*(\_ \,\widehat{\otimes}\, D)$ for a suitable tensor product and suitable auxiliary algebras D, are indeed homology theories (homotopy invariant, half-exact, with long exact sequences).

Exercise 8.12. The *mod-m Moore space* is a CW-complex X with three cells defined by attaching a 2-cell to $\mathbb{S}^1$ by a map $\mathbb{S}^1 \to \mathbb{S}^1$ of degree m. Let $x_0 \in X$ be the 0-cell in X.

Show that the above definition of $K_*(\_; \mathbb{Z}/m)$ using the mapping cone of the unital map $\mathbb{C} \to \mathbb{M}_m(\mathbb{C})$ agrees with the more classical choice

$$K_*(A; \mathbb{Z}/m) := K_*\big(C_0(X \setminus \{x_0\}, A)\big).$$

8.2   Algebraic dual K-theory

Definition 8.13. Let A be a local Banach algebra, and let $DK^j(A)$ (D for dual) be the set of commutative diagrams

$$
\begin{array}{ccc}
K_j(A;\mathbb{Q}) & \xrightarrow{\;\varrho_*\;} & K_j(A;\mathbb{Q}/\mathbb{Z}) \\
\downarrow & & \downarrow \\
\mathbb{Q} & \xrightarrow{\;\varrho\;} & \mathbb{Q}/\mathbb{Z},
\end{array}
$$

where $\varrho\colon \mathbb{Q} \to \mathbb{Q}/\mathbb{Z}$ is the quotient map and the induced map ϱ_* is as in (8.10). Then $DK^*(A)$ can be made into an Abelian group, a subgroup of

$$
\mathrm{Hom}_{\mathbb{Z}}(K_j(A;\mathbb{Q}),\mathbb{Q}) \oplus \mathrm{Hom}_{\mathbb{Z}}(K_j(A;\mathbb{Q}/\mathbb{Z}),\mathbb{Q}/\mathbb{Z}).
$$

(This definition may be found in [79].)

Theorem 8.14. DK^* *is a cohomology theory on local Banach algebras and satisfies Bott periodicity and a Universal Coefficient Theorem natural exact sequence*

$$
0 \to \mathrm{Ext}^1_{\mathbb{Z}}(K_{j-1}(A),\mathbb{Z}) \to DK^j(A) \to \mathrm{Hom}_{\mathbb{Z}}(K_j(A),\mathbb{Z}) \to 0. \tag{8.15}
$$

Proof. Clearly DK^* is a contravariant homotopy functor with Bott periodicity. The UCT map

$$
DK^j(A) \twoheadrightarrow \mathrm{Hom}_{\mathbb{Z}}(K_j(A),\mathbb{Z})
$$

comes from chasing the commutative diagram with exact rows

$$
\begin{array}{ccccccc}
K_j(A) & \xrightarrow{\;\iota\;} & K_j(A;\mathbb{Q}) & \xrightarrow{\;\varrho_*\;} & K_j(A;\mathbb{Q}/\mathbb{Z}) & \xrightarrow{\;\partial\;} & K_{j-1}(A) \\
\downarrow & & \downarrow & & \downarrow & & \\
\mathbb{Z} & \rightarrowtail & \mathbb{Q} & \xrightarrow{\;\varrho\;} & \mathbb{Q}/\mathbb{Z} & \longrightarrow & 0.
\end{array} \tag{8.16}
$$

We go through the details. An element of $DK^j(A)$ corresponds to a pair of maps $\alpha\colon K_j(A;\mathbb{Q}) \to \mathbb{Q}$ and $\beta\colon K_j(A;\mathbb{Q}/\mathbb{Z}) \to \mathbb{Q}/\mathbb{Z}$ giving a commutative square in the middle of (8.16). Composing α with the canonical map $\iota\colon K_j(A) \to K_j(A;\mathbb{Q})$ gives a map $K_j(A) \to \mathbb{Q}$, which takes its values in $\mathbb{Z}$ because

$$
\varrho \circ (\alpha \circ \iota) = (\varrho \circ \alpha) \circ \iota = (\beta \circ \varrho_*) \circ \iota = \beta \circ (\varrho_* \circ \iota) = 0.
$$

Thus we get a map $DK^j(A) \to \mathrm{Hom}(K_j(A),\mathbb{Z})$.

We claim that this map is surjective. Given $\gamma\colon K_j(A) \to \mathbb{Z}$, we tensor γ with $\mathbb{Q}$ to get $\alpha\colon K_j(A;\mathbb{Q}) \to \mathbb{Q}$. This determines $\beta\colon K_j(A;\mathbb{Q}/\mathbb{Z}) \to \mathbb{Q}/\mathbb{Z}$ on the image of ϱ_*. The extension to a map on all of $K_j(A;\mathbb{Q}/\mathbb{Z})$ is possible because the target group $\mathbb{Q}/\mathbb{Z}$ is injective as a $\mathbb{Z}$-module.

The same diagram also gives the left side of the UCT exact sequence once we remember that $\mathrm{Ext}^1_{\mathbb{Z}}(\mathrm{K}_{j-1}(A), \mathbb{Z})$ is the cokernel of the map

$$\mathrm{Hom}_{\mathbb{Z}}(\mathrm{K}_{j-1}(A), \mathbb{Q}) \to \mathrm{Hom}_{\mathbb{Z}}(\mathrm{K}_{j-1}(A), \mathbb{Q}/\mathbb{Z}).$$

Indeed, suppose an element of $\mathrm{DK}^j(A)$, given by

$$
\begin{array}{ccc}
\mathrm{K}_j(A; \mathbb{Q}) & \xrightarrow{\varrho_*} & \mathrm{K}_j(A; \mathbb{Q}/\mathbb{Z}) \\
\alpha \downarrow & & \downarrow \beta \\
\mathbb{Q} & \xrightarrow{\varrho} & \mathbb{Q}/\mathbb{Z},
\end{array}
$$

goes to 0 in $\mathrm{Hom}(\mathrm{K}_j(A), \mathbb{Z})$. This means $\alpha \circ \iota = 0$, so that α vanishes on $\mathrm{im}\,\iota = \ker \varrho_*$, and factors through $\mathrm{im}\,\varrho_* \subseteq \mathrm{K}_j(A; \mathbb{Q}/\mathbb{Z})$. But $\mathrm{K}_j(A; \mathbb{Q}/\mathbb{Z})$ is a torsion group and $\mathbb{Q}$ is torsion-free, so that $\alpha = 0$. Thus β vanishes on $\mathrm{im}\,\varrho_* = \ker \partial$, and β factors through $\mathrm{im}\,\partial \subseteq \mathrm{K}_{j-1}(A)$. Since $\mathbb{Q}/\mathbb{Z}$ is $\mathbb{Z}$-injective, we can extend the map $\mathrm{im}\,\partial \to \mathbb{Q}/\mathbb{Z}$ to a map $\delta\colon \mathrm{K}_{j-1}(A) \to \mathbb{Q}/\mathbb{Z}$. We claim that β only depends on the image of δ in $\mathrm{Ext}^1_{\mathbb{Z}}(\mathrm{K}_{j-1}(A), \mathbb{Z}) = \mathrm{coker}\big(\varrho_*\colon \mathrm{Hom}(\mathrm{K}_{j-1}(A), \mathbb{Q}) \to \mathrm{Hom}(\mathrm{K}_{j-1}(A), \mathbb{Q}/\mathbb{Z})\big)$. Indeed, suppose we add to δ something that factors as $\mathrm{K}_{j-1}(A) \to \mathbb{Q} \xrightarrow{\varrho} \mathbb{Q}/\mathbb{Z}$. This has no effect on β because β is defined on the torsion group $\mathrm{K}_j(A; \mathbb{Q}/\mathbb{Z})$, and is thus unaffected by something factoring through the torsion-free group $\mathbb{Q}$. Thus the kernel of $\mathrm{DK}^j(A) \to \mathrm{Hom}(\mathrm{K}_j(A), \mathbb{Z})$ comes from $\mathrm{Ext}^1_{\mathbb{Z}}(\mathrm{K}_{j-1}(A), \mathbb{Z})$. The same calculation shows that any element of $\mathrm{Ext}^1_{\mathbb{Z}}(\mathrm{K}_{j-1}(A), \mathbb{Z})$ gives rise to an element of $\mathrm{DK}^j(A)$ of the special form

$$
\begin{array}{ccc}
\mathrm{K}_j(A; \mathbb{Q}) & \xrightarrow{\varrho_*} & \mathrm{K}_j(A; \mathbb{Q}/\mathbb{Z}) \\
0 \downarrow & & \downarrow \beta \\
\mathbb{Q} & \xrightarrow{\varrho} & \mathbb{Q}/\mathbb{Z},
\end{array}
$$

with β factoring through $\mathrm{K}_j(A; \mathbb{Q}/\mathbb{Z}) \xrightarrow{\partial} \mathrm{K}_{j-1}(A)$.

To complete the proof of the UCT, we just need to see that the induced map $\mathrm{Ext}^1_{\mathbb{Z}}(\mathrm{K}_{j-1}(A), \mathbb{Z}) \to \mathrm{DK}^j(A)$ is injective. If an element of Ext represented by $\delta\colon \mathrm{K}_{j-1}(A) \to \mathbb{Q}/\mathbb{Z}$ yields $\beta = 0$, then δ vanishes on the image of $\partial\colon \mathrm{K}_j(A; \mathbb{Q}/\mathbb{Z}) \to \mathrm{K}_{j-1}(A)$, which is the same as the kernel of the map $\mathrm{K}_{j-1}(A) \to \mathrm{K}_{j-1}(A; \mathbb{Q}) = \mathrm{K}_{j-1}(A) \otimes_{\mathbb{Z}} \mathbb{Q}$. This kernel is clearly the torsion subgroup of $\mathrm{K}_{j-1}(A)$. But if a map $\delta\colon \mathrm{K}_{j-1}(A) \to \mathbb{Q}/\mathbb{Z}$ vanishes on the torsion subgroup of $\mathrm{K}_{j-1}(A)$, then it comes from a map $\mathrm{K}_{j-1}(A) \to \mathbb{Q}$, and thus represents 0 in Ext. This completes the proof of the UCT.

We show that DK^* comes with long exact sequences. Here we use that $\mathbb{Q}$ and $\mathbb{Q}/\mathbb{Z}$ are divisible and hence injective as $\mathbb{Z}$-modules. It suffices (as in the case of ordinary topological K-theory) to prove split-exactness and middle-exactness. Split-exactness is immediate from split-exactness of K-theory with coefficients and exactness of the functors $\mathrm{Hom}(\llcorner, \mathbb{Q})$ and $\mathrm{Hom}(\llcorner, \mathbb{Q}/\mathbb{Z})$.

To prove middle-exactness, let

$$A \overset{\phi}{\rightarrowtail} B \overset{\psi}{\twoheadrightarrow} C$$

be an extension of local Banach algebras. We get a commuting diagram with long exact rows

$$
\begin{array}{ccccccc}
\cdots \overset{\partial}{\longrightarrow} & K_j(A;\mathbb{Q}) & \overset{\phi_*}{\longrightarrow} & K_j(B;\mathbb{Q}) & \overset{\psi_*}{\longrightarrow} & K_j(C;\mathbb{Q}) & \overset{\partial}{\longrightarrow} \cdots \\
& \downarrow{\varrho_*} & & \downarrow{\varrho_*} & & \downarrow{\varrho_*} & \\
\cdots \overset{\partial}{\longrightarrow} & K_j(A;\mathbb{Q}/\mathbb{Z}) & \overset{\phi_*}{\longrightarrow} & K_j(B;\mathbb{Q}/\mathbb{Z}) & \overset{\psi_*}{\longrightarrow} & K_j(C;\mathbb{Q}/\mathbb{Z}) & \overset{\partial}{\longrightarrow} \cdots,
\end{array}
\tag{8.17}
$$

which induces

$$\mathrm{DK}^j(C) \overset{\psi^*}{\longrightarrow} \mathrm{DK}^j(B) \overset{\phi^*}{\longrightarrow} \mathrm{DK}^j(A).$$

We must show that this is exact in the middle at $\mathrm{DK}^j(B)$. Obviously $\phi^* \circ \psi^* = (\psi \circ \phi)^* = 0$. Suppose we are given an element of $\mathrm{DK}^j(B)$, given by

$$
\begin{array}{ccc}
K_j(B;\mathbb{Q}) & \overset{\varrho_*}{\longrightarrow} & K_j(B;\mathbb{Q}/\mathbb{Z}) \\
\alpha \downarrow & & \downarrow \beta \\
\mathbb{Q} & \overset{\varrho}{\longrightarrow} & \mathbb{Q}/\mathbb{Z},
\end{array}
$$

which goes to 0 in $\mathrm{DK}^j(A)$ under ϕ^*. Since the functor $\mathrm{Hom}(\llcorner,\mathbb{Q})$ is exact, the sequence

$$\mathrm{Hom}(K_j(C;\mathbb{Q}),\mathbb{Q}) \overset{(\psi_*)^*}{\longrightarrow} \mathrm{Hom}(K_j(B;\mathbb{Q}),\mathbb{Q}) \overset{(\phi_*)^*}{\longrightarrow} \mathrm{Hom}(K_j(A;\mathbb{Q}),\mathbb{Q})$$

is exact. Thus α comes from an element $\alpha' \in \mathrm{Hom}(K_j(C;\mathbb{Q}),\mathbb{Q})$. Similarly, β comes from an element $\beta' \in \mathrm{Hom}(K_j(C;\mathbb{Q}/\mathbb{Z}),\mathbb{Q}/\mathbb{Z})$. It remains to arrange $\beta' \circ \varrho_* = \varrho \circ \alpha'$. The difference between these, $\beta' \circ \varrho_* - \varrho \circ \alpha'$, is a map $K_j(C;\mathbb{Q}) \to \mathbb{Q}/\mathbb{Z}$ whose image under ψ^* is 0. The exactness of the functor $\mathrm{Hom}(\llcorner;\mathbb{Q}/\mathbb{Z})$ yields an exact sequence

$$\mathrm{Hom}(K_{j-1}(A;\mathbb{Q}),\mathbb{Q}/\mathbb{Z}) \overset{\partial^*}{\longrightarrow} \mathrm{Hom}(K_j(C;\mathbb{Q}),\mathbb{Q}/\mathbb{Z}) \overset{(\psi_*)^*}{\longrightarrow} \mathrm{Hom}(K_j(B;\mathbb{Q}),\mathbb{Q}/\mathbb{Z}).$$

Hence $\beta' \circ \varrho_* - \varrho \circ \alpha'$ factors as a map $K_j(C;\mathbb{Q}) \overset{\partial}{\to} K_{j-1}(A;\mathbb{Q}) \to \mathbb{Q}/\mathbb{Z}$. Add to α' a composite $\gamma \colon K_j(C;\mathbb{Q}) \overset{\partial}{\to} K_{j-1}(A;\mathbb{Q}) \to \mathbb{Q}$ with $\varrho \circ \gamma = \beta' \circ \varrho_* - \varrho \circ \alpha'$. Then $\alpha' + \gamma$ still maps to α, but now we have $\beta' \circ \varrho_* = \varrho \circ (\alpha' + \gamma)$. This concludes the proof of exactness. $\square$

8.3 Homotopy-theoretic KK-theory

Homotopy-theoretic KK is a bivariant theory that is hard to locate in the literature, but that was constructed independently by a number of people, including the author of this chapter (J. Rosenberg) and Stephan Stolz (see for example [24]). We will be brief about this since formal definitions require a lot of machinery. If A and B are local Banach algebras, the K-groups of A and B are homotopy groups of *spectra* $\mathbf{K}(A)$ and $\mathbf{K}(B)$, in fact of $\mathbf{K}$-*module spectra*, where $\mathbf{K} = \mathbf{K}(\mathbb{C})$ is the spectrum of complex K-theory.

Here we are dealing with spectra in the sense of algebraic topology — we are not talking about operator theory or Banach algebra theory, where the word has a completely different meaning. Good references for the theory of spectra are [1, Part III] and [81]. The category of $\mathbf{K}$-module spectra is studied by Bousfield [11]. Roughly speaking, spectra are generalised spaces that give concrete representations of generalised homology theories.

In a suitable category of $\mathbf{K}$-module spectra, we can define

$$\mathbf{KK}(A, B) = \mathrm{Hom}_{\mathbf{K}}\big(\mathbf{K}(A), \mathbf{K}(B)\big).$$

This is itself a $\mathbf{K}$-module spectrum, so that it has homotopy groups satisfying Bott periodicity. These are the *homotopy-theoretic* KK-*groups* of A and B, $\mathrm{HKK}_*(A, B)$. Properties of the category of $\mathbf{K}$-module spectra yield a UCT exact sequence

$$\mathrm{Ext}^1_{\mathbb{Z}}\big(\mathrm{K}_{*-1}(A), \mathrm{K}_*(B)\big) \rightarrowtail \mathrm{HKK}_*(A, B) \twoheadrightarrow \mathrm{Hom}_{\mathbb{Z}}\big(\mathrm{K}_*(A), \mathrm{K}_*(B)\big). \tag{8.18}$$

It is fairly easy to see that all the other bivariant K-theories we are discussing have natural transformations to HKK, which in good cases are isomorphisms. To construct the natural transformation, we need that a class in the bivariant K-theory yields a map of spectra $\mathbf{K}(A) \to \mathbf{K}(B)$ making the following diagram commute:

$$
\begin{array}{ccc}
\mathbf{K}(\mathbb{C}) \wedge \mathbf{K}(A) & \longrightarrow & \mathbf{K}(\mathbb{C}) \wedge \mathbf{K}(B) \\
\downarrow{\scriptstyle \mu_A} & & \downarrow{\scriptstyle \mu_B} \\
\mathbf{K}(A) & \longrightarrow & \mathbf{K}(B).
\end{array}
$$

Here μ is the natural multiplication map for $\mathbf{K}$-module spectra. This gives a way to prove a UCT in many situations.

8.4 Brown–Douglas–Fillmore extension theory

Of great historical importance, because of its connection with the Weyl–von Neumann Theorem, is the extension theory by Brown, Douglas, and Fillmore developed in [18, 22, 23, 46], often called BDF Theory (for short).

Definition 8.19. Let $\mathcal{L} = \mathcal{L}(\mathcal{H})$ be the algebra of bounded operators on an infinite-dimensional separable Hilbert space $\mathcal{H}$, and let $\mathcal{Q} = \mathcal{L}/\mathcal{K}$ be the *Calkin algebra*. If A is a separable C^*-algebra, an *extension of A by $\mathcal{K}$* is a C^*-algebra E containing $\mathcal{K}$ as an ideal together with a fixed $*$-isomorphism $E/\mathcal{K} \xrightarrow{\cong} A$. The extension is called *essential* if no element of E commutes with $\mathcal{K}$; equivalently, E embeds in $\mathcal{L} = \mathcal{M}(\mathcal{K})$ (with $\mathcal{K}$ going to itself).

Any extension of A by $\mathcal{K}$ is a pullback

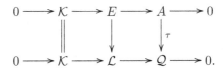

Thus we think of *-homomorphisms $\tau\colon A \to \mathcal{Q}$ as extensions; the essential extensions are those for which τ is injective. An extension *splits* if and only if τ lifts to a *-homomorphism $\tau\colon A \to \mathcal{L}$.

Two extensions are considered equivalent if they differ by conjugation via unitaries in $\mathcal{L}$. We can add extensions via

$$A \xrightarrow{\tau_1 \oplus \tau_2} \mathcal{Q} \oplus \mathcal{Q} \longrightarrow \mathcal{Q}(\mathcal{H} \oplus \mathcal{H}) \xrightarrow{\cong} \mathcal{Q}.$$

The result is well-defined modulo unitary conjugation, and makes classes of extensions into an Abelian semigroup (in general without unit). After dividing out by the *split* extensions (this is unnecessary, by a result of Voiculescu [124], if A is non-unital), we get an Abelian monoid $\mathrm{Ext}(A)$.

Recall that a linear map between C^*-algebras $f\colon A \to B$ is called *completely positive* if, for each $n \in \mathbb{N}$, the induced map $f_n = f \otimes 1_{\mathbb{M}_n}\colon \mathbb{M}_n(A) \to \mathbb{M}_n(B)$ is *positive*, that is, sends positive elements to positive elements. Besides *-homomorphisms, which obviously have this property, the obvious examples are compressions to a corner. In other words, if $B = pAp$, where p is a (self-adjoint) projection in the multiplier algebra of A, then $a \mapsto pap$ is readily seen to be completely positive, and is unital if A is unital.

Theorem 8.20 (Arveson [3], Choi–Effros [25]). *An extension $\tau\colon A \to \mathcal{Q}$ is invertible in $\mathrm{Ext}(A)$ if and only if it has a completely positive lifting $A \to \mathcal{L}$. The liftable extensions form a group, and if A is nuclear, this group is all of $\mathrm{Ext}(A)$.*

Partial sketch of proof. Suppose that the extension $\tau\colon A \to \mathcal{Q}$ is invertible in $\mathrm{Ext}(A)$. This means that there is some other extension τ' such that $\tau \oplus \tau'$ is split, or in other words, lifts to a homomorphism $\varphi\colon A \to \mathcal{L}(\mathcal{H})$. Let us be careful about the Hilbert spaces; say that $\tau\colon A \to \mathcal{Q}(\mathcal{H}_1)$ and $\tau'\colon A \to \mathcal{Q}(\mathcal{H}_2)$. Then $\tau \oplus \tau'\colon A \to \mathcal{Q}(\mathcal{H}_1 \oplus \mathcal{H}_2)$. Write $\mathcal{H} = \mathcal{H}_1 \oplus \mathcal{H}_2$, so that $\mathcal{H}_1 = p\mathcal{H}$ for p the orthogonal projection killing $\mathcal{H}_2$. Then φ, the lifting of $\tau \oplus \tau'$, followed by compression into $\mathcal{L}(\mathcal{H}_1) = p\mathcal{L}(\mathcal{H})p$, is a completely positive lifting of τ.

The other direction of the first statement follows from Stinespring's Dilation Theorem [117]. This asserts that if A is a unital C^*-algebra and $f\colon A \to \mathcal{L}(\mathcal{H}_1)$ is unital and completely positive, then there is a unital *-homomorphism $\varphi\colon A \to \mathcal{L}(\mathcal{H})$, where $\mathcal{H}$ is a larger Hilbert space, with $\mathcal{H}_1 = p\mathcal{H}$ for some self-adjoint projection p, so that $f(a) = p\varphi(a)p$ for all $a \in A$. Assuming this result and the unitality hypotheses (which are easy to remove), the assumption that τ has a completely positive lifting f gives us a representation φ such that τ lifts to one corner of φ. Let τ' be the image in the Calkin algebra of the compression to the opposite corner of φ: $\tau'(a) = (1-p)\varphi(a)(1-p)$ on $\mathcal{H}_2 = (1-p)\mathcal{H}$, modulo $\mathcal{K}(\mathcal{H}_2)$. Then by construction, $\tau \oplus \tau'$ lifts to a $*$-homomorphism $\varphi\colon A \to \mathcal{L}(\mathcal{H})$, and so τ is invertible in Ext A. Thus extensions are invertible in Ext A if and only if they are liftable. It follows that a sum of liftable extensions is a sum of invertible extensions, hence is invertible (since we are in an Abelian monoid), hence is liftable, so the liftable extensions form a group.

The rest of the theorem involves the Choi–Effros theory of nuclearity, and we omit the proof. $\qquad\square$

Incidentally, the condition in Theorem 8.20 for invertibility of an element of Ext(A) is not automatic. Joel Anderson [2] has constructed a separable C^*-algebra for which Ext(A) is *not* a group.

Theorem 8.21 (O'Donovan [94], Salinas [111]). Ext *is homotopy-invariant on quasi-diagonal C^*-algebras.*

It is easy to construct a natural transformation Ext $\to$ DK^1. Given an extension $\mathcal{K} \rightarrowtail E \twoheadrightarrow A$, tensor the extension with nuclear C^*-algebras C that satisfy $K_1(C) = 0$ and $K_0(C) = \mathbb{Q}$ or $\mathbb{Q}/\mathbb{Z}$, respectively. Such algebras were constructed above: U in (8.3) and SC_∞ in Definition 8.9. Then use the connecting map $K_1(A \otimes C) \xrightarrow{\partial} K_0(\mathcal{K} \otimes C) \cong K_0(C)$ in the long exact K-theory sequences for the tensored extensions to define an element of DK^1. In favourable circumstances, for instance, if A is a type I C^*-algebra, this natural map Ext $\to$ DK^1 is an isomorphism. This special case of the Universal Coefficient Theorem is due to Lawrence Brown [17, 19, 21].

Exercise 8.22. This exercise deals with the original motivation for BDF Theory: the classification of essentially normal operators. Let $\mathcal{H}$ be a separable Hilbert space. An operator $T \in \mathcal{L}(\mathcal{H})$ is *essentially normal* if $TT^* - T^*T \in \mathcal{K}(\mathcal{H})$. If $\pi\colon \mathcal{L}(\mathcal{H}) \to \mathcal{Q}(\mathcal{H})$ is the projection map, this is equivalent to $\pi(T)$ being a normal element of the C^*-algebra $\mathcal{Q}(\mathcal{H})$. The unital commutative C^*-subalgebra of $\mathcal{Q}(\mathcal{H})$ generated by $\pi(T)$ is isomorphic to $C(X)$, where X is the *essential spectrum* of T, that is, the spectrum of $\pi(T)$. It is known that X is the closure of the set obtained from the spectrum of T by removing all eigenvalues of finite multiplicity.

The following is the original problem treated by Brown, Douglas, and Fillmore: Given an essentially normal operator T, when is there a compact operator K for which $T + K$ is normal?

1. Show that the problem is non-trivial: the unilateral shift is essentially normal but not of the form normal + compact.

2. Show that the map $C(X) \to \mathcal{Q}(\mathcal{H})$ given by $z \mapsto \pi(T)$, where $z\colon X \hookrightarrow \mathbb{C}$ is the usual inclusion of the spectrum of $\pi(T)$ into the complex numbers, is an

element of $\operatorname{Ext} C(X)$; it is trivial (represents 0 in this group) if and only if T is of the form normal + compact.

3. Show that every element of $\operatorname{Ext} C(X)$ is invertible (liftable), so that $\operatorname{Ext} C(X)$ is a group. This is a lot easier than the general Choi–Effros Theorem since $C(X)$ is generated by the single normal element z.

4. Show that every element of $\operatorname{Ext} C(X)$ corresponds to an essentially normal operator with essential spectrum X, and thus that understanding $\operatorname{Ext} C(X)$ is equivalent to classifying essentially normal operators with essential spectrum X modulo compact operators.

8.5   Bivariant K-theories for C^*-algebras

We adapt our bivariant K-theories to the realm of C^*-algebras. The result agrees with Kasparov's bivariant K-theory for separable C^*-algebras because both theories enjoy the same universal property. We explicitly describe the natural transformation from Kasparov theory to our new theory. Generalising this construction, we arrive at the notion of abstract Kasparov module for bornological algebras. This is useful for translating constructions from Kasparov theory to bornological algebras.

8.5.1   Adapting our machinery

We work in the category of C^*-algebras with $*$-homomorphisms as morphisms; we allow non-separable C^*-algebras here, although Kasparov's definition only works in the separable case. We assume that the reader knows about some technical notions like completely positive maps (see section 8.4 above), which are explained in [10].

Let A and B be C^*-algebras. A *(continuous) homotopy* between two morphisms $f_0, f_1 \colon A \to B$ is, of course, a $*$-homomorphism $f \colon A \to C([0,1], B)$ with $\operatorname{ev}_t \circ f = f_t$ for $t = 0, 1$. It is important here that $C([0,1], B)$ is again a C^*-algebra. Thus we redefine $B[0,1] := C([0,1], B)$ in our new context. The *suspension* $SB = B(0,1)$ and the *cone* $CB = B(0,1]$ are redefined accordingly. As usual, they fit into a *cone extension* $CB \rightarrowtail SB \twoheadrightarrow B$.

Homotopy defines an equivalence relation on the space of $*$-homomorphisms $A \to B$. We let $\langle A, B \rangle$ be the set of equivalence classes and $\langle f \rangle$ the class of $f \colon A \to B$ in $\langle A, B \rangle$. Concatenation turns $\langle A, SB \rangle$ into a group for all A, B; the different group structures on $\langle A, S^n B \rangle$ for $n \geq 1$ agree and are Abelian (compare §6.1).

There are at least two useful exterior products in our category. We choose the *minimal C^*-tensor product* $A \widehat{\otimes}_{\min} B$ because this is most commonly used in connection with Kasparov's theory. The *maximal C^*-tensor product* works equally well, and it depends on the situation which one is preferable. Fortunately, both tensor products agree for nuclear C^*-algebras; we denote the tensor product by $\widehat{\otimes}_{C^*}$ in such cases. For example, the C^*-algebras $C([0,1])$ and $\mathcal{K}(\ell^2\mathbb{N})$ are nuclear. We have $C([0,1]) \widehat{\otimes}_{C^*} A \cong A[0,1]$ and $\mathcal{K}(\ell^2\mathbb{N}) \widehat{\otimes}_{C^*} A \cong \mathcal{K}_{C^*}(A)$. The latter is our choice of stabilisation.

Definition 8.23. An extension of C^*-algebras is called *cpc-split* if it has a completely positive contractive linear section; the letters "cpc" stand for *completely positive contractive*, of course.

This is the class of extensions for which Kasparov's theory is known to have long exact sequences in both variables. By Theorem 8.20, an extension $\mathcal{K}_{C^*} \rightarrowtail E \twoheadrightarrow A$ is cpc-split if and only if its class in the semigroup $\mathrm{Ext}(A)$ has an inverse.

Example 8.24. The cone extension $SB \rightarrowtail CB \twoheadrightarrow B$ for a C^*-algebra is cpc-split with section $(\sigma b)(t) := t \cdot b$. Pull-backs of cpc-split extensions remain cpc-split. Thus the mapping cone extension $SB \rightarrowtail C(f) \twoheadrightarrow A$ for a $*$-homomorphism $f\colon A \to B$ and the extension $I \rightarrowtail C(\pi) \twoheadrightarrow CQ$ for a cpc-split extension $I \rightarrowtail E \twoheadrightarrow Q$ are cpc-split.

The following tensor algebra construction is adapted to cpc-split extensions.

Definition 8.25. Let A be a C^*-algebra; its *cpc-tensor algebra* is a C^*-algebra $T_{\mathrm{cpc}}A$ with a cpc linear map $\sigma_A\colon A \to T_{\mathrm{cpc}}A$ that is universal in the sense that any cpc linear map $A \to B$ into a C^*-algebra B factors uniquely through σ_A.

This universal property determines $T_{\mathrm{cpc}}A$ and σ_A uniquely up to natural isomorphism. In order to construct $T_{\mathrm{cpc}}A$, we start with the bornological algebra TA. It carries a unique algebra involution such that $\sigma_A(a^*) = \sigma_A(a)^*$ for all $a \in A$. Call a C^*-semi-norm on TA *good* if the map σ_A is a completely positive contraction with respect to it. It is easy to see that the supremum of a family of good C^*-semi-norms is again good; hence there is a maximal good C^*-semi-norm. We let $T_{\mathrm{cpc}}A$ be the completion of TA for this maximal good C^*-semi-norm and let $\sigma_A\colon A \to T_{\mathrm{cpc}}A$ be the obvious map. This satisfies the universal property of Definition 8.25.

The identity map $A \to A$ is cpc and hence induces a natural $*$-homomorphism $T_{\mathrm{cpc}}A \to A$. Let $J_{\mathrm{cpc}}A \subseteq T_{\mathrm{cpc}}A$ be its kernel. We get a cpc-split extension $J_{\mathrm{cpc}}A \rightarrowtail T_{\mathrm{cpc}}A \twoheadrightarrow A$ with natural cpc section $\sigma_A\colon A \to T_{\mathrm{cpc}}A$.

The same arguments as in Chapter 6 show that this cpc-tensor algebra extension is universal among cpc-split extensions. That is, any cpc-split extension has a *classifying map*, which is unique up to homotopy (compare Definition 6.16 and Lemma 6.17). Moreover, the tensor algebra extension is functorial, and J_{cpc} is a homotopy functor in the sense that it descends to a map $\langle A, B \rangle \to \langle J_{\mathrm{cpc}}A, J_{\mathrm{cpc}}B \rangle$ (compare Lemma 6.19). Here we use that the functor $A \mapsto A[0,1]$ preserves cpc-split extensions; this is so because it is functorial for cpc linear maps. More generally, we get canonical maps

$$\kappa_{A,B}\colon J(A \widehat{\otimes}_{\min} B) \to (JA) \widehat{\otimes}_{\min} B$$

as in Definition 6.22 because $\widehat{\otimes}_{\min}$ is exact on cpc-split extensions.

The *cone extension* $SB \rightarrowtail CB \twoheadrightarrow B$ is cpc-split by the section $b \mapsto t \otimes b$, where $t \in \mathbb{C}(0,1]$ denotes the identical function on $[0,1]$. Hence it has a classifying map $J_{\mathrm{cpc}}B \to SB$. More generally, we get a natural operator

$$\Lambda \colon \langle A, B \rangle \to \langle J_{\mathrm{cpc}}A, SB \rangle$$

as in Definition 6.23.

Now we use our new suspension, cone, and tensor algebra functors as in §6.3 to define the *suspension-stable homotopy category of C^*-algebras* $\Sigma\mathrm{Ho}^{C^*}$ and its product $\#$. The same arguments as for $\Sigma\mathrm{Ho}$ show that $\#$ is well-defined and associative.

We define the *mapping cone* of a $*$-homomorphism $f \colon A \to B$ as

$$C(f) := \{(a,b) \in A \oplus C(B) \mid f(a) = b(1)\}.$$

There are natural maps $S(B) \to C(f) \to A \to B$, so that we can define mapping cone triangles in $\Sigma\mathrm{Ho}^{C^*}$ as in Definition 6.46. Literally the same arguments as in §6.4 show that $\Sigma\mathrm{Ho}^{C^*}$ is a triangulated category (the treatment of the Octahedral Axiom is postponed to §13.2); some proofs simplify because continuous homotopies are easier to manipulate than smooth homotopies. As a consequence, we get Puppe exact sequences for $\Sigma\mathrm{Ho}^{C^*}$ in both variables.

Any cpc-split extension gives rise to an extension triangle using its classifying map. The same argument as in the proof of Theorem 6.63 shows that such extension triangles are exact. Hence the category $\Sigma\mathrm{Ho}^{C^*}$ has long exact sequences for cpc-split extensions in both variables as in Theorem 6.63. Furthermore, the canonical functor from the category of C^*-algebras to $\Sigma\mathrm{Ho}^{C^*}$ is split-exact, so that quasi-homomorphisms induce morphisms in $\Sigma\mathrm{Ho}^{C^*}$ as in §3.1.1. We also get Mayer–Vietoris exact sequences for pull-backs of extensions as in Corollary 6.67; here we use that the pull-back of a cpc-split extension is again cpc-split. The theory $\Sigma\mathrm{Ho}^{C^*}$ also enjoys a universal property: it is the universal triangulated homology theory for C^*-algebras. Here the definition of a (triangulated) homology theory is adapted to use continuous homotopies and cpc-split extensions, of course. Again the proof carries over literally.

The C^*-algebraic stabilisation behaves like the smooth stabilisation in the following respects:

- there are natural isomorphisms $\mathcal{K}_{C^*}\mathcal{K}_{C^*}(A) \cong \mathcal{K}_{C^*}(A)$;

- if $V \colon \ell^2\mathbb{N} \to \ell^2\mathbb{N}$ is an isometry, then the resulting inner endomorphism $\mathrm{Ad}_{V,V^*} \colon \mathcal{K}_{C^*}(A) \to \mathcal{K}_{C^*}(A)$ is homotopic to the identity map;

- the stabilisation homomorphism $\mathcal{K}_{C^*}(A) \to \mathcal{K}_{C^*}\mathcal{K}_{C^*}(A)$ is a homotopy equivalence.

The proofs are similar to those for the smooth stabilisation. We also use that $\mathcal{H} \cong \ell^2(\mathbb{N})$ for any separable Hilbert space $\mathcal{H}$.

It follows that the functor $A \mapsto F\big(\mathcal{K}_{C^*}(A)\big)$ is $\mathcal{K}_{C^*}$-stable and $\mathbb{M}_2$-stable for any homotopy functor F (compare Lemma 3.38).

Definition 8.26. We let

$$\mathrm{kk}^{C^*}(A, B) := \Sigma\mathrm{Ho}^{C^*}\big(\mathcal{K}_{C^*}(A), \mathcal{K}_{C^*}(B)\big)$$

as in Definition 7.19, and we let $\mathbf{kk}^{C^*}$ be the canonical functor from the category of C^*-algebras (or from $\Sigma\mathrm{Ho}$) to kk^{C^*}.

The following theorem summarises the properties of kk^{C^*}.

Theorem 8.27. *The category* kk^{C^*} *is triangulated, and the functor*

$$\mathbf{kk}^{C^*} : \Sigma\mathrm{Ho}^{C^*} \to \mathrm{kk}^{C^*}$$

is exact. The functor $\mathbf{kk}^{C^*}$ *is a homotopy functor,* $\mathcal{K}_{C^*}$*-stable, half-exact for cpc-split extensions, split-exact, and satisfies Bott periodicity.*

Let F be any functor from the category of C^*-algebras to an additive category that is homotopy invariant, $\mathcal{K}_{C^*}$-stable, and half-exact for cpc-split extensions. Then F factors uniquely through $\mathbf{kk}^{C^*}$.

If F has these properties and is a functor to a triangulated category $\mathfrak{T}$, then the resulting functor $\mathrm{kk}^{C^*} \to \mathfrak{T}$ is exact if and only if F maps mapping cone triangles to exact triangles in $\mathfrak{T}$.

Let F_1 and F_2 be functors with the above properties, so that they descend to functors $\bar{F}_1$ and $\bar{F}_2$ on kk^{C^*}. If $\Phi\colon F_1 \to F_2$ is a natural transformation, then Φ remains natural with respect to morphisms in kk^{C^*}, that is, Φ is a natural transformation $\bar{F}_1 \to \bar{F}_2$.

The proof is literally the same as for $\mathrm{kk}^?$, see §7.2–7.3.

We also get Pimsner–Voiculescu exact sequences for crossed products by automorphisms. For Bott periodicity and the Pimsner–Voiculescu sequence, we use the C^*-algebraic variant of the crossed Toeplitz extension, of course.

Theorem 8.28. *There is a natural isomorphism* $\mathrm{kk}^{C^*}(A, B) \cong \mathrm{KK}(A, B)$ *for all separable C^*-algebras A and B. Here KK denotes Kasparov's bivariant K-theory.*

Even more, we still have $\mathrm{kk}^{C^*}(A, B) \cong \mathrm{KK}(A, B)$ if A is separable and B arbitrary.

Proof. If A and B are separable, then so are the C^*-algebras $J_{\mathrm{cpc}}^k \mathcal{K}_{C^*}(A)$ and $S^k \mathcal{K}_{C^*}(B)$ that arise in our definition of $\mathrm{kk}^{C^*}(A, B)$. Therefore, we may restrict attention to separable C^*-algebras in our construction of kk^{C^*}. This implies that the restriction of kk^{C^*} to separable C^*-algebras still enjoys an analogous universal property for functors defined on the category of separable C^*-algebras.

It is known that Kasparov's bivariant K-theory is the universal split-exact $\mathcal{K}_{C^*}$-stable homotopy functor for separable C^*-algebras (see [59]). Hence we get a natural transformation $KK \to kk^{C^*}$ because kk^{C^*} has these properties. Moreover, KK is known to be half-exact for cpc-split extensions. Therefore, we also get a natural transformation $kk^{C^*} \to KK$.

The natural transformations $kk^{C^*} \to KK \to kk^{C^*}$ and $KK \to kk^{C^*} \to KK$ act identically on $*$-homomorphisms by construction. By the uniqueness parts of the universal properties, they act identically on the bivariant K-theories. That is, we have isomorphisms $kk^{C^*}(A, B) \cong KK(A, B)$. $\square$

We usually write KK for kk^{C^*} in the following, unless we want to emphasise the different definitions of these two theories.

Remark 8.29. Although kk^{C^*} is defined for inseparable C^*-algebras, it does not seem the right generalisation of Kasparov theory to this realm because of a technical problem: kk^{C^*} has no reason to be compatible with direct sums (compare §6.3.1). We only know that its restriction to separable C^*-algebras has this property because it holds for Kasparov theory.

8.5.2   Another variant related to E-theory

Now we modify our construction so that it recovers the E-theory of Alain Connes and Nigel Higson, which is originally defined in [30]. We use the same category of algebras, the same notion of homotopy, and the same stabilisation functor, but we modify the class of extensions, allowing *all* C^*-algebra extensions this time; this forces us to use another tensor algebra. In addition, we now use the maximal C^*-tensor product as exterior product because it is exact for all extensions, unlike the minimal C^*-tensor product.

The tensor algebra extension is supposed to be universal for all extensions; this determines it uniquely up to homotopy equivalence. In order to actually construct a tensor algebra with the required universal property, we examine what kinds of sections C^*-algebra extensions admit.

Let $K \rightarrowtail E \twoheadrightarrow Q$ be an extension of C^*-algebras and let $D \subseteq Q$ be some dense subset of the open unit ball; we can take the whole open unit ball, but if Q is separable then we may want to choose a dense sequence instead. Let $A^*(D)$ be the free $*$-algebra with one generator for each element of D. There is a maximal C^*-semi-norm on $A^*(D)$ for which all generators have norm at most 1. We let $T'Q$ be the completion of $A^*(D)$ with respect to this C^*-semi-norm. Since the map $E \to Q$ is a quotient mapping, we may lift elements of D to the closed unit ball of E. This induces a $*$-homomorphism $A^*(D) \to E$, which extends to the completion $T'Q$. Thus $T'Q$ has the required universal property.

Using this new tensor algebra, we can now repeat the arguments above and construct another bivariant K-theory for C^*-algebras. This theory has very similar properties. The only difference is that it is half-exact for all extensions, not just for the cpc-split extensions. Thus its universal property is different.

This new bivariant K-theory agrees with E-theory for separable C^*-algebras. The proof is almost the same as for Theorem 8.28, so that we omit further details.

8.5.3 Comparison with Kasparov's definition

Much work in bivariant K-theory is done in the context of Kasparov theory. If we want to translate it to bornological algebras, we must first extend Kasparov's definition of KK. The right notion here seems to be that of an *abstract Kasparov module*; simpler notions like Fredholm modules and spectral triples are also in use, but they have some deficiencies. Abstract Kasparov modules give rise to elements of $\mathrm{kk}^?$. Hence they can be used to translate constructions from Kasparov theory to $\mathrm{kk}^?$. Along the way, we also describe the natural isomorphism $\mathrm{KK}(A, B) \cong \mathrm{kk}^{C^*}(A, B)$ for separable C^*-algebras A and B that we have obtained in Theorem 8.28.

Fredholm modules and spectral triples

There are two closely related ways to define the groups $\mathrm{KK}_*(A, B)$ for two C^*-algebras A and B; one uses bounded, the other unbounded operators. We first describe both of them for $B = \mathbb{C}$. The resulting $\mathbb{Z}/2$-graded group $\mathrm{K}^*(A) := \mathrm{KK}_*(A, \mathbb{C})$ is also called the K-*homology* of A.

The groups $\mathrm{KK}_*(A, \mathbb{C})$ are generated by certain cycles, which are called Fredholm modules in the bounded picture and spectral triples in the unbounded picture. Depending on whether $* = 0, 1$, we are dealing with even or odd Fredholm modules and spectral triples. We define these notions right away in the generality where A is a bornological algebra. If A is a C^*-algebra, then we merely replace bounded homomorphisms by $*$-homomorphisms in the following definitions.

Definition 8.30. Let A be a bornological algebra. An *even Fredholm module* over A consists of a pair (φ, F), where φ is a bounded homomorphism from A into the algebra $\mathcal{L}(\mathcal{H})$ of bounded operators on a $\mathbb{Z}/2$-graded Hilbert space $\mathcal{H} = \mathcal{H}_+ \oplus \mathcal{H}_-$ and F is a self-adjoint element of $\mathcal{L}(\mathcal{H})$ such that φ is even, F is odd, and such that for all $x \in A$ the following operators are compact:

$$\varphi(x)(1 - F^2), \quad [\varphi(x), F]. \tag{8.31}$$

In the direct sum decomposition $\mathcal{H} = \mathcal{H}_+ \oplus \mathcal{H}_-$, the operators F and φ correspond to block matrices

$$F = \begin{pmatrix} 0 & v \\ v^* & 0 \end{pmatrix} \qquad \varphi = \begin{pmatrix} \alpha & 0 \\ 0 & \bar{\alpha} \end{pmatrix}. \tag{8.32}$$

In most examples, $1 - F^2$ is compact, so that F is a Fredholm operator. This is the source of the name "Fredholm module".

Definition 8.33. An *odd Fredholm module* over A is a pair (φ, F), where φ is a bounded homomorphism from A to the algebra $\mathcal{L}(\mathcal{H})$ of bounded operators on a Hilbert space $\mathcal{H}$ (which is this time trivially graded) and F is a self-adjoint element of $\mathcal{L}(\mathcal{H})$ such that $\varphi(x)(1 - F^2)$ and $[\varphi(x), F]$ are compact as in (8.31).

The only difference between even and odd Fredholm modules is the additional grading in the even case.

It is often required that F should satisfy $F^2 = 1$; we require less in (8.31), but we can achieve $F^2 = 1$ using functional calculus. First we have to double the Hilbert space and consider $\hat{\mathcal{H}} := \mathcal{H} \oplus \mathcal{H}^{\mathrm{op}}$, where $\mathcal{H}^{\mathrm{op}}$ means the same Hilbert space; but in the even case, we use the opposite grading on $\mathcal{H}^{\mathrm{op}}$. We let $\hat{\varphi} := \varphi \oplus 0$ and

$$\hat{F} := \begin{pmatrix} F & \sqrt{1 - F^2} \\ \sqrt{1 - F^2} & -F \end{pmatrix}.$$

It is easy to check that $\hat{F}^2 = 1$ and that $(\hat{\varphi}, \hat{F})$ is a Fredholm module of the same parity as (φ, F).

When we are dealing with bornological algebras, we want to replace the C^*-algebra $\mathcal{K}(\mathcal{H})$ by a Schatten ideal. Thus we often restrict attention to Fredholm modules satisfying the following additional requirement:

Definition 8.34. A Fredholm module (φ, F) is called *p-summable* for some $p \in \mathbb{R}_{\geq 1}$ if $\varphi(x)(1 - F^2)$ and $[\varphi(x), F]$ in (8.31) even belong to the Schatten ideal $\mathscr{L}^p(\mathcal{H})$.

Example 8.35. We are going to construct an important 1-summable odd Fredholm module for $A = C^\infty(\mathbb{T})$.

Let $\mathcal{H} = L^2(\mathbb{T})$; we identify $\mathcal{H} \cong \ell^2(\mathbb{Z})$ via Fourier transform. Equivalently, we equip $\mathcal{H}$ with the orthonormal basis $e_n = \exp(2\pi i n t)$ for $n \in \mathbb{Z}$. The representation $\varphi \colon A \to \mathcal{L}(\mathcal{H})$ is given by pointwise multiplication on $L^2(\mathbb{T})$, which becomes convolution on $\ell^2(\mathbb{Z})$. We let $F(e_n) = e_n$ for $n \geq 0$ and $F(e_n) = -e_n$ for $n < 0$. This defines a self-adjoint operator on $\mathcal{H}$ with $F^2 = \mathrm{id}_{\mathcal{H}}$.

It is evident that $[F, \varphi(z)]$ is a finite-rank operator, where $z \colon \mathbb{T} \to \mathbb{C}$ is the identical inclusion. More generally, $[F, \varphi(a)] \in \mathscr{L}^1(\mathcal{H})$ for all $a \in C^\infty(\mathbb{T})$. As a consequence, (φ, F) is a 1-summable odd Fredholm module over A. If we replace A by $C(\mathbb{T})$, then we still have $[F, \varphi(a)] \in \mathcal{K}(\mathcal{H})$ for all $a \in C(\mathbb{T})$, and φ is a $*$-representation. Hence we get a Fredholm module over $C(\mathbb{T})$.

As we shall see, we can get Fredholm modules for $A = C^\infty(M)$ for a closed smooth manifold M from elliptic pseudo-differential operators on M of order 0. In practice, it is much easier to write down elliptic differential operators of order 1. We can also describe K-homology using such unbounded operators:

Definition 8.36. An *odd p-summable spectral triple* is a triple $(\varphi, \mathcal{H}, D)$ consisting of a Hilbert space $\mathcal{H}$, a bounded homomorphism $\varphi \colon A \to \mathcal{L}(\mathcal{H})$, and a self-adjoint (unbounded) operator D on $\mathcal{H}$ such that

$$[\varphi(x), D] \in \mathcal{L}(\mathcal{H}), \qquad \varphi(x)(1 + D^2)^{-1/2} \in \mathscr{L}^p(\mathcal{H}) \tag{8.37}$$

for all $x \in A$; in addition, the resulting maps $A \to \mathcal{L}(\mathcal{H})$, $x \mapsto [\varphi(x), D]$, and $A \to \mathscr{L}^p(\mathcal{H})$, $x \mapsto \varphi(x)(1 + D^2)^{-1/2}$, are required to be bounded.

An *even p-summable spectral triple* is a triple $(\varphi, \mathcal{H}, D)$ where $\mathcal{H} = \mathcal{H}_+ \oplus \mathcal{H}_-$ is a $\mathbb{Z}/2$-graded Hilbert space, φ and D are as above and, in addition, φ is even and D is odd; thus we have block matrix decompositions as in (8.32).

A p-summable spectral triple yields a p-summable Fredholm module (φ, F) by

$$F := \frac{D}{\sqrt{1 + D^2}}. \tag{8.38}$$

If D is invertible, we may replace this by the sign $D \cdot |D|^{-1}$ of D.

It is much harder to pass, conversely, from a Fredholm module to a spectral triple. Thus spectral triples contain more information than Fredholm modules. This information is quite crucial for noncommutative geometry, but we shall not use it here.

Example 8.39. Let $A = C^\infty(\mathbb{T})$ and $\mathcal{H} = L^2(\mathbb{T})$ as in Example 8.35, and let D act on $\mathcal{H}$ by $Df := \frac{1}{2\pi i} \frac{d}{dt} f$. In the orthonormal basis $e_n = \exp(2\pi i n t)$, we find $D(e_n) = n e_n$. Hence D is an unbounded self-adjoint operator (with suitably chosen domain). The operator $(1 + D^2)^{-1/2}$ is the diagonal operator $e_n \mapsto (1 + n^2)^{-1/2} e_n$; since this sequence grows like $1/n$, the operator $(1 + D^2)^{-1/2}$ is compact and belongs to $\mathscr{L}^p(\mathcal{H})$ for all $p > 1$ but not for $p = 1$. It is easy to check that $[D, \varphi(a)]$ is bounded for all $a \in A$. Hence $(\varphi, \mathcal{H}, D)$ is a spectral triple that is p-summable for all $p > 1$.

The bounded operator F associated to D by our general recipe is the diagonal operator $e_n \mapsto n (1 + n^2)^{-1/2} e_n$. This operator is a compact perturbation of the operator F in Example 8.35 (compare Definition 8.43 below).

Kasparov modules over C^*-algebras

Let A and B be C^*-algebras. A *Kasparov A, B-module* is defined like a Fredholm module over A, except that the Hilbert space $\mathcal{H}$ is replaced by a Hilbert B-module and $\mathcal{L}(\mathcal{H})$ and $\mathcal{K}(\mathcal{H})$ are replaced by the C^*-algebras of adjointable and compact operators. More generally, we may use an arbitrary unital C^*-algebra L containing a closed ideal $K \subseteq L$ that is *stably isomorphic* to B; this means that it comes equipped with an isomorphism $\mathcal{K}_{C^*}(K) \cong \mathcal{K}_{C^*}(B)$ (whose equivalence class up to inner automorphisms is part of the data). It is well-known that two separable C^*-algebras are stably isomorphic if and only if they are Morita–Rieffel equivalent, if and only if $K \cong \mathcal{K}(\mathcal{H}_B)$ for some Hilbert B-module $\mathcal{H}_B$.

Definition 8.40. An *odd Kasparov A, B-module* is a pair (φ, F), where φ is a $*$-homomorphism $A \to L$ and $F \in L$ is self-adjoint, such that $\varphi(x)(1 - F^2) \in K$ and $[\varphi(x), F] \in K$ for all $x \in A$; here L is a unital C^*-algebra and $K \subseteq L$ is an ideal that is stably isomorphic to B.

Even Kasparov A, B-modules are defined similarly; we add a *grading operator* $\varepsilon \in L$ satisfying $\varepsilon = \varepsilon^*$ and $\varepsilon^2 = 1$ to our data and require ε to commute with $\varphi(x)$ for all $x \in A$ and anti-commute with F, that is, $\varepsilon F = -F\varepsilon$. In addition, we require $\varphi(x)(1 - F^2) \in K$ and $[\varphi(x), F] \in K$ for all $x \in A$ as before.

A Kasparov $A, \mathbb{C}$-module is nothing but a Fredholm module in the sense of Definitions 8.30 and 8.33.

A *homotopy* between two Kasparov A, B-modules is a Kasparov $A, B[0, 1]$-module with appropriate restrictions at 0 and 1. This defines an equivalence relation on the sets of even and odd Kasparov A, B-modules. The sets of equivalence classes are the Kasparov groups $\mathrm{KK}_0(A, B)$ and $\mathrm{KK}_1(A, B)$.

Now we associate elements of $\mathrm{kk}_*^{C^*}(A, B)$ to even and odd Kasparov A, B-modules. This provides a natural transformation $\mathrm{KK}_*(A, B) \to \mathrm{kk}_*^{C^*}(A, B)$.

We begin with the even case. First, we modify the Kasparov module to satisfy $F^2 = 1$; this is done as in the case of Fredholm modules. Secondly, let $P := \frac{1}{2}(1+\varepsilon)$ and $P^\perp := 1 - P = \frac{1}{2}(1 - \varepsilon)$, then $P, P^\perp \in L$ are complementary projections that commute with $\varphi(A)$; we get two $*$-homomorphisms

$$\alpha, \bar{\alpha} \colon A \to L, \qquad \alpha(a) := P\varphi(a)P, \quad \bar{\alpha}(a) := P^\perp \varphi(a)P^\perp.$$

Since $F^2 = 1$ and $F = F^*$, we have another $*$-homomorphism

$$\mathrm{Ad}_F \circ \bar{\alpha} \colon A \to L, \qquad a \mapsto F\bar{\alpha}(a)F.$$

The condition $[\varphi(x), F] \in K$ yields $\mathrm{Ad}_F \circ \bar{\alpha}(a) - \alpha(a) \in K$ for all $a \in A$. Hence we get a quasi-homomorphism

$$(\alpha, \mathrm{Ad}_F \circ \bar{\alpha}) \colon A \rightrightarrows L \rhd K.$$

This defines an element in $\mathrm{kk}_0^{C^*}(A, K)$ by split-exactness. The stability of kk^{C^*} yields an isomorphism in $\mathrm{kk}_0^{C^*}(K, B)$. Composing these two ingredients, we get the desired element

$$\mathrm{KK}(\varphi, F, \varepsilon) \in \mathrm{kk}_0^{C^*}(A, B).$$

Since $\mathrm{KK}_0(A, B)$ is defined using homotopy classes of even Kasparov A, B-modules, we get a map $\mathrm{KK}_0(A, B) \to \mathrm{kk}_0^{C^*}(A, B)$. An abstract nonsense argument shows that this reproduces the natural isomorphism of Theorem 8.28.

Next we discuss the odd case, which is slightly simpler. Here we do not need the additional condition $F^2 = 1$. We let $P = \frac{1}{2}(1 + F)$. Let $q \colon L \to L/K$ be the quotient map. The conditions for a Kasparov module imply that the map

$$\psi \colon A \to L/K, \qquad a \mapsto q(P\varphi(a)P),$$

is a $*$-homomorphism. Hence we get a singular morphism-extension

$$
\begin{array}{c}
A \\
\downarrow{\psi} \\
K \rightarrowtail L \xrightarrow{\ q\ } L/K.
\end{array}
$$

It is singular because the extension $K \rightarrowtail L \twoheadrightarrow L/K$ need not be cpc-split. Since $P\varphi(\_)P$ is a completely positive lifting, we may proceed as in Lemma 6.26 and associate a classifying map $J_{\mathrm{cpc}}A \to K$ to it, which yields a class in $\mathrm{kk}_1^{C^*}(A, K)$. Combining this with the isomorphism in $\mathrm{kk}_0^{C^*}(K, B)$, we get the desired element

$$\mathrm{KK}(\varphi, F) \in \mathrm{kk}_1^{C^*}(A, B).$$

Passage to bornological algebras

The constructions in §8.5.3 can be carried over to the setting of bornological algebras. There is only one step that does not carry over literally: in order to achieve
$F^2 = 1$ in the even case, we have used functional calculus for the function $\sqrt{1 - x^2}$;
even in a local Banach algebra, where the holomorphic functional calculus is available, this only makes sense if we know something about the spectrum of F^2. Since
this extra information does not come for free, we have to add a hypothesis to the
extent that $\sqrt{1 - F^2}$ exists. This is the point of the following definition:

Definition 8.41. Let A, K, and L be bornological algebras. Assume that L is
unital and that K is an ideal in L (Definition 3.1). An *abstract even Kasparov*
(A, K)-*module relative to* L is a triple $(\alpha, \bar{\alpha}, U)$ where

- α and $\bar{\alpha}$ are bounded homomorphisms $A \to L$;

- U is an invertible element in L;

- $U\bar{\alpha}(x) - \alpha(x)U \in K$ for all $x \in A$, and the resulting map $A \to K$ is bounded.

We have omitted the stable isomorphism between K and B for simplicity.
Many applications use the ideal $K = \mathscr{L}^p(\mathcal{H})$ in $L = \mathcal{L}(\mathcal{H})$.

An even Kasparov module yields a quasi-homomorphism

$$(\alpha, \mathrm{Ad}_U \circ \bar{\alpha}) \colon A \rightrightarrows L \rhd K,$$

where Ad_U denotes conjugation by U. Since the functor $\mathrm{kk}_0^?(A, \_)$ is split-exact,
this yields a class in $\mathrm{kk}_0^?(A, K)$—which we denote by $\mathrm{kk}(\alpha, \bar{\alpha}, U)$.

Conversely, any quasi-homomorphism $\varphi_+, \varphi_- \colon A \rightrightarrows L \rhd K$ comes from an
abstract even Kasparov module $(\varphi_+, \varphi_-, \mathrm{id})$. Thus an abstract even Kasparov
module is essentially the same thing as a quasi-homomorphism, and the opera-
tor U is redundant. Definition 8.45 is meaningful nevertheless because, in most
applications, the homomorphisms $\alpha, \bar{\alpha}$ do not carry much information and the
operator U is the most crucial ingredient of the construction.

If $(\varphi, F, \varepsilon)$ is an even Kasparov A, B-module over C^*-algebras A and B as in
§8.5.3, then we get an abstract even Kasparov A, K-module by setting

$$U := \varepsilon\sqrt{1 - F^2} + F, \qquad \alpha := P\varphi, \qquad \bar{\alpha} := P^\perp\varphi,$$

where we use $P := \frac{1}{2}(1 + \varepsilon)$, $P^\perp := 1 - P = \frac{1}{2}(1 - \varepsilon)$.

Exercise 8.42. Check that $U^2 = 1$ and that $\mathrm{KK}(\alpha, \bar{\alpha}, U)$ agrees with the construction in §8.5.3 where we double the Hilbert modules to achieve that $F^2 = 1$. Notice
that U does not commute with ε, so that (φ, U) is not a Kasparov module in the
sense of §8.5.3.

We can still carry out the above construction for bornological algebras whenever $1 - F^2$ has a square-root in L. The easiest case of this is $F^2 = 1$, where we
may simply take $U := F$.

Whereas in the C^*-algebra setting, all elements of KK come from Kasparov modules, this is no longer the case for bornological algebras. Another issue is what equivalence relation to put on Kasparov modules. One of Kasparov's main results is that all reasonable equivalence relations agree in the C^*-algebra setting; the easiest to work with is usually homotopy. In the bornological context, smooth homotopy is a good substitute. But we do not expect abstract Kasparov modules that define the same class in $\mathrm{kk}^?$ to be smoothly homotopic.

The following definition contains some finer equivalence relations on abstract Kasparov modules.

Definition 8.43. Let $(\alpha, \bar{\alpha}, U)$ be an abstract even Kasparov module for A, K relative to L. A *compact perturbation* of $(\alpha, \bar{\alpha}, U)$ is a triple $(\alpha, \bar{\alpha}, U')$ where $\alpha(a) \cdot (U' - U) \in K$ and $(U' - U) \cdot \bar{\alpha}(a) \in K$ for all $a \in A$.

We call $(\alpha, \bar{\alpha}, U)$ *degenerate* if $\alpha = \mathrm{Ad}_U \circ \bar{\alpha}$.

Exercise 8.44. If $(\alpha, \bar{\alpha}, U')$ is a compact perturbation of $(\alpha, \bar{\alpha}, U)$, then

$$\mathrm{kk}(\alpha, \bar{\alpha}, U) = \mathrm{kk}(\alpha, \bar{\alpha}, U').$$

Hint: work with $\alpha \oplus 0, \bar{\alpha} \oplus 0 \colon A \to \mathbb{M}_2(L)$ and use that $U \oplus U'$ and $U' \oplus U$ are smoothly homotopic via rotations.

If two abstract even Kasparov modules differ by addition of degenerate ones, then they define the same class in kk.

Finally, we come to abstract odd Kasparov modules:

Definition 8.45. Let A, K, and L be as in Definition 8.41. An *abstract odd Kasparov (A, K)-module relative to L* is a pair (φ, P) where

- φ is a bounded homomorphism $A \to L$;

- $P \in L$ is such that $[P, \varphi(x)] \in K$ and $\varphi(x)(P - P^2) \in K$ for all $x \in A$.

This is equivalent to Definition 8.40 with the substitutions $P := \frac{1}{2}(F + 1)$, $F = 2P - 1$. Whereas F is almost an involution, P is almost a projection.

There is a good reason to use the operator F instead: this unifies the even and odd theories. The parallels between these two cases are less apparent when we use even and odd abstract Kasparov modules.

An odd Kasparov module yields a singular morphism-extension

$$K \rightarrowtail L \xrightarrow{\;\;\psi\;\;} L/K$$

with $\psi(x) := P\varphi(x)P$ (it is irrelevant whether or not $P^2 = P$). This yields a classifying map in $\mathrm{kk}^?_{-1}(A, K)$ by Lemma 6.26, which we denote by $\mathrm{kk}(\varphi, P)$.

Example 8.46. We consider the Fredholm module over $A = C^\infty(\mathbb{T})$ defined in Example 8.35. We take $L = \mathcal{L}(\ell^2(\mathbb{Z}))$ and $K = \mathscr{L}^1(\ell^2(\mathbb{Z}))$. Define $\varphi\colon A \to L$ and $F \in L$ as above, and let $P := \frac{1}{2}(1+F)$. Then $Pe_n = e_n$ for $n \geq 0$ and $Pe_n = 0$ for $n < 0$, that is, P is the orthogonal projection onto $\ell^2(\mathbb{N}) \subseteq \ell^2(\mathbb{Z})$. We claim that the extension that we get from this abstract odd Kasparov module is, essentially, the familiar *Toeplitz extension*. The main point is the following observation: if $z \in C^\infty(\mathbb{T})$ is the identical function, then PzP is the unilateral shift on $\ell^2(\mathbb{N})$, extended by 0 to $\ell^2(\mathbb{Z})$.

There is a variant of the Toeplitz extension $\mathscr{L}^1(\ell^2(\mathbb{N})) \rightarrowtail T \twoheadrightarrow C^\infty(\mathbb{T})$ where we extend $\mathcal{K}_{\mathscr{S}}$ to $\mathscr{L}^1$ and use the same rules for the multiplication. We can further enlarge the kernel to $\mathscr{L}^1(\ell^2(\mathbb{Z})) \cong \mathbb{M}_2(\mathscr{L}^1(\ell^2(\mathbb{N})))$ by putting $C^\infty(\mathbb{T})$ in one corner of $\mathbb{M}_2$. The resulting extension is exactly the one that we get from our abstract odd Kasparov module.

Since the maps $\mathcal{K}_{\mathscr{S}} \to \mathscr{L}^1(\ell^2(\mathbb{N})) \to \mathbb{M}_2(\mathscr{L}^1(\ell^2\mathbb{N}))$ are invertible in $\mathrm{kk}^{\mathscr{L}}$, our variations on the Toeplitz extension have no effect in $\mathrm{kk}^{\mathscr{L}}$.

Comparison with K-theory

A basic feature of Kasparov theory is the natural isomorphism $\mathrm{KK}_*(\mathbb{C}, A) \cong \mathrm{K}_*(A)$ for all separable C^*-algebras A. The isomorphism

$$\mathrm{K}_0(A) \to \mathrm{KK}_0(\mathbb{C}, A)$$

is easy to construct. Recall that elements of $\mathrm{K}_0(A)$ are represented by pairs of projections (e_+, e_-) in $\mathbb{M}_n(A^+)$ such that $e_+ - e_- \in \mathbb{M}_n(A)$. This data gives rise to an even Kasparov module with underlying Hilbert module $\mathcal{H} := A^n \oplus (A^n)^{\mathrm{op}}$, $*$-homomorphism $\mathbb{C} \to \mathcal{L}(\mathcal{H})$, $1 \mapsto e_+ \oplus e_-$, and $F = \left(\begin{smallmatrix} 0 & 1 \\ 1 & 0 \end{smallmatrix}\right)$. (There are various other representatives; for instance, we may cut down $\mathcal{H}$ to the range of $e_+ \oplus e_-$.) It is not hard to see that this construction yields an isomorphism $\mathrm{K}_0(A) \to \mathrm{KK}_0(\mathbb{C}, A)$.

Using suspensions, we conclude that $\mathrm{K}_1(A) \cong \mathrm{KK}_1(\mathbb{C}, A)$. It is remarkable that this map is harder to write down explicitly. Any odd Kasparov cycle with underlying Hilbert module $\mathcal{H}$ over A gives rise to an extension $\mathcal{K}(\mathcal{H}) \rightarrowtail E \twoheadrightarrow \mathbb{C}$. If A is unital, then we need an infinitely generated Hilbert module $\mathcal{H}$ in order to have room for non-trivial extensions. Therefore, we need a Fredholm operator on a large Hilbert module.

For odd KK-theory, what is easy to write down is a natural isomorphism

$$\mathrm{K}_1(A) \xrightarrow{\cong} \mathrm{KK}_0(C_0(\mathbb{R}), A).$$

We need a significant part of Bott periodicity to go from here to $\mathrm{KK}_1(\mathbb{C}, A)$. Given a unitary $u \in \mathrm{Gl}_n(A)$, we view the associated functional calculus as a map $C(\mathbb{S}^1) \to \mathbb{M}_n(A^+)$, whose restriction to $C_0(\mathbb{R}) \cong C_0(\mathbb{S}^1 \setminus \{1\})$ is a $*$-homomorphism into $\mathbb{M}_n(A)$. This $*$-homomorphism defines a class in $\mathrm{KK}_0(C_0(\mathbb{R}), A)$.

8.5.4   Some remarks on the Kasparov product

One of Kasparov's main achievements is the construction of an associative product

$$\mathrm{KK}_*(A, D) \times \mathrm{KK}_*(D, B) \to \mathrm{KK}_*(A, B).$$

This is more difficult than the construction of the product in $\Sigma\mathrm{Ho}$ and $\mathrm{kk}^?$. From our point of view, the crucial point of Kasparov's construction is that the composition of two quasi-homomorphisms $\varphi_\pm \colon A \rightrightarrows \tilde{D} \triangleright D$ and $\psi_\pm \colon D \rightrightarrows \tilde{B} \triangleright B$—which is comparatively easy to define in $\mathrm{kk}^{C^*}(A, B)$—can again be represented by a quasi-homomorphism from A to B (actually, we must replace B by a stabilisation here). Since this uses special features of C^*-algebras, we do not expect this to work out for Banach algebras or bornological algebras.

In order to compute Kasparov products, we need a sufficient criterion for a quasi-homomorphism to represent the product of two quasi-homomorphisms. We only formulate this in terms of Kasparov modules, as usual in the literature. It seems likely that a similar sufficient condition characterises Kasparov products for abstract Kasparov modules over bornological algebras. Since we have not yet investigated this issue, we mostly limit our discussion to the case of C^*-algebras.

A universal algebra related to quasi-homomorphisms

We want to classify quasi-homomorphisms $A \rightrightarrows D \triangleright B$ by homomorphisms $qA \to B$ for a suitable universal algebra qA; this construction is analogous to the construction of classifying maps $JA \to I$ for extensions $I \rightarrowtail E \twoheadrightarrow A$.

First, we need *free products* of bornological algebras and C^*-algebras. We will not distinguish between these two parallel cases in our notation.

Definition 8.47. The free product of two algebras A and B is defined by the universal property

$$\mathrm{Hom}(A * B, D) \cong \mathrm{Hom}(A, D) \times \mathrm{Hom}(B, D),$$

where Hom denotes morphisms in the categories of bornological or C^*-algebras. That is, $A * B$ is the coproduct of A and B in the appropriate category. It comes equipped with two canonical maps

$$i_A \colon A \to A * B, \qquad i_B \colon B \to A * B.$$

The free product of bornological algebras can be described explicitly. The underlying bornological vector space is the direct sum of all alternating tensor products $\cdots \hat{\otimes}\, A \,\hat{\otimes}\, B \,\hat{\otimes}\, A \,\hat{\otimes}\, B \,\hat{\otimes}\, \cdots$, which may begin and end in A or B. The product is defined by concatenation of tensors, followed by multiplication in A or B if two factors in the same algebra meet. The embeddings i_A, i_B identify A and B with the corresponding direct summands in $A * B$. Thus a monomial $a_1 \otimes b_1 \otimes a_2 \otimes b_2 \otimes \cdots \otimes a_n \otimes b_n$ corresponds to the product

$$i_A(a_1) \cdot i_B(b_1) \cdot i_A(a_2) \cdot i_B(b_2) \cdots i_A(a_n) \cdot i_B(b_n).$$

In the C^*-algebra case, this bornological free product carries a unique involution extending the involutions on A and B. The C^*-algebraic free product is the completion of this $*$-algebra for the maximal C^*-seminorm, which exists because

$$\|a_1 \otimes b_1 \otimes a_2 \otimes b_2 \otimes \cdots \otimes a_n \otimes b_n\| \leq \|a_1\| \cdot \|b_1\| \cdot \|a_2\| \cdot \|b_2\| \cdots \|a_n\| \cdot \|b_n\|$$

holds for any C^*-seminorm.

The universal property provides a natural map $\varphi \colon A * B \to A \oplus B$ whose compositions with i_A and i_B are the coordinate inclusions in $A \oplus B$.

Definition 8.48. We let $QA := A * A$. The pair of homomorphisms $(\mathrm{id}_A, \mathrm{id}_A)$ induces a natural map $\pi_A \colon QA \to A$. We let $qA := \ker \pi_A \subseteq QA$. Let $\varrho_A \colon qA \to A$ be the restriction of the map $QA \to A$ induced by the pair $(\mathrm{id}_A, 0)$.

The free product $QA = A * A$ comes equipped with two canonical maps $i_1, i_2 \colon A \to QA$. We have $\pi_A \circ i_1 = \pi_A \circ i_2 = \mathrm{id}_A$, that is, $\pi_A \colon QA \to A$ is a split surjection with two sections $i_1, i_2 \colon A \to QA$. The difference $i_1 - i_2$ maps A into qA. Thus we get a special quasi-homomorphism $i_1, i_2 \colon A \rightrightarrows QA \rhd qA$. It is universal in the following sense: if $(f_\pm) \colon A \rightrightarrows D \rhd B$ is any quasi-homomorphism, then there is a commuting diagram

$$
\begin{array}{ccccc}
A & \overset{i_1}{\underset{i_2}{\rightrightarrows}} & QA & \rhd & qA \\
\| & & \downarrow & & \downarrow \\
A & \overset{f_+}{\underset{f_-}{\rightrightarrows}} & D & \rhd & B.
\end{array}
$$

The map $QA \to D$ is induced by the pair of maps (f_+, f_-); it restricts to a map $qA \to B$. In the bornological case, this restriction is a bounded map $qA \to B$ because $f_+ - f_- \colon A \to B$ and the multiplication map $D \times B \to B$ are bounded and the map

$$QA \,\widehat{\otimes}\, A \to qA, \qquad x \otimes a \mapsto x \cdot \big(i_1(a) - i_2(a)\big)$$

has a bounded linear section. The map $qA \to B$ above is called the *classifying map of the quasi-homomorphism*.

Proposition 8.49. *Let F be a functor on the category of C^*-algebras (or bornological algebras) that is $\mathbb{M}_2$-stable and (smoothly) homotopy invariant. Then the natural map $A * B \to A \oplus B$ induces an isomorphism $F(A * B) \to F(A \oplus B)$. If F is additive as well, then $F(A * B) \cong F(A) \oplus F(B)$.*

Proof. The map

$$i_A \oplus i_B \colon A \oplus B \to \mathbb{M}_2(A * B), \qquad (a, b) \mapsto \begin{pmatrix} i_A(a) & 0 \\ 0 & i_B(b) \end{pmatrix}$$

induces a map $F(A \oplus B) \to F\big(\mathbb{M}_2(A * B)\big) \cong F(A * B)$ by $\mathbb{M}_2$-stability. We claim that this map is inverse to $F(\varphi)$. It is easy to see that $\varphi \circ (i_A \oplus i_B)$ differs from the stabilisation homomorphism $A \oplus B \to \mathbb{M}_2(A \oplus B)$ by an inner endomor-phism of $\mathbb{M}_2(A \oplus B)$. Hence this composition induces the identity map on F by Proposition 3.16. We claim that the other composite map $A * B \to \mathbb{M}_2(A * B)$ is (smoothly) homotopic to the stabilisation homomorphism. By the universal prop-erty of $A*B$, it suffices to construct (smooth) homotopies of maps $A \to \mathbb{M}_2(A*B)$, $B \to \mathbb{M}_2(A * B)$ *separately.* We take a constant homotopy on A; on B, we take a rotation homotopy $B \to \mathbb{M}_2(B)[0,1]$, composed with the canonical embedding $\mathbb{M}_2(B)[0,1] \to \mathbb{M}_2(A * B)[0,1]$. $\qquad\square$

Proposition 8.50. *If F is split-exact, $\mathbb{M}_2$-stable, and smoothly homotopy invariant, then the map $\varrho_A \colon qA \to A$ induces an isomorphism $F(qA) \cong F(A)$. Its inverse is the map induced by the universal quasi-homomorphism $A \rightrightarrows QA \triangleright qA$.*

Proof. Apply F to the morphism of extensions

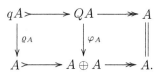

The vertical map $F(QA) \to F(A \oplus A)$ is an isomorphism by Proposition 8.49. Since F is split-exact, the Snake Lemma shows that the map $F(\varrho_A) \colon F(qA) \to F(A)$ is invertible as well. Let $\tilde{F}(i_1, i_2) \colon F(A) \to F(qA)$ be the map associated to the universal quasi-homomorphism. Its composition with ϱ_A is the map associated to the quasi-homomorphism $i_1, i_2 \colon A \to A \oplus A \triangleright A$. Obviously, the latter induces the identity map on $F(A)$. This yields the last assertion. $\qquad\square$

Since the functor $\mathbf{kk}^?$ (or $\mathbf{kk}^{C^*}$) has all the properties required in Proposi-tion 8.50, it follows that ϱ_A becomes invertible in $\mathrm{kk}^?(qA, A)$.

The algebra qA plays a crucial role in the proof of the universal property of Kasparov theory. The main idea of [34, 35] is that there is a natural bijection between $\mathrm{KK}_0(A, B)$ and the set of homotopy classes of $*$-homomorphisms $qA \to \mathcal{K}_{C^*}(B)$ for all separable C^*-algebras A and B. There are similar descriptions for $\mathrm{KK}_1(A, B)$ and in the $\mathbb{Z}/2$-graded case [57, 58, 129]. A concise treatment of this topic can also be found in [83].

These descriptions of Kasparov theory by universal algebras are forerunners of our construction of bivariant K-theories in Chapters 6 and 7. The approach in [129] is particularly close to ours. We are forced to use different universal al-gebras because the Kasparov product for quasi-homomorphisms does not work in general.

Now we use our universal algebra qA to discuss the Kasparov product. The universal quasi-homomorphisms $A \rightrightarrows QA \triangleright qA$ and $qA \rightrightarrows Q(qA) \triangleright q(qA)$ may be viewed as elements of $\mathrm{KK}(A, qA)$ and $\mathrm{KK}(qA, qqA)$. Their Kasparov product is then an element of $\mathrm{KK}(A, qqA)$. Unravelling the construction, we see that this ele-ment is represented by a $*$-homomorphism $\alpha_A \colon qA \to \mathbb{M}_2(qqA)$. The associativity of the Kasparov product implies that the compositions of α_A with $\varrho_{qA}\colon qqA \to qA$ and $\mathrm{M}_2(\varrho_{qA})$ are homotopic to the stabilisation homomorphisms for qqA and qA, respectively. The map α_A is constructed more directly in [35].

No matter how we construct α_A, we need special features of separable C^*-algebras: Kasparov's Technical Theorem in Kasparov's approach, and a derivation lifting theorem by Gert K. Pedersen in Cuntz's approach. Hence neither construction has a chance to work for more general algebras.

We can use the map α_A to compose more general quasi-homomorphisms

$$\varphi_\pm\colon A \rightrightarrows \tilde{D} \rhd D, \qquad \psi_\pm\colon D \rightrightarrows \tilde{B} \rhd B.$$

Let $\Phi\colon qA \to D$ and $\Psi\colon qD \to B$ be their classifying maps. We get a composite map $\Psi \circ q(\Phi)\colon qqA \to B$. Finally, $\mathrm{M}_2(\Psi) \circ \mathrm{M}_2 q(\Phi) \circ \alpha_A\colon qA \to \mathrm{M}_2(B)$ is the classifying map of a quasi-homomorphism from A to B; it translates the Kasparov product in the notation of quasi-homomorphisms.

Computation of Kasparov products

The description of the Kasparov product above is not constructive. Even if we already know what the Kasparov product should be, we cannot use it to verify our guess. This can be achieved by an axiomatic description of Kasparov products due to Alain Connes and Georges Skandalis [31, Theorem A.3].

For more general algebras than separable C^*-algebras, we no longer expect a Kasparov product to exist: we must pass from quasi-homomorphisms to extensions of higher length. Nevertheless, it should be possible in nice cases to represent the composition of two quasi-homomorphisms again by a quasi-homomorphism. We discuss a sufficient criterion for this here, but only in the C^*-algebra case. It is possible to trace explicitly which properties of C^*-algebras are needed, so that the criterion extends to bornological algebras; but the requirements get rather technical, so that it is not clear at the moment how useful this extension is.

Our construction of the Kasparov product is the same as in [10, §18]. We differ from [10] by relating it to the ideas of §8.5.4. We also get a construction of the maps $\alpha_A\colon qA \to \mathrm{M}_2(qqA)$ needed there. To simplify the exposition, we mostly ignore stabilisations. The following discussion focuses on the C^*-algebraic case.

Let $\varphi_1, \bar{\varphi}_1\colon A \to \mathcal{M}(D)$ and $U_1 \in \mathcal{M}(D)$ define an abstract even Kasparov A, D-module, and let $\varphi_2, \bar{\varphi}_2\colon D \to \mathcal{M}(B)$ and $U_2 \in \mathcal{M}(B)$ define an abstract even Kasparov D, B-module (Definition 8.41). Here $\mathcal{M}(D)$ and $\mathcal{M}(B)$ denote the multiplier algebras of D and B, respectively; we put

$$\Phi_1 := \begin{pmatrix} \varphi_1 & 0 \\ 0 & \bar{\varphi}_1 \end{pmatrix}, \qquad F_1 := \begin{pmatrix} 0 & U_1 \\ U_1^{-1} & 0 \end{pmatrix}$$

and similarly for Φ_2 and F_2. Since we are in the C^*-algebraic case, we assume U_1 and U_2 to be unitary and Φ_1 and Φ_2 to be *-homomorphisms. By design, F_j is odd and self-adjoint and satisfies $F_j^2 = 1$ for $j = 1, 2$. We also have $[F_1, \Phi_1(A)] \subseteq D$ and $[F_2, \Phi_2(D)] \subseteq B$, so that we have even Kasparov modules as in Definition 8.40.

Let $K := \mathbb{M}_4(B)$ and $L := \mathcal{M}(K)$; by design, K is an ideal in L. We equip K and L with the grading operators

$$\varepsilon_1 := \begin{pmatrix} 1 & 0 & 0 & 0 \\ 0 & -1 & 0 & 0 \\ 0 & 0 & 1 & 0 \\ 0 & 0 & 0 & -1 \end{pmatrix}, \qquad \varepsilon_2 := \begin{pmatrix} 1 & 0 & 0 & 0 \\ 0 & 1 & 0 & 0 \\ 0 & 0 & -1 & 0 \\ 0 & 0 & 0 & -1 \end{pmatrix}, \qquad \varepsilon := \varepsilon_1 \varepsilon_2.$$

Now we *assume* that Φ_2 extends to a unital homomorphism $\Phi_2 \colon \mathcal{M}(D) \to \mathbb{M}_2(\mathcal{M}(B))$, which we again write as $\Phi_2 = \varphi_2 \oplus \bar{\varphi}_2$. This comes for free if Φ_2 is essential, that is, $\Phi_2(D) \cdot \mathbb{M}_2(B)$ is dense in $\mathbb{M}_2(B)$. It is known that any Kasparov module is homotopic to an essential one; this step involves stabilisations.

By assumption, we get a composite $*$-homomorphism

$$\hat{\Phi} := \begin{pmatrix} \varphi_2 \circ \varphi_1 & 0 & 0 & 0 \\ 0 & \varphi_2 \circ \bar{\varphi}_1 & 0 & 0 \\ 0 & 0 & \bar{\varphi}_2 \circ \varphi_1 & 0 \\ 0 & 0 & 0 & \bar{\varphi}_2 \circ \bar{\varphi}_1 \end{pmatrix}$$

and self-adjoint odd operators

$$\hat{F}_1 := \Phi_2(F_1) = \begin{pmatrix} 0 & \varphi_2(U_1) & 0 & 0 \\ \varphi_2(U_1^{-1}) & 0 & 0 & 0 \\ 0 & 0 & 0 & \bar{\varphi}_2(U_1) \\ 0 & 0 & \bar{\varphi}_2(U_1^{-1}) & 0 \end{pmatrix},$$

$$\hat{F}_2 := (F_2 \oplus F_2)\varepsilon_1 = \begin{pmatrix} 0 & 0 & U_2 & 0 \\ 0 & 0 & 0 & -U_2 \\ U_2^{-1} & 0 & 0 & 0 \\ 0 & -U_2^{-1} & 0 & 0 \end{pmatrix}.$$

These data $\hat{\Phi}, \hat{F}_1, \hat{F}_2$ have the following properties:

- $\hat{\Phi}$ is even and $\hat{F}_1$ and $\hat{F}_2$ are odd with respect to ε;

- $(\hat{F}_1 - \hat{F}_1^*) \cdot \hat{\Phi}(A) \subseteq K$ and $(\hat{F}_1^2 - 1) \cdot \hat{\Phi}(A) \subseteq K$, and similarly for $\hat{F}_2$;

- the graded commutators

$$\big[\hat{F}_2, [\hat{F}_1, \hat{\Phi}(a)]\big] = \hat{F}_2 \cdot [\hat{F}_1, \hat{\Phi}(a)] + [\hat{F}_1, \hat{\Phi}(a)]\hat{F}_2$$

  belong to K for all $a \in A$;

- $[\hat{F}_1, \hat{\Phi}(a)] \cdot [\hat{F}_2, \hat{\Phi}(b)] \in K$ for all $a, b \in A$.

Definition 8.51. We call a triple $(\hat{\Phi}, \hat{F}_1, \hat{F}_2)$ with these properties a *double Kasparov module*; the bigrading given by ε_1 and ε_2 or, equivalently, the direct sum decomposition $K = B \oplus B^{\mathrm{op}} \oplus B^{\mathrm{op}} \oplus B$ is also part of the data, but always suppressed from the notation.

We have intentionally left out another additional property: in the above situation, we have $[\hat{F}_1, \hat{\Phi}(a)] \cdot [\hat{F}_1, \hat{F}_2] \in K$ for all $a \in A$ (here $[\hat{F}_1, \hat{F}_2] = \hat{F}_1\hat{F}_2 + \hat{F}_2\hat{F}_1$); this property is neither needed nor preserved by the following constructions.

Lemma 8.52. *Let $(\hat{\Phi}, \hat{F}_1, \hat{F}_2)$ be a double Kasparov module with $\hat{F}_1^2 = 1 = \hat{F}_2^2$ and $\hat{F}_1^* = \hat{F}_1$ and $\hat{F}_2^* = \hat{F}_2$; let $P_{00}: K \to B$ be the orthogonal projection onto the subspace that is even with respect to both gradings $\varepsilon_1, \varepsilon_2$. Then the maps $P_{00}\hat{\Phi}$, $P_{00}\operatorname{Ad}(\hat{F}_1)\hat{\Phi}$, $P_{00}\operatorname{Ad}(\hat{F}_2)\hat{\Phi}$, and $P_{00}\operatorname{Ad}(\hat{F}_1\hat{F}_2)\hat{\Phi}$ induce a *-homomorphism $q(qA) \to K$ called its* classifying map.

The proof below explains how to construct this classifying map.

Proof. The pairs of maps $(\hat{\Phi}, \operatorname{Ad}(\hat{F}_1)\hat{\Phi})$ and $(\operatorname{Ad}(\hat{F}_2)\hat{\Phi}, \operatorname{Ad}(\hat{F}_2\hat{F}_1)\hat{\Phi})$ define two *-homomorphisms $\Phi_0', \Phi_1': qA \to L$, which we combine to a *-homomorphism $\Phi'': q(qA) \to L$. The point is that the range of Φ'' is contained in K. Equivalently, (Φ_0', Φ_1') is a quasi-homomorphism $A \rightrightarrows L \triangleright K$. Since the range of Φ'' commutes with P_{00}, we may compress it to a *-homomorphism $q(qA) \to B$.

We must check that $[\hat{F}_2, \Phi_0'(x)] \in K$ for all $x \in qA$. The products of the form

$$\hat{\Phi}(a_0)[\hat{F}_1, \hat{\Phi}(a_1)] \cdots [\hat{F}_1, \hat{\Phi}(a_n)], \qquad [\hat{F}_1, \hat{\Phi}(a_1)] \cdots [\hat{F}_1, \hat{\Phi}(a_n)]$$

with $a_0, \ldots, a_n \in A$ span a dense subspace of $\Phi_0'(qA)$. Using the Leibniz rule and the requirements $[\hat{F}_2, \hat{\Phi}(a_0)] \cdot [\hat{F}_1, \hat{\Phi}(a_1)] \in K$ and $\big[\hat{F}_2, [\hat{F}_1, \hat{\Phi}(a_j)]\big] \in K$ for all $j \in \{1, \ldots, n\}$, we find that commutators of such products with $\hat{F}_2$ always belong to B, as desired. $\qquad\square$

A *-homomorphism $q(qA) \to B$ yields a class in $\mathrm{KK}(A, B)$ because $qA \cong A$ in KK. This class is the product of the classes in $\mathrm{KK}(A, D)$ and $\mathrm{KK}(D, B)$ if our double Kasparov module arises from two such Kasparov modules as above.

To construct the Kasparov product, we have to simplify a double Kasparov module to a Kasparov module in the usual sense. This uses the following notion:

Definition 8.53. A double Kasparov module is *simple* if $[\hat{F}_1, \hat{\Phi}(A)] \subseteq K$.

Lemma 8.54. *Let $(\hat{\Phi}, \hat{F}_1, \hat{F}_2)$ be a simple double Kasparov module. Then $(\hat{\Phi}, \hat{F}_1)$ is a Kasparov module in its own right. The classes in $\mathrm{KK}_0(A, B)$ associated to $(\hat{\Phi}, \hat{F}_1, \hat{F}_2)$ and $(\hat{\Phi}, \hat{F}_1)$ agree. Hence they induce the same map $H(A) \to H(B)$ for any split-exact $\mathbb{M}_2$-stable functor H.*

Proof. Let $\Phi_0': qA \to \mathcal{M}(B)$ be the *-homomorphism defined in the proof of Lemma 8.52. Since the Kasparov module is simple, $\Phi_0'(qA) \subseteq B$. Thus our quasi-homomorphism $q(qA) \to B$ comes from a pair of *-homomorphisms $qA \to B$, so that we can simplify it using (3.7). Replacing the sums in (3.7) by orthogonal direct sums as in Exercise 3.17, we get the quasi-homomorphism attached to the Kasparov module $(\hat{\Phi}, \hat{F}_1)$. $\qquad\square$

Our task is to find a homotopy $\hat{F}_{1,t}$ between $\hat{F}_1 = \hat{F}_{1,0}$ and another operator $\hat{F}_{1,1}$ such that $(\hat{\Phi}, \hat{F}_{1,1}, \hat{F}_2)$ is simple and $(\hat{\Phi}, \hat{F}_{1,t}, \hat{F}_2)$ is a double Kasparov module over $A, B[0,1]$. Then $(\hat{\Phi}, \hat{F}_{1,1})$ is the desired Kasparov product by Lemma 8.54.

Lemma 8.55. *Let $M, N \in L$ be positive, even operators with*

- $M + N = 1$;
- *the commutators $[M, \hat{F}_1]$, $[M, \hat{F}_2]$, and $[M, \hat{\Phi}(a)]$ for $a \in A$ belong to K;*
- $M \cdot [\hat{F}_1, \hat{\Phi}(a)]$ *(equivalently, $[\hat{F}_1, \hat{\Phi}(a)] \cdot M$) belongs to B for all $a \in A$;*
- $N \cdot [\hat{F}_2, \hat{\Phi}(a)]$ *and $N \cdot [\hat{F}_1, \hat{F}_2]$ (equivalently, $[\hat{F}_2, \hat{\Phi}(a)] \cdot N$ and $[\hat{F}_1, \hat{F}_2] \cdot N$), belong to B for all $a \in A$.*

Put

$$\hat{F}_{1,t} := \sqrt[4]{1 - tN} \cdot \hat{F}_1 \cdot \sqrt[4]{1 - tN} + \sqrt[4]{tN} \cdot \hat{F}_2 \cdot \sqrt[4]{tN}$$

for $t \in [0, 1]$, so that

$$\hat{F}_{1,1} = \sqrt[4]{M} \cdot \hat{F}_1 \cdot \sqrt[4]{M} + \sqrt[4]{N} \cdot \hat{F}_2 \cdot \sqrt[4]{N}.$$

Then $(\hat{F}_{1,t}, \hat{F}_2)_{t \in [0,1]}$ is a double Kasparov module over $A, B[0, 1]$ and $\hat{F}_{1,1}$ is simple.

Here $[\hat{F}_1, \hat{F}_2]$ is the graded commutator $\hat{F}_1 \hat{F}_2 + \hat{F}_2 \hat{F}_1$.

Proof. It is clear that $F_{1,t}$ is odd and self-adjoint. Since M and N commute with $\hat{F}_1, \hat{F}_2, \hat{\Phi}(A)$ modulo K, the same holds for their roots. Hence we have

$$F_{1,t}^2 \hat{\Phi}(a) \equiv (1 - tN)\hat{F}_1^2 \hat{\Phi}(a) + tN\hat{F}_2^2 \hat{\Phi}(a) + \sqrt{1 - tN}\sqrt{tN}[\hat{F}_1, \hat{F}_2]\hat{\Phi}(a) \equiv \hat{\Phi}(a)$$

modulo B. Still computing modulo B, we have

$$[\hat{F}_{1,t}, \hat{\Phi}(a)] \equiv \sqrt{1 - tN}[\hat{F}_1, \hat{\Phi}(a)] + \sqrt{tN}[\hat{F}_2, \hat{\Phi}(a)] \equiv \sqrt{1 - tN}[\hat{F}_1, \hat{\Phi}(a)];$$

this vanishes modulo B if $t = 1$, so that $(\hat{\Phi}, \hat{F}_{1,1}, \hat{F}_2)$ is simple; we also get

$$[\hat{F}_2, [\hat{F}_{1,t}, \hat{\Phi}(a)]] \equiv [\hat{F}_2, [\hat{F}_1, \hat{\Phi}(a)]] \equiv 0 \bmod B,$$
$$[\hat{F}_{1,t}, \hat{\Phi}(a)] \cdot [\hat{F}_2, \hat{\Phi}(a)] \equiv [\hat{F}_1, \hat{\Phi}(a)] \cdot [\hat{F}_2, \hat{\Phi}(a)] \equiv 0 \bmod B.$$

Thus we get a double Kasparov module over $A, B[0, 1]$.   □

If A and D are separable C^*-algebras, then the existence of operators M, N as in Lemma 8.55 is ensured by Kasparov's Technical Theorem (see [10]). Hence we can always apply this result to construct a Kasparov product. For the universal quasi-homomorphisms $A \rightrightarrows QA \triangleright qA$ and $qA \rightrightarrows Q(qA) \triangleright q(qA)$, it yields a homomorphism $A \to \mathbb{M}_2(qqA)$ as needed in §8.5.4.

In order to carry this over to bornological algebras, we drop all those assumptions on M and N that involve positivity and require instead the existence of various square roots. Concerning the commutator conditions, we should require these for the roots of $1 - tN$ and N that we need because now they are no longer constructed from $1 - tN$ and N by some kind of functional calculus. It still remains to explore some examples to see whether the above construction is useful for bornological algebras.

8.6   Equivariant bivariant K-theories

Let G be a locally compact group. Recall that a G-C^*-algebra is a C^*-algebra equipped with a continuous action of G by $*$-automorphisms. Similarly, a *bornological G-algebra* is a bornological algebra equipped with a *smooth* action of G by automorphisms. Smooth group actions on bornological vector spaces are defined in [85]. We want to construct equivariant versions of the theories KK and kk$^?$ for G-C^*-algebras and bornological G-algebras, respectively. Of course, the morphisms in these categories are the equivariant $*$-homomorphisms and the equivariant bounded algebra homomorphisms, respectively.

We mainly need KK^G for our description of the Baum–Connes assembly map in terms of localisation of categories in §13.1.2. We omit most details because the construction of KK^G is parallel to the construction of kk^{C^*} in §8.5.1.

Since the tensor algebra, suspension, and mapping cone constructions are natural, they inherit group actions by functoriality. One checks easily that these induced group actions are again smooth or continuous if we start with smooth or continuous actions. Hence we get corresponding constructions for G-C^*-algebras and bornological G-algebras. Moreover, the natural maps that are needed to define the product and verify its properties are natural and therefore compatible with a group action. Hence it is easy to take into account the group action in the definitions of $\Sigma\mathrm{Ho}$ and $\Sigma\mathrm{Ho}^{C^*}$: simply restrict to G-equivariant homomorphisms (and homotopies) everywhere. This does not change anything substantial.

Additional care is necessary in connection with classifying maps and extension triangles. We call an extension of bornological G-algebras *G-equivariantly semi-split* if it has a G-equivariant bounded linear section. In the definition of a morphism-extension, we have to assume that the extension is G-equivariantly semi-split because the universal property of the tensor algebra only works for such extensions. If G is compact, then any semi-split extension is G-equivariantly semi-split because we can ensure equivariance by averaging over the group; this fails for non-compact groups.

Fortunately, all extensions that we need to define the theory have *natural* bounded linear sections, where naturality holds in the formal sense that implies G-equivariance. Therefore, we can construct the categories $\Sigma\mathrm{Ho}^G$ and $\Sigma\mathrm{Ho}^{G,C^*}$ exactly as above, and these categories are again triangulated. They have extension triangles for G-equivariantly semi-split and G-equivariantly cpc-split extensions, respectively.

Next we discuss the choice of stabilisation by compact operators. We can, of course, use the same stabilisations as in the non-equivariant case. But it is better to consider algebras of compact operators with non-trivial G-action. We first discuss this in the easier C^*-algebraic case.

Definition 8.56. A functor F defined on the category of G-C^*-algebras is called *equivariantly stable* if the natural embeddings

$$\mathcal{K}(\mathcal{H}_1) \mathbin{\widehat{\otimes}}_{C^*} A \longrightarrow \mathcal{K}(\mathcal{H}_1 \oplus \mathcal{H}_2) \mathbin{\widehat{\otimes}}_{C^*} A \longleftarrow \mathcal{K}(\mathcal{H}_1) \mathbin{\widehat{\otimes}}_{C^*} A$$

induce isomorphisms on F for any pair of G-Hilbert spaces $\mathcal{H}_1, \mathcal{H}_2$; here $\mathcal{K}(\dots)$ is equipped with the induced action and the tensor products are equipped with the diagonal action.

Lemma 8.57. *Let F be a functor defined on the category of G-C^*-algebras. Suppose that F is homotopy invariant and stable with respect to $\mathcal{K}(\ell^2\mathbb{N})$, that is, the stabilisation homomorphism $A \to \mathcal{K}(\ell^2\mathbb{N}) \widehat{\otimes}_{C^*} A$ induces an isomorphism on F for all A. Then the functor $A \mapsto F(\mathcal{K}(L^2G) \widehat{\otimes}_{C^*} A)$ is equivariantly stable.*

Proof. The left regular representation on the Hilbert space $L^2(G)$ has the property that, for any G-Hilbert space $\mathcal{H}$, the diagonal representation and the left regular representation on the first factor on $L^2(G) \overline{\otimes} \mathcal{H}$ are unitarily equivalent. The intertwining unitary acts by $Uf(g) := g \cdot f(g)$. Therefore, up to isomorphism $\mathcal{K}(L^2G) \widehat{\otimes}_{C^*} \mathcal{K}(\mathcal{H}) \cong \mathcal{K}(L^2G \overline{\otimes} \mathcal{H})$ is independent of the group action on $\mathcal{H}$. Hence equivariant stability reduces to ordinary stability.   □

This motivates choosing the stabilisation

$$\mathcal{K}(A) := \mathcal{K}(L^2G \overline{\otimes} \ell^2\mathbb{N}) \widehat{\otimes}_{C^*} A.$$

Exercise 8.58. Combine Lemma 8.57 with our previous non-equivariant stability results to show that $A \mapsto F(\mathcal{K}(A))$ is equivariantly stable for any functor F.

Now we have all ingredients to define the bivariant K-theory:

$$\mathrm{kk}^{G,C^*}(A, B) := \Sigma\mathrm{Ho}^{G,C^*}(\mathcal{K}(A), \mathcal{K}(B)).$$

It is important here to stabilise both A and B. It is evident that this defines a category kk^{G,C^*} and that we have a functor $\Sigma\mathrm{Ho}^{G,C^*} \to \mathrm{kk}^{G,C^*}$. This functor is equivariantly stable by Exercise 8.58, so that we get isomorphisms

$$\mathcal{K}(A) := \mathcal{K}(L^2G \overline{\otimes} \ell^2\mathbb{N}) \widehat{\otimes}_{C^*} A \cong \mathcal{K}(L^2G \overline{\otimes} \ell^2\mathbb{N} \oplus \mathbb{C}) \widehat{\otimes}_{C^*} A \cong A$$

for all A. Using this isomorphism $A \cong \mathcal{K}(A)$, we show that kk^{G,C^*} is a triangulated category. It has extension triangles for G-equivariantly cpc-split extensions, and it is universal for functors that are stable, homotopy invariant, and split-exact for equivariantly cpc-split extensions. We omit the straightforward proof. Since Kasparov's equivariant theory can be characterised by a similar universal property (see [83, 120]), the proof of Theorem 8.28 yields

$$\mathrm{KK}^G(A, B) \cong \mathrm{kk}^{G,C^*}(A, B)$$

for all separable G-C^*-algebras A, B.

If we work with bornological algebras, then a good substitute for $\mathcal{K}(L^2G)$ is the dense isoradial subalgebra $C_c^\infty(G \times G) \subseteq \mathcal{K}(L^2G)$ of compactly supported smooth integral kernels on G. Thus we put

$$\mathcal{K}(A) := C_c^\infty(G \times G) \widehat{\otimes} \mathcal{K}_{\mathscr{S}} \widehat{\otimes} \mathscr{L}^1 \widehat{\otimes} A.$$

We do not discuss the resulting theory any further here.

Chapter 9

Algebras of continuous trace, twisted K-theory

9.1 Algebras of continuous trace

In this chapter, we only deal with C^*-algebras. Since the only stabilisation we need is the C^*-stabilisation, we denote $\mathcal{K}(\ell^2(\mathbb{N}))$ by $\mathcal{K}$. We begin with a few facts about the structure of C^*-algebras. Fortunately, the algebras we will be dealing with are quite close to being tensor products of commutative algebras with $\mathcal{K}$, so the amount of structure theory we need is rather minimal.

Definition 9.1. Let A be a (complex) C^*-algebra. The *primitive ideal space* Prim A is the set of ideals in A which are kernels of irreducible $*$-representations on Hilbert spaces. This space carries a natural topology, called the *hull-kernel topology* or *Jacobson topology*, in which the closed sets are the sets of the form $\{J \in \text{Prim}\, A \mid J \supseteq K\}$ for some $K \triangleleft A$. Such a closed set can also be identified with the primitive ideal space of a quotient of A,

$$\{J \in \text{Prim}\, A \mid J \supseteq K\} \equiv \text{Prim}(A/K).$$

The topology on Prim A is always T_0 (that is, given two distinct primitive ideals J_0 and J_1, there is always a closed set that contains one and not the other). But it need not be T_1 (that is, singletons may not be closed), and it certainly need not be Hausdorff.

The *dual space* of A, also known sometimes as the spectrum of A, is defined to be $\widehat{A}$, the set of unitary equivalence classes of irreducible $*$-representations on Hilbert spaces. By definition, the kernel $J = \ker \pi$ of any irreducible representation π of A is a primitive ideal, which is unchanged if we replace π by a unitarily equivalent representation. So we have a natural surjective map $[\pi] \mapsto \ker \pi$ from $\widehat{A}$ onto Prim A. We give $\widehat{A}$ the topology pulled back from the topology of Prim A

under this map. This topology, called the *Fell topology*, will be T_0 if and only if the map $\hat{A} \to \text{Prim}\, A$ is bijective because the topology cannot distinguish irreducible representations with the same kernel.

Definition 9.2. A C^*-algebra A is called *liminary* (*liminaire* in French — the term *liminal* is also used) if, for each irreducible *-representation π of A on a Hilbert space $\mathcal{H}$ (which of course can depend on π), $\pi(A)$ is *elementary*, that is, equal to $\mathcal{K}(\mathcal{H})$. This is automatic if all irreducible *-representations are finite-dimensional because every operator on a finite-dimensional Hilbert space is compact, and a C^*-subalgebra of the operators on a finite-dimensional Hilbert space $\mathcal{H}$ acts irreducibly if and only if it consists of *all* linear operators on $\mathcal{H}$.

We need to remind the reader about some basic properties of elementary C^*-algebras, which we will use for various purposes later.

Proposition 9.3. *Let $\mathcal{H}$ be a Hilbert space, and let $\mathcal{K} = \mathcal{K}(\mathcal{H})$ be the algebra of compact operators on $\mathcal{H}$. Then every irreducible *-representation of $\mathcal{K}$ is unitarily equivalent to the standard representation of $\mathcal{K}$ on $\mathcal{H}$, and every *-automorphism of $\mathcal{K}$ is given by conjugation by a unitary operator on $\mathcal{H}$. The *-automorphism group of $\mathcal{K}$ can be identified with the topological group $PU(\mathcal{H}) := U(\mathcal{H})/\mathbb{T}$, the projective unitary group of $\mathcal{H}$, with the quotient topology from the strong operator topology on $U(\mathcal{H})$.*

Proof. To begin with, we claim that the dual space of $\mathcal{K}(\mathcal{H})$ is the space $\mathscr{L}^1(\mathcal{H})$ of trace-class operators on $\mathcal{H}$. The dual pairing is given by the trace:

$$\langle a, b \rangle = \text{tr}(ab) \qquad \forall a \in \mathcal{K}(\mathcal{H}),\ b \in \mathscr{L}^1(\mathcal{H}).$$

To prove this, we start with the standard fact from linear algebra that this is true if $\mathcal{H}$ is finite-dimensional. Then we recall that finite-rank operators are dense in $\mathcal{K}$ to deduce this in general. Thus the *states* of $\mathcal{K}$, that is, the positive linear functionals of norm 1, can be identified with trace-class operators $b \in \mathscr{L}^1(\mathcal{H})$ such that $\text{tr}(ab) \geq 0$ for $a \geq 0$ and $a \mapsto \text{tr}(ab)$ has norm 1. These are precisely the *positive* trace-class operators $b \geq 0$ with trace norm 1, that is, with $\text{tr}(b) = 1$. (For positive operators, there is no difference between the trace and the trace norm.)

The *pure states* of any C^*-algebra A are the extreme points in the convex set of all states. (If A is also unital, then the states can also be described as positive linear functionals φ with $\varphi(1) = 1$, so that the state space is compact in the weak-* topology of the dual space A^* of A.) Pure states are precisely the states that give rise to irreducible *-representations via the GNS (Gelfand–Naimark–Segal) construction (see for example [44, Chapter II, §5]). Now any positive trace-class operator of trace 1 is unitarily equivalent to a diagonal operator, with the diagonal entries summing to 1. Clearly such an operator is a convex combination of other such operators *unless* there is only one non-zero diagonal entry, that is, the operator is a rank-one projection. Hence the pure states of $\mathcal{K}$ are precisely those of the form $a \mapsto \text{tr}(ae)$, where e is a rank-one projection in $\mathcal{H}$. Let π denote the standard representation of $\mathcal{K}(\mathcal{H})$ on $\mathcal{H}$. Then if e is the rank-one projection onto the span of a unit vector $\xi \in \mathcal{H}$, the pure state $a \mapsto \text{tr}(ae)$ coincides with the *vector state* $a \mapsto \langle a\xi, \xi \rangle = \langle \pi(a)\xi, \xi \rangle$. Hence all pure states of $\mathcal{K}$ are vector states of π. If η is another unit vector in $\mathcal{H}$, then the corresponding vector state $\langle \pi(a)\eta, \eta \rangle$ can be written also as $\langle u\pi(a)u^*\xi, \xi \rangle$, provided that u is a unitary operator with $u^*\xi = \eta$. This shows that all irreducible $*$-representations are unitarily equivalent to the standard representation.

Finally, suppose φ is a $*$-automorphism of $\mathcal{K}$. We have just seen that if π is the standard representation of $\mathcal{K}$ on $\mathcal{H}$, then $\pi \circ \varphi$ is unitarily equivalent to π. Thus we get a unitary operator $u \in \mathcal{L}(\mathcal{H})$ with $\pi \circ \varphi(a) = u\pi(a)u^*$ for all $a \in \mathcal{K}(\mathcal{H})$. Since π is just the identity map $\mathcal{K}(\mathcal{H}) \to \mathcal{K}(\mathcal{H}) \subseteq \mathcal{L}(\mathcal{H})$, this means $\varphi(a) = uau^*$ for all $a \in \mathcal{K}(\mathcal{H})$. Hence every $*$-automorphism of $\mathcal{K}$ is given by conjugation by a unitary operator on $\mathcal{H}$. Two unitary operators induce the same automorphism if and only if they differ by a scalar in $\mathbb{T}$. So $\text{Aut}\,\mathcal{K} \cong PU(\mathcal{H})$.

The natural topology on the automorphism group of a C^*-algebra is the topology of pointwise convergence. When the algebra is $A = C_0(X)$, X locally compact Hausdorff, then $\text{Aut}\,A = \text{Homeo}\,X$ and the topology of pointwise convergence on $\text{Aut}\,C_0(X)$ is the same as the compact-open topology on the homeomorphism group. But $u_\alpha a u_\alpha^* \to uau^*$ for all $a \in \mathcal{K}$ if and only if this happens for a of rank one, which is the same as saying that $\mathbb{C} \cdot u_\alpha \xi \to \mathbb{C} \cdot u\xi$ for all unit vectors ξ. Hence the topology on $\text{Aut}\,\mathcal{K} \cong PU(\mathcal{H})$ is induced from the strong operator topology on $U(\mathcal{H})$. $\qquad \square$

Lemma 9.4. *Let A be a C^*-algebra, and let π be an irreducible $*$-representation of A on a Hilbert space $\mathcal{H}$ with $\pi(A) \cap \mathcal{K}(\mathcal{H}) \neq 0$. Then $\pi(A) \supseteq \mathcal{K}(\mathcal{H})$.*

Proof. Let $J = \pi^{-1}\big(\mathcal{K}(\mathcal{H})\big)$. Then J is a non-zero closed ideal in A and $0 \neq \pi(J) \subseteq \mathcal{K}(\mathcal{H})$. Since π is irreducible and J is an ideal in A, so is $\pi|_J$. Now $\pi(J)$ contains a non-zero compact self-adjoint operator. By the spectral theorem for such operators, it contains a finite-rank projection. Let n be the minimal rank of a non-zero finite-rank projection in $\pi(J)$. If $n > 1$, then it is easy to see that $\pi(J)$ is not irreducible. Thus $\pi(A)$ contains a rank-one projection. By irreducibility, it contains *all* rank-one projections, and hence all compact self-adjoint operators (since these can be approximated by finite linear combinations of rank-one projections). The result follows. $\qquad \square$

Proposition 9.5. *A liminary C^*-algebra A has a T_1 dual space $\widehat{A}$ — that is, points in $\widehat{A}$ are closed. Conversely, if A is a separable C^*-algebra and $\widehat{A}$ is T_1, then A is liminary.*

Proof. Suppose A is liminary. Then for every primitive ideal J of A, $A/J \cong \mathcal{K}(\mathcal{H}) \subseteq \mathcal{L}(\mathcal{H})$ for some Hilbert space $\mathcal{H}$. If J' were another proper ideal of A with $J \subsetneq J'$, then J'/J would be a proper non-zero two-sided ideal of A/J, contradicting the simplicity of $\mathcal{K}(\mathcal{H})$. Thus every primitive ideal of A is maximal, which means that every point in $\text{Prim}\,A$ is closed, that is, $\text{Prim}\,A$ is a T_1 space. Furthermore, $\widehat{A} \to \text{Prim}\,A$ is bijective because, for any primitive ideal J,

$A/J \cong \mathcal{K}(\mathcal{H})$ for some $\mathcal{H}$, and Proposition 9.3 asserts that $\mathcal{K}(\mathcal{H})$ has a unique irreducible $*$-representation up to unitary equivalence.

The converse direction is much deeper since it uses the difficult theorem that a separable C^*-algebra for which the map $\widehat{A} \to \operatorname{Prim} A$ is a bijection has the property that the image of each of its irreducible $*$-representations contains the compact operators (see [44, §9.1]). The result is then immediate since, if the image of some irreducible $*$-representation were to strictly contain the compact operators, then its kernel could not be maximal.   □

We now want to focus on a special class of liminary C^*-algebras that are particularly close to being of the special form $C_0(X) \widehat{\otimes}_{C^*} \mathcal{K}(\mathcal{H})$. In particular, they have a dual space which is Hausdorff (T_2), not just T_1.

Definition 9.6 (See [44, §10.6]). Let X be a locally compact Hausdorff space. An *algebra of continuous trace* over X is a C^*-algebra A with dual space $\widehat{A} = X$, such that for each $x_0 \in X$, there is an element $a \in A$ such that $x(a)$ is a rank-one projection for each x in a neighbourhood of x_0 (*Fell's condition*).

Notice that while x is only a unitary *equivalence class* of representations, the notion of $x(a)$ being a rank-one projection makes perfect sense.

Such algebras were studied by Fell and Dixmier–Douady, and are algebras of sections of continuous fields of elementary C^*-algebras. The term *continuous trace* is explained by the following:

Proposition 9.7. *Let A be a C^*-algebra. Then A is of continuous trace in the sense of Definition 9.6 if and only if the set of $x \in A$ for which the map $\pi \mapsto \operatorname{tr}(\pi(x)\pi(x)^*)$ is finite and continuous on $\widehat{A}$ is dense in A.*

Proof. First suppose A is of continuous trace in the sense of Definition 9.6, that is, A satisfies Fell's condition. Then $X = \widehat{A}$ is Hausdorff, and for each $x_0 \in X$, there is an element $a \in A$ such that $x(a)$ is a rank-one projection for each x in a neighbourhood of x_0. Since $X = \widehat{A}$ is Hausdorff, A is liminary by Proposition 9.5. (We do not need to use the difficult part of this proposition, since Fell's condition implies that for each $x \in X$, the image of A under the corresponding representation contains a non-zero compact operator, hence contains *all* compact operators by Lemma 9.4, hence consists *exactly* of the compact operators since $\operatorname{Prim} A$ is T_1 and thus the kernel of the representation is a maximal ideal.) Let $\mathfrak{n}$ be the set of $x \in A$ such that the map $\pi \mapsto \operatorname{tr}(\pi(x)\pi(x)^*)$ is finite and continuous on $\widehat{A}$. Then $\mathfrak{n} = \mathfrak{n}^*$ since $\operatorname{tr}(\pi(x^*)\pi(x)) = \operatorname{tr}(\pi(x)\pi(x)^*)$. If $x, y \in \mathfrak{n}$ and $a \in A$, then $aa^* \leq \|a\|^2$ and

$$2xx^* + 2yy^* - (x+y)(x+y)^* = xx^* + yy^* - xy^* - yx^* = (x-y)(x-y)^* \geq 0,$$

so that $(x+y)(x+y)^* \leq 2xx^* + 2yy^*$ and

$$(xa)(xa)^* = x(aa^*)x^* \leq \|a\|^2 xx^*.$$

Thus $\pi \mapsto \mathrm{tr}(\pi(x+y)\pi(x+y)^*)$ and $\pi \mapsto \mathrm{tr}(\pi(xa)\pi(xa)^*)$ are finite on $\widehat{A}$. They are also continuous, since the trace is always lower semi-continuous and

$$-\,\mathrm{tr}\big(\pi(xa)\pi(xa)^*\big) = -\|a\|^2\,\mathrm{tr}\big(\pi(x)\pi(x)^*\big) + \mathrm{tr}\big(\pi(x(\|a\|^2 - aa^*)x^*)\big)$$

is the sum of a continuous function and a lower semi-continuous function, hence is also lower semi-continuous, and similarly with $-\,\mathrm{tr}(\pi(x+y)\pi(x+y)^*)$. Thus $\mathfrak{n}$ is a self-adjoint left ideal, hence a two-sided ideal, in A. We need to show that $\mathfrak{n}$ is dense in A. For this it suffices to show that $\pi(\mathfrak{n})$ is dense in $\pi(A)$ for each $\pi \in \widehat{A}$. Let $x_0 = [\pi] \in \widehat{A}$, where the square brackets denote the unitary equivalence class of π, and let a be as in Fell's condition. Thus $a(x)$ is a rank-one projection for x in a neighbourhood N of x_0. Multiplying by a function which is 1 in a smaller neighbourhood of x_0 and 0 outside N, we can guarantee that $a \in \mathfrak{n}$. Thus $\pi(\mathfrak{n})$ is non-zero. Since $\pi(A) = \mathcal{K}(\mathcal{H})$, we conclude that $\pi(\mathfrak{n})$ is dense in $\pi(A)$ by Lemma 9.4. This completes one direction of the proof.

For the converse, retain the same notation and assume that $\mathfrak{n}$ is dense in $\widehat{A}$. Since $\pi(\mathfrak{n})$ consists of Hilbert–Schmidt operators and is dense in $\pi(A)$ for each $\pi \in \widehat{A}$, A is liminary. To show that $\widehat{A}$ is Hausdorff, suppose $x = [\pi] \neq y = [\sigma]$ in $\widehat{A}$, and choose $a \in \mathfrak{n}$ with $\pi(a) \neq 0$, $b \in A$ with $\pi(ab) \neq 0$, $\sigma(b) = 0$. (This is possible since $\ker \sigma$ cannot be contained in $\ker \pi$ since A is liminary, and $\pi(\ker \sigma)$ is thus all of $\pi(A)$ by Lemma 9.4.) Since $\mathfrak{n}$ is an ideal (see the reasoning above), $ab \in \mathfrak{n}$ and $[\varrho] \mapsto \mathrm{tr}(\varrho(ab)\varrho(ab)^*)$ is a continuous function which is non-zero at $[\pi]$ and zero at $[\sigma]$. Thus $\widehat{A}$ is Hausdorff. It remains to verify Fell's condition. Let π be an irreducible $*$-representation of A, and choose $a \in \mathfrak{n}$ with $\pi(a) \neq 0$. Then $[\sigma] \mapsto \mathrm{tr}(\sigma(aa^*))$ is a continuous map $\widehat{A} \to [0, \infty)$ and strictly positive at $[\pi]$. We may assume some positive number $r \in \mathrm{spec}\,\pi(aa^*)$ has multiplicity 1, since (as in the proof of Lemma 9.4) if all spectral projections had rank > 1 for all $a \in \mathfrak{n}$, that would contradict the irreducibility of π. Since the spectrum of a positive self-adjoint compact operator is discrete, except perhaps for 0, there is a small interval containing r for which the corresponding spectral projection has rank 1. Then if f is a suitable non-negative real-valued function supported near r and with $f \equiv 1$ on a very small neighbourhood of r, $\sigma(f(aa^*))$ is a rank-one projection for σ in a neighbourhood of π. This verifies Fell's condition. $\qquad\square$

For simplicity, we assume henceforth that X is second countable (or equivalently, that $C_0(X)$ is separable) and consider only separable C^*-algebras. As far as K-theory is concerned, it is no loss of generality to *stabilise*, that is, to tensor with $\mathcal{K} = \mathcal{K}(\mathcal{H})$, for $\mathcal{H}$ a fixed separable, infinite-dimensional Hilbert space, such as $\ell^2(\mathbb{N})$. Since $\mathcal{K} \widehat{\otimes}_{C^*} \mathcal{K} \cong \mathcal{K}$, looking only at stable algebras is the same as restricting to algebras A with $A \cong A \widehat{\otimes}_{C^*} \mathcal{K}$.

Lemma 9.8 (Dixmier–Douady). *Let $\mathcal{H}$ be a separable infinite-dimensional Hilbert space, and let $U = U(\mathcal{H})$ be its unitary group, viewed as a topological group in the strong operator topology. Then U is contractible.*

Proof. We may take $\mathcal{H} = L^2([0,1])$. For $0 \leq t \leq 1$, define $V_t \colon \mathcal{H} \to \mathcal{H}$ by $V_t(f)(s) := \sqrt{t}f(ts)$ and $P_t \colon \mathcal{H} \to \mathcal{H}$ by $P_t(f)(s) := f(s)$ for $0 \leq s \leq t$ and $P_t(f)(s) = 0$ for $s > t$. Then P_t is an orthogonal projection and V_t is a partial isometry with domain projection P_t, annihilating $\{f \mid f \equiv 0 \text{ on } [0,t]\}$ and mapping its orthogonal complement

$$\mathcal{H}_t = \{f \mid f \equiv 0 \text{ on } [t,1]\} = P_t \mathcal{H} \cong L^2([0,t])$$

isometrically onto $\mathcal{H}$. It is clear that V_t and $P_t = V_t^* V_t$ vary continuously with t in the strong operator topology. For $u \in U(\mathcal{H})$, let $H_t(u) = 1 - P_t + V_t^* u V_t$ for $0 \leq t \leq 1$. Clearly $H_0(u) = 1$ and $H_1(u) = u$. Then H is a contraction of U in the strong operator topology (a homotopy from the identity to the map $U \to 1$).   □

Theorem 9.9 (Dixmier–Douady). *Any stable separable algebra A of continuous trace over a second-countable locally compact Hausdorff space X is isomorphic to $\Gamma_0(X, \mathcal{A})$, the sections vanishing at infinity of a locally trivial bundle of algebras over X, with fibres $\mathcal{K}$ and structure group $\mathrm{Aut}(\mathcal{K}) = PU = U/\mathbb{T}$. Classes of such bundles are in natural bijection with the Čech cohomology group $H^3(X, \mathbb{Z})$. The 3-cohomology class $\delta(A)$ attached to (the stabilisation of) a continuous-trace algebra A is called its* Dixmier–Douady class.

Proof. The proof of local triviality is best done using other results from C^*-algebra theory, and we will just sketch it. (For more detailed but slightly different proofs, see [44, Chapter X] or [104].) Suppose A is of continuous trace with $\widehat{A} = X$. Since the conclusion is local, it is enough to show that for each $x_0 \in X$, there is a compact neighbourhood K of x_0 in X such that the quotient of A defined by K is isomorphic to $C(K, \mathcal{K})$. By Fell's condition, we can choose a compact neighbourhood K of x_0 and an element p of A such that $x(p)$ is a rank-one projection for each $x \in K$. Without loss of generality, replace X by K. Then pAp is a *corner* of A, that is, it is the cut-down of A by a projection,[1] and this corner is *full*, that is, ApA is dense in A. The latter follows from the fact that for a rank-one projection e in $\mathcal{K}$, $\mathcal{K}e\mathcal{K}$ is dense in $\mathcal{K}$, and from the fact that A is liminary. Since we assume A to be stable, Brown's Stable Isomorphism Theorem [20] yields $A \cong pAp \otimes_{C^*} \mathcal{K}$. Since $x(p)$ is a rank-one projection for each $x \in K$, pAp is commutative and hence isomorphic to $C(K)$. Thus A is isomorphic to $C(K, \mathcal{K})$ as desired.

   Now we explain the last part. By Lemma 9.8, U (in the strong operator topology) is contractible, and $\mathbb{T} \cong \mathbb{S}^1$ acts freely on it. Thus PU has the homotopy type of a *classifying space* $B\mathbb{T} = K(\mathbb{Z}, 2)$, and BPU has the homotopy type of $K(\mathbb{Z}, 3)$.[2] In other words, BPU has exactly one non-zero homotopy group, π_3.

---

[1]The name "corner" comes from the fact that we can view elements of A as matrices $\begin{pmatrix} pap & pa(1-p) \\ (1-p)ap & (1-p)a(1-p) \end{pmatrix}$.

[2]Any topological group G acts freely on some weakly contractible space EG [88]. The quotient $BG = EG/G$ is called a classifying space for G. If G has the homotopy type of a CW-complex, EG may be chosen contractible and BG has the homotopy type of a CW-complex, and the homotopy type of BG is independent of the choice of EG. The main use of classifying spaces is

Principal PU-bundles over X are thus classified by

$$[X, BPU] = [X, K(\mathbb{Z}, 3)] = H^3(X, \mathbb{Z}).$$

Given a principal PU-bundle $PU \to E \to X$, we can form the associated bundle $E \times_{PU} \mathcal{K}$, where PU acts on $\mathcal{K}$ by automorphisms (see Proposition 9.3). This is now a locally trivial bundle of algebras, with fibres $\mathcal{K}$ and structure group PU, and its algebra of sections vanishing at infinity is *locally* isomorphic to $C_0(X, \mathcal{K})$, hence satisfies Fell's condition and is a stable continuous-trace algebra over X. In the other direction, given a stable continuous-trace algebra A over X, it comes from a locally trivial bundle with structure group PU, hence is determined by a homotopy class of maps $X \to BPU = K(\mathbb{Z}, 3)$. □

Definition 9.10. The group $H^3(X, \mathbb{Z})$ can also be described as the *Brauer group* of $C_0(X)$, that is, the group of algebras of continuous trace over X modulo Morita equivalence over X. The group operation then corresponds to tensor product over X. More precisely, if A and B are algebras of continuous trace over X, we define $A \otimes_X B$ to be the largest C^*-algebra whose irreducible $*$-representations are generated by $\pi_1(A) \otimes \pi_2(B)$ on $\mathcal{H}_1 \otimes \mathcal{H}_2$, where π_1 is an irreducible $*$-representation of A on $\mathcal{H}_1$ and π_2 is an irreducible $*$-representation of B on $\mathcal{H}_2$, both corresponding to the same point in X.

Proposition 9.11 (P. Green [51, 96, 104]). *Let X be a second-countable locally compact Hausdorff space, and let A and B be stable algebras of continuous trace over X. Then $A \otimes_X B$ is also a stable continuous-trace algebra over X, and the Dixmier–Douady class $\delta(A \otimes_X B)$ of $A \otimes_X B$ is given by $\delta(A) + \delta(B)$. The Dixmier–Douady class of the* opposite algebra A^{op} *is given by $\delta(A^{\mathrm{op}}) = -\delta(A)$, so that $A \otimes_X A^{\mathrm{op}} \cong C_0(X, \mathcal{K})$.*

Proof. It is clear from the definition that if $A = \Gamma_0(X, \mathcal{A})$ and $B = \Gamma_0(X, \mathcal{B})$, where $\mathcal{A}$ and $\mathcal{B}$ are locally trivial bundles of C^*-algebras over X with fibres isomorphic to $\mathcal{K}$, then $A \otimes_X B \cong \Gamma_0(X, \mathcal{A} \otimes \mathcal{B})$, where $\mathcal{A} \otimes \mathcal{B}$ is the locally trivial bundle with fibre $\mathcal{A}_x \widehat{\otimes}_{C^*} \mathcal{B}_x$ over $x \in X$. The pairing $(\mathcal{A}, \mathcal{B}) \mapsto \mathcal{A} \otimes \mathcal{B}$ on bundles of algebras corresponds to a pairing on the corresponding principal PU-bundles, coming from a map $BPU \times BPU \to BPU$ or $K(\mathbb{Z}, 3) \times K(\mathbb{Z}, 3) \to K(\mathbb{Z}, 3)$ which by the universal property of Eilenberg–Mac Lane spaces is determined up to homotopy by a class in

$$H^3(K(\mathbb{Z}, 3) \times K(\mathbb{Z}, 3), \mathbb{Z}) \cong \mathbb{Z} \oplus \mathbb{Z}.$$

This class obviously corresponds to $(1, 1) \in \mathbb{Z} \oplus \mathbb{Z}$, since if either $\mathcal{A}$ or $\mathcal{B}$ is the trivial $\mathcal{K}$-bundle, tensoring with it (over X) has no effect. It follows that $\delta(A \otimes_X B) = \delta(A) + \delta(B)$.

---

for classifying principal G-bundles. Any principal G-bundle over a paracompact base space X is pulled back from the "universal G-bundle" $EG \to BG$, via a map $X \to BG$. The homotopy class of this "classifying map" is uniquely determined, and in this way, one gets a natural bijection between isomorphism classes of principal G-bundles over X and homotopy classes $[X, BG]$ of continuous maps $X \to BG$.

Similarly, the map $A \mapsto A^{\mathrm{op}}$ comes from a similar map $\mathcal{A} \mapsto \mathcal{A}^{\mathrm{op}}$ of bundles and an involutive map $BPU \to BPU$ determined up to homotopy by a class in $H^3(K(\mathbb{Z},3),\mathbb{Z}) \cong \mathbb{Z}$. We claim this class is given by the element -1, from which the formula $\delta(A^{\mathrm{op}}) = -\delta(A)$ follows. All this follows from the fact that if $u \in U(\mathcal{H})$, $a \in \mathcal{K}$, and a^{op} is the corresponding element of $\mathcal{K}^{\mathrm{op}} \cong \mathcal{K}$, then $\left((\mathrm{Ad}_u)(a)\right)^{\mathrm{op}} = (uau^*)^{\mathrm{op}} = (u^*)^{\mathrm{op}}a^{\mathrm{op}}u^{\mathrm{op}} = \left(\mathrm{Ad}(u^*)^{\mathrm{op}}\right)(a^{\mathrm{op}})$, while $u \mapsto u^*$ induces multiplication by -1 on $\pi_3(BPU) \cong \mathbb{Z}$. $\square$

Exercise 9.12 (P. Green). Use Proposition 9.11 to construct an example of a separable C^*-algebra A not isomorphic to its opposite algebra A^{op}. (Hint: It suffices to find a compact space X with a class $\delta \in H^3(X,\mathbb{Z})$ such that there is no homeomorphism $X \to X$ sending δ to its negative.)

For X a finite CW-complex, Serre and Grothendieck had earlier studied the Brauer group of $C(X)$ in the purely algebraic sense, that is, the group of algebras of sections of bundles of matrix algebras over X, modulo algebraic Morita equivalence over X. Translated into our language, their result is:

Theorem 9.13 (Serre, Grothendieck [55]). *Let X be a finite CW-complex. Then an element of the Brauer group $H^3(X,\mathbb{Z})$ of continuous-trace algebras over X is represented by a bundle of finite-dimensional matrix algebras if and only if the class is torsion.*

Proof. Since $\mathrm{Aut}\,\mathbb{M}_n(\mathbb{C}) \cong PU(n) = U(n)/\mathbb{T}$, in the same way that $\mathrm{Aut}\,\mathcal{K} \cong PU = U(\mathcal{H})/\mathbb{T}$, we see that bundles of n-dimensional matrix algebras arise from principal $PU(n)$-bundles over X, and are classified by $[X, BPU(n)]$. Stabilisation via tensoring with $\mathcal{K}$ gives us a map $BPU(n) \to BPU$, which is induced by the map of topological groups $\mathrm{Ad}_u \mapsto \mathrm{Ad}(u \otimes 1) \colon \mathrm{Aut}\,\mathbb{M}_n(\mathbb{C}) \to \mathrm{Aut}\left(\mathbb{M}_n(\mathbb{C}) \otimes \mathcal{K}\right) \cong \mathrm{Aut}\,\mathcal{K}$. The map $BPU(n) \to BPU \cong K(\mathbb{Z},3)$ is determined up to homotopy by a class in $H^3(BPU(n),\mathbb{Z})$. Since $\pi_k\left(BPU(n)\right) \cong \pi_{k-1}\left(PU(n)\right)$ and we know $\pi_0\left(PU(n)\right) = 0$, $\pi_1\left(PU(n)\right) \cong \mathbb{Z}/n$, and $\pi_2\left(PU(n)\right) = 0$, it follows from the Hurewicz Theorem that the first non-trivial homology group of $BPU(n)$ is $H_2(BPU(n),\mathbb{Z}) \cong \mathbb{Z}/n$, and that $H^3(BPU(n),\mathbb{Z}) = \mathbb{Z}/n$. The map $BPU(n) \to BPU$ is easily seen to correspond to the usual generator 1 of this group. So if $\delta \in H^3(X,\mathbb{Z}) \cong [X, BPU]$, the stable continuous-trace algebra with Dixmier–Douady class δ comes from a locally trivial $\mathbb{M}_n(\mathbb{C})$-bundle if and only of we have a factorisation of the classifying map

$$\begin{array}{ccc} & & BPU(n) \\ & \nearrow & \downarrow \\ X & \xrightarrow{\ \delta\ } & K(\mathbb{Z},3). \end{array} \qquad (9.14)$$

Existence of such a factorisation (9.14) obviously implies that $\delta \in H^3(X,\mathbb{Z})$ factors through $H^3(BPU(n),\mathbb{Z}) \cong \mathbb{Z}/n$, and so implies that δ is n-torsion. (So far we have not used the finiteness of X.)

For the other direction, suppose that X is a finite CW-complex and that we are given a torsion class $\delta \in H^3(X, \mathbb{Z})$. We must show we have a factorisation (9.14) for sufficiently large n (chosen to be a multiple of the order of δ). The idea of Serre [27, undated letter of Serre from "Wednesday afternoon," 1964–65] is to compute the homotopy groups of the homotopy limit $BPU(\infty) = \varinjlim BPU(n)$ for the maps induced by

$$\varphi \mapsto \varphi \otimes 1 \colon PU(n) \cong \operatorname{Aut} \mathbb{M}_n(\mathbb{C})$$
$$\mapsto \operatorname{Aut} \mathbb{M}_n(\mathbb{C}) \otimes \mathbb{M}_k(\mathbb{C}) \cong \operatorname{Aut} \mathbb{M}_{nk}(\mathbb{C}) \cong PU(nk),$$

where one takes the homotopy limit over the positive integers, partially ordered by divisibility. (Alternatively, one can just take the limit of the sequence

$$BPU(2) \to BPU(2^2 \cdot 3) \to BPU(2^3 \cdot 3^2 \cdot 5) \to \cdots$$

of $BPU(n)$'s for n's in which every prime occurs as a factor infinitely often.) Now $\pi_2\big(BPU(\infty)\big)$ is easy to compute; it's just $\mathbb{Q}/\mathbb{Z}$, since as we saw above, $\pi_2\big(BPU(n)\big) \cong \mathbb{Z}/n$, and the map $\pi_2\big(BPU(n)\big) \to \pi_2\big(BPU(nk)\big)$ corresponds to the inclusion of a cyclic group of order n into a cyclic group of order nk. Next, we observe that $\pi_{2j+1}\big(BPU(\infty)\big)$ vanishes for all j; this follows from the facts that all $BPU(n)$ are simply connected and that $\pi_{2j+1}\big(BPU(n)\big) = \pi_{2j+1}\big(BU(n)\big)$ vanishes for any fixed $j > 1$ once n is sufficiently large (by Bott periodicity, reformulated). Finally, we claim that $\pi_{2j}\big(BPU(\infty)\big) \cong \mathbb{Q}$ for $j > 1$. This is again a consequence of Bott periodicity: $\pi_{2j}\big(BPU(n)\big) \cong \pi_{2j}\big(BU(n)\big)$ for $j > 1$, and this is $\cong \mathbb{Z}$ for sufficiently large n (compared to j). So we only need the map

$$\mathbb{Z} \cong \pi_{2j}\big(BPU(n)\big) \cong \pi_{2j}\big(BU(n)\big) \to \pi_{2j}\big(BU(nk)\big) \cong \pi_{2j}\big(BPU(nk)\big) \cong \mathbb{Z}$$

for n sufficiently large. This map is detected by the map on the jth Chern class $c_j \in H^{2j}$ induced by tensor product with a trivial bundle of rank k, and this is multiplication by k. (Just as an example, the map $BU(2) \to BU(2k)$ sends c_2 to $kc_2 + k(k-1)c_1^2/2$, so the induced map on π_4 is multiplication by k.) Passing to the limit, we get $\pi_{2j}\big(BPU(\infty)\big) \cong \mathbb{Q}$.

Finally, we can finish the proof. Because of the long exact sequence

$$\cdots \to H^2(X, \mathbb{Q}) \to H^2(X, \mathbb{Q}/\mathbb{Z}) \to H^3(X, \mathbb{Z}) \to H^3(X, \mathbb{Q}) \to \cdots,$$

a torsion class in $H^3(X, \mathbb{Z})$ comes by the Bockstein homomorphism from a map $X \to K(\mathbb{Q}/\mathbb{Z}, 2)$. Consider the Postnikov tower of $BPU(\infty)$. We write $BPU(\infty)$ as a principal fibration over $K(\mathbb{Q}/\mathbb{Z}, 2)$ whose fibre F has $\pi_k(F) \cong \mathbb{Q}$ if $k \geq 4$ is even, and 0 otherwise. The fibration must be trivial because the rational cohomology of $K(\mathbb{Q}/\mathbb{Z}, 2)$ vanishes. Thus $BPU(\infty)$ splits up to homotopy as $K(\mathbb{Q}/\mathbb{Z}, 2) \times F$. In particular, any map $X \to K(\mathbb{Q}/\mathbb{Z}, 2)$ factors through $BPU(\infty)$. Finally, since $BPU(\infty)$ is defined as a homotopy limit and X is assumed finite, any map $X \to K(\mathbb{Q}/\mathbb{Z}, 2)$ factors through $BPU(n)$ for n sufficiently large. $\qquad\square$

Exercise 9.15. Finish the details of the proof above, by verifying that the map $BU(n) \to BU(nk)$ induces multiplication by k on π_{2n}.

9.2   Twisted K-theory

Definition 9.16. The *twisted* K-*theory* $K_\delta^{-*}(X)$ of a (locally compact) space X with respect to a cohomology class $\delta \in H^3(X, \mathbb{Z})$ is the K-theory of the stable continuous-trace algebra $CT(X, \delta)$ with Dixmier–Douady class δ.

Recall that $CT(X, \delta)$ is *locally* isomorphic to $C_0(X, \mathcal{K})$, but is globally twisted as prescribed by δ. This is somewhat analogous to the *twisted cohomology* (or cohomology with local coefficients) attached to a flat line bundle. (For more details about twisted cohomology, see [14, 125].) Twisted K-theory was first introduced by Karoubi and Donovan in [45]. Their treatment was more general in one sense because they also treated the real case and considered $\mathbb{Z}/2$-graded algebras, but more specific in another sense because they only considered bundles of finite-dimensional matrix algebras, which by Theorem 9.13 amounts to requiring the Dixmier–Douady class to be torsion. The present point of view may be found, for instance, in [5, 108].

Proposition 9.17. *Twisted* K-*theory is 2-periodic and comes with a cup-product* $K_\delta^{-*}(X) \otimes K_\varrho^{-*}(X) \to K_{\delta+\varrho}^{-*}(X)$. *Twisted* K-*theory for the trivial twist,* $K_0^{-*}(X)$, *is just usual* K-*theory with compact supports* $K^{-*}(X)$.

Proof. The last statement is clear since, by definition,

$$K_0^{-*}(X) = K_*\big(CT(X, 0)\big) = K_*\big(C_0(X, \mathcal{K})\big) = K_*\big(C_0(X) \widehat{\otimes}_{C^*} \mathcal{K}\big)$$
$$= K_*\big(C_0(X)\big) = K^{-*}(X)$$

(stabilising has no effect on K-theory). Periodicity of period 2 follows from Bott periodicity for the K-theory of (local) Banach algebras (Theorem 4.7). The cup-product is induced by the tensor product over X: as indicated in Proposition 9.11,

$$CT(C, \delta) \otimes_X CT(C, \varrho) \cong CT(C, \delta + \varrho). \qquad \square$$

Example 9.18 ([107]). Let $X = \mathbb{S}^3$, so that $H^3(X) \cong \mathbb{Z}$. Thus we have a stable continuous-trace algebra over X for each integer m. It can be obtained by glueing together two copies of $C(D^3, \mathcal{K})$ via a map $\mathbb{S}^2 \to \mathrm{Aut}(\mathcal{K}) = PU$ of degree m. If $m \neq 0$, then

$$K_m^{-*}(\mathbb{S}^3) = K_*\big(CT(\mathbb{S}^3, \delta_m)\big) = \begin{cases} 0, & * \text{ even,} \\ \mathbb{Z}/m, & * \text{ odd.} \end{cases}$$

Exercise 9.19. Complete the calculation of $K_m^{-*}(\mathbb{S}^3)$.

Exercise 9.20. (difficult) Use the last exercise and the Atiyah–Hirzebruch spectral sequence (the spectral sequence induced by the skeletal filtration) to show that if X is a finite CW-complex and $\delta \in H^3(X, \mathbb{Z})$, then there is a spectral sequence

$$H^p\big(X, K_q(\mathbb{C})\big) \Longrightarrow K_\delta^{p+q \bmod 2}(X),$$

in which the first non-trivial differential is $d_3 = \_ \cup \delta + \mathrm{Sq}^3$.

Solution Hints 9.21. Since Exercise 9.20 is a bit difficult, we give some details on how to get started. Let $X^{(j)}$ be the j-skeleton of X, so that $H_*(X^{(j)}, X^{(j-1)})$ is concentrated in degree j and can be identified with $C_j(X)$, the cellular j-chains of X. Each $X^{(j)}$ is closed in X, so that we have extensions

$$CT(X^{(j)} \setminus X^{(j-1)}, \delta) \rightarrowtail CT(X^{(j)}, \delta) \twoheadrightarrow CT(X^{(j-1)}, \delta).$$

We get a filtration of $CT(X, \delta)$ by ideals, with subquotients $C_0(X^{(j)} \setminus X^{(j-1)}) \widehat{\otimes}_{C_*} \mathcal{K}$ having K-theory groups $C^j(X) \otimes K^*(\mathcal{K})$ (concentrated in even or odd degree, depending on the parity of j). As in §4.3.1, we get a spectral sequence converging to $K_\delta^{p+q \bmod 2}(X)$ with

$$E_1^{p,q} = C^p(X, K_q(\mathbb{C}))$$

1. Check that d_1 is the usual cellular cochain differential, so that E_2 is as claimed.

2. Check that d_2 vanishes, simply because many of the groups in the sequence vanish.

3. Check that d_3 is given by a universal formula involving δ and cohomology operations on integral cohomology raising degree by 3 and commuting with suspension.

4. It is known that there is only one non-trivial cohomology operation on integral cohomology raising degree by 3 and commuting with suspension, namely, the Steenrod operation Sq^3. Hence the number of possibilities for d_3 is quite limited, and it suffices to check a few examples such as spheres.

Chapter 10

Crossed products by $\mathbb{R}$ and Connes' Thom Isomorphism

In this chapter, we deal mainly with C^*-algebras, although we sometimes use certain dense subalgebras. We define C^*-algebraic crossed products in greater detail than in Chapter 5 and discuss Pontrjagin Duality and Takesaki–Takai Duality for Abelian locally compact groups. These are used to compute the K-theory for crossed products by $\mathbb{R}$. The result is closely related to the Pimsner–Voiculescu sequence. A recommended source for further reading is [126].

10.1 Crossed products and Takai Duality

Definition 10.1. Let A be a C^*-algebra and let α be an action of a locally compact group G on A (by $*$-automorphisms). Let $\Delta_G \colon G \to \mathbb{R}_+^\times$ be the modular function of G. (The reader not familiar with this need not worry about it, since we will mostly be interested in the case where G is Abelian, in which case $\Delta_G \equiv 1$.) The C^*-*crossed product* of A by G (via the action α), denoted $A \rtimes_\alpha G$, is the completion of $C_c(G, A)$ in the universal C^*-norm, with convolution multiplication determined by the formal relation $g \cdot a \cdot g^{-1} = \alpha_g(a)$.

More precisely, we turn $C_c(G, A)$ into an involutive algebra by

$$(f \star g)(s) = \int_G f(s)\alpha_s\big(g(s^{-1}t)\big)\,\mathrm{d}t, \qquad f^*(s) = \alpha_s\big(f(s^{-1})^*\big) \cdot \Delta_G(s)^{-1}.$$

Here $\mathrm{d}t$ is a left Haar measure on G. Then $A \rtimes_\alpha G$ is the completion of this algebra for the largest C^*-algebra norm dominated by the L^1-norm

$$\|f\|_{L^1} := \int_G \|f(s)\|_A\,\mathrm{d}s.$$

In particular, if $A = \mathbb{C}$, then α must be trivial, and $A \rtimes_\alpha G$ is just the completion of $L^1(G)$ in the largest C^*-algebra norm, and is called the *group C^*-algebra*, denoted $C^*(G)$.

The crossed product can also be described in another way: it is the universal C^*-algebra for *covariant pairs* (π, σ), where π is a (strongly continuous) unitary representation of G and σ is a $*$-representation of A, both on the same Hilbert space $\mathcal{H}$ and satisfying the compatibility relation $\pi(g)\sigma(a)\pi(g)^{-1} = \sigma(\alpha_g(a))$. The *integrated form* of a covariant pair (π, σ) is a $*$-representation of $C_c(G, A)$ defined by $f \mapsto \int_G \pi(s)\sigma(f(s))\,ds$; this extends to a $*$-representation of $A \rtimes_\alpha G$. Conversely, any $*$-representation of $A \rtimes_\alpha G$ is of this form for some covariant pair (π, σ), which we can recover by

$$\pi(g) = \lim_{\alpha,\beta} L_g(f_\beta) \cdot u_\alpha, \qquad \sigma(a) = \lim_\beta f_\beta \cdot a, \qquad (10.2)$$

where L_g denotes left translation by g, (u_α) is a bounded approximate identity for A and f_β is an approximate identity in $C_c(G)$ (consisting of non-negative functions of integral 1 becoming more and more concentrated near the identity element of G).

Exercise 10.3. Let A be a C^*-algebra and let α be a strongly continuous action of a locally compact group G on A. Check that the integrated form of a covariant pair (π, σ) for (A, G, α) is a $*$-representation of the twisted convolution algebra $C_c(G, A)$ that is bounded with respect to the L^1-norm and hence extends to a $*$-representation of $A \rtimes_\alpha G$.

Exercise 10.4. Conversely, check that (10.2) associates a covariant pair to a $*$-representation of $A \rtimes_\alpha G$.

The definition of C^*-algebra crossed product is a bit easier to understand, and is easier to reconcile with the original occurrence of crossed products in algebra,[1] if we use the *multiplier algebra* $\mathcal{M}(A)$ of a C^*-algebra A. Recall that this is the largest unital C^*-algebra that contains A as an essential two-sided ideal. In general, $A \rtimes_\alpha G$ contains copies of neither A nor G. But both of them naturally embed into $\mathcal{M}(A \rtimes_\alpha G)$, with G embedding into the unitary group of $\mathcal{M}(A \rtimes_\alpha G)$; and these embeddings satisfy the basic commutation identity

$$g \cdot a \cdot g^{-1} = \alpha_g(a) \qquad \forall g \in G, a \in A,$$

which is the "hallmark" of the crossed product.

From a $*$-representation of $A \rtimes_\alpha G$, we get the corresponding covariant pair by first extending the representation to the multiplier algebra and then restricting to these copies of G and A. Furthermore, the homomorphism $G \to U(\mathcal{M}(A \rtimes_\alpha G))$ induces a $*$-homomorphism $\varphi \colon C^*(G) \to \mathcal{M}(A \rtimes_\alpha G)$, and products $\varphi(f) \cdot a$ with

---

[1] Crossed products can be traced back to [42], where the key equation $ji = \theta(i)j$ appears as equation (4) on the first page. As indicated in a footnote, most of this paper was actually written in 1906.

$f \in C^*(G)$ and $a \in A$ lie in the crossed product $A \rtimes_\alpha G$ itself (not just in its multiplier algebra), and generate the crossed product (see [52, §1]); because of the commutation rule, we can equally well consider products $a \cdot \varphi(f)$ with a on the left here.

When G is locally compact Abelian, there is a duality theory for crossed products, generalising Pontrjagin Duality, and culminating in the Takai Duality Theorem.

Definition 10.5. Let G be a locally compact Abelian group. Its *Pontrjagin dual group* is $\widehat{G} = \mathrm{Hom}(G, \mathbb{T})$, where Hom denotes the space of continuous group homomorphisms. This is again a locally compact Abelian group with respect to pointwise product of homomorphisms and the compact-open topology.

Theorem 10.6 (Pontrjagin). *Let G be a locally compact Abelian group and let $\widehat{G}$ be its Pontrjagin dual. Then the Pontrjagin dual of $\widehat{G}$ is naturally identified with G itself. Furthermore, if H is a closed subgroup of G, then $\widehat{H}$ may naturally be identified with $\widehat{G}/H^\perp$, where $H^\perp$ is the annihilator of H inside $\widehat{G}$. A locally compact Abelian group G is discrete if and only if its Pontrjagin dual is compact, connected if and only if its Pontrjagin dual is torsion-free.*

Exercise 10.7. Prove Theorem 10.6. This is mostly elementary once you observe that homomorphisms $G \to \mathbb{T}$ separate points.

Definition 10.8. Let G be a locally compact Abelian group with Pontrjagin dual group $\widehat{G}$. Let α be an action of G on a C^*-algebra A by $*$-automorphisms. The *dual action* $\widehat{\alpha}$ of $\widehat{G}$ on $A \rtimes_\alpha G$ is defined by extending the action on the dense subalgebra $C_c(G, A)$ given by

$$\big(\widehat{\alpha}(\gamma)(f)\big)(s) = f(s)\langle \gamma, s \rangle,$$

where $\langle \gamma, s \rangle$ denotes the dual pairing between $s \in G$ and $\gamma \in \widehat{G}$.

We also recall the following classical fact:

Lemma 10.9. *If G is a locally compact Abelian group with dual group $\widehat{G}$, then the Fourier transform provides an isomorphism from $C^*(G)$ onto $C_0(\widehat{G})$.*

Proof. By definition, $C^*(G)$ is the completion of $L^1(G)$ in the greatest C^*-algebra norm. It is a commutative C^*-algebra because G is Abelian. Thus it is isomorphic to $C_0(X)$ for some locally compact topological space X. But the Fourier transform is an injective algebra $*$-homomorphism from $L^1(G)$ to a dense subalgebra of $C_0(\widehat{G})$. The result follows. $\qquad\square$

Theorem 10.10 (Takai). *Let A be a C^*-algebra and let α be an action of a locally compact Abelian group G on A. Then*

$$(A \rtimes_\alpha G) \rtimes_{\widehat{\alpha}} \widehat{G} \cong A \,\widehat{\otimes}_{C^*}\, \mathcal{K}\big(L^2(G)\big).$$

Furthermore, the isomorphism can be chosen so that the double dual action $\widehat{\widehat{\alpha}}$ is conjugate to $\alpha \otimes \mathrm{Ad}_L$, where $\mathrm{Ad}_{L(s)}$ denotes conjugation by left translation $L(s)$ by s on $L^2(G)$.

Proof. There are basically two parts to the proof: the proof of the Stone–von Neumann–Mackey Theorem, which is the special case $A = \mathbb{C}$, and a somewhat formal argument reducing everything down to this special case. Slightly different versions of the proof, written out in slightly greater detail, may be found in [98, §7.9] and [126, Chapter 7].

We begin with the special case $A = \mathbb{C}$. The algebra $(\mathbb{C} \rtimes G) \rtimes_{\widehat{\alpha}} \widehat{G} = C^*(G) \rtimes \widehat{G}$ has a natural $*$-representation on $L^2(\widehat{G})$, corresponding to the covariant pair (π, σ), where π is the left regular representation of $\widehat{G}$, defined by $\pi(\gamma)f(\gamma') = f(\gamma' - \gamma)$, and σ is the $*$-representation of $C^*(G) \cong C_0(\widehat{G})$ on $L^2(\widehat{G})$ given by pointwise multiplication. The integrated form of the representation is an action of a certain completion of $C_c(\widehat{G}, C_0(\widehat{G}))$ on $L^2(\widehat{G})$, which acts by the formula

$$f \cdot \xi(s) = \int_{\widehat{G}} f(t, s - t)\xi(s - t)\,\mathrm{d}t = \int_{\widehat{G}} f(s - t, t)\xi(t)\,\mathrm{d}t. \tag{10.11}$$

This is the usual form for an integral operator with continuous kernel, and if the kernel function lies in L^2 (in particular, if it has compact support on $\widehat{G} \times \widehat{G}$), then the operator lies in the Schatten class $\mathscr{L}^2$ of Hilbert–Schmidt operators. Thus (10.11) shows that the image of the representation contains all Hilbert–Schmidt operators, and that these are norm-dense in the image. Thus the image of the representation is precisely $\mathcal{K}(L^2(\widehat{G})) \cong \mathcal{K}(L^2(G))$ (the Fourier transform gives an isometry from $L^2(G)$ onto $L^2(\widehat{G})$). We need to show that this representation is faithful, so that we have captured the structure of the *entire* crossed product. There are several ways to do this. The simplest is to note that the proof so far already shows that a dense subalgebra of the crossed product is isomorphic to a dense subalgebra of the Hilbert–Schmidt operators, which admits only one C^*-norm, the norm of the compact operators. Thus $(\mathbb{C} \rtimes G) \rtimes_{\widehat{\alpha}} \widehat{G} \cong \mathcal{K}(L^2(G))$.

Now we reduce the general case to this by a somewhat formal trick. Without loss of generality we may assume A is unital. (Otherwise, we can always adjoin an identity.) Then $(A \rtimes_\alpha G) \rtimes_{\widehat{\alpha}} \widehat{G}$ contains a copy of $(\mathbb{C} \rtimes G) \rtimes_{\widehat{\alpha}} \widehat{G} = C^*(G) \rtimes \widehat{G}$, which is isomorphic to $\mathcal{K} = \mathcal{K}(L^2(G))$. We know the double crossed product $(A \rtimes_\alpha G) \rtimes_{\widehat{\alpha}} \widehat{G}$ is generated (inside its multiplier algebra) by products of the form $a \cdot b \cdot c$, where $a \in A$, $b \in C^*(\widehat{G})$, and $c \in C^*(G)$. The products $b \cdot c$ generate $\mathcal{K}$. Furthermore, A and the copy of $C^*(\widehat{G})$ commute, since by definition of the dual action, $\widehat{G}$ acts on $C^*(G)$ but not on A. Hence the products $a \cdot b$ generate the crossed product $A \rtimes \widehat{G}$ for the trivial action of $\widehat{G}$ on A, which is nothing but $A \widehat{\otimes}_{C^*} C^*(\widehat{G}) = A \widehat{\otimes}_{C^*} C_0(G) = C_0(G, A)$; the double crossed product is generated by products of $C_0(G, A)$ with elements of $C^*(G)$ and can be rewritten as $C_0(G, A) \rtimes G$. The action of G on $C_0(G, A)$ is the tensor product action of the translation on G and the original action α of G on A (because the tensor factor

$C_0(G)$ comes from $C^*(\widehat{G})$, which certainly does *not* commute with $C^*(G)$, as the two together generate $\mathcal{K}$). The automorphism Φ of $C_0(G, A)$ defined by

$$\big(\Phi(f)\big)(s) = \alpha_{-s}f(s)$$

intertwines this action with the tensor product action of translation on G and the *trivial* action on A:

$$(L \otimes 1)_s\big(\Phi(f)\big)(t) = \big(\Phi(f)\big)(t - s) = \alpha_{s-t}f(t - s), \quad \text{while}$$
$$\Phi\big((L \otimes \alpha)_s(f)\big)(t) = \alpha_{-t}(L \otimes \alpha)_s(f)(t) = \alpha_{-t}\alpha_s\big(f(t - s)\big) = \alpha_{s-t}f(t - s).$$

The upshot is that

$$(A \rtimes_\alpha G) \rtimes_{\widehat{\alpha}} \widehat{G} \cong C_0(G, A) \rtimes_{L \otimes \alpha} G$$
$$\xrightarrow[\cong]{\Phi} C_0(G, A) \rtimes_{L \otimes 1} G = A \,\widehat{\otimes}_{C^*} (C_0(G) \rtimes G) = A \,\widehat{\otimes}_{C^*} \mathcal{K}.$$

This completes the proof of the isomorphism. We leave it to the reader to check the assertion about the double dual action. (Just follow it through the isomorphism.) $\qquad\square$

10.2 Connes' Thom Isomorphism Theorem

Theorem 10.12 (Connes). *Let A be a C^*-algebra and let α be an action of $\mathbb{R}$ on A. Then there is a natural isomorphism*

$$\phi \colon \mathrm{K}_*(A) \to \mathrm{K}_{*+1}(A \rtimes_\alpha \mathbb{R}).$$

Thus the K-theory of $A \rtimes_\alpha \mathbb{R}$ is independent of the action α.

We will sketch two proofs, Connes' original one [28] and a modification of one due to Rieffel [105]. In both cases there are two steps: the construction of ϕ and the proof that it is an isomorphism.

10.2.1 Connes' original proof

Connes' original proof relies on the following 2×2 matrix trick:

Lemma 10.13 (Connes). *Let α be an action of a locally compact group G on a C^*-algebra A, and let u be a unitary cocycle for G; that is, u is a strictly continuous map $G \to U\big(\mathcal{M}(A)\big)$ that satisfies the cocycle relation $u_{gh} = u_g\alpha_g(u_h)$. Then there is an action of G on $\mathbb{M}_2(A)$ restricting to α on one corner and to α' on the other corner. Here $\alpha'_g = \mathrm{Ad}(u_g) \circ \alpha_g$.*

Proof. The cocycle condition guarantees that α' is an action. Simply define β on $\mathbb{M}_2(A)$ by the formula:

$$\beta_g \begin{pmatrix} a & b \\ c & d \end{pmatrix} = \begin{pmatrix} \alpha_g(a) & \alpha_g(b)u_g^* \\ u_g\alpha_g(c) & u_g\alpha_g(d)u_g^* \end{pmatrix}$$

and check that it works.   □

Definition 10.14. The actions α and α' related as in Lemma 10.13 are called *exterior equivalent*.

Exercise 10.15. Let α and α' be exterior equivalent actions of a locally compact group G on a C^*-algebra A. Prove that $A \rtimes_\alpha G$ and $A \rtimes_{\alpha'} G$ are $*$-isomorphic. Construct an isomorphism that acts identically on the natural copies of A in the multiplier algebras of $A \rtimes_\alpha G$ and $A \rtimes_{\alpha'} G$.

In many ways, the most satisfying proof of Theorem 10.12 is the original one by Connes. This depends on the following lemma:

Lemma 10.16 (Connes). *Let α be an action of $\mathbb{R}$ on a C^*-algebra A, and let e be a projection in A which is a smooth vector for α. Then there is an exterior equivalent action α' of $\mathbb{R}$ on A that fixes e.*

Proof. The fact that e is α-smooth means that it lies in the domain of the derivation δ which is the infinitesimal generator of α. Write δ formally as i ad H, where H is an unbounded self-adjoint multiplier of A. Then replace H by

$$H' = eHe + (1-e)H(1-e) = H + \mathrm{i}[\delta(e), e],$$

which commutes with e. Define α'_t by $\mathrm{Ad}\big(e^{\mathrm{i}tH'}\big)$, defined by expanding the series, and check that it works.

In order to show that α'_t is exterior equivalent to α_t, we define

$$u_t := \exp(\mathrm{i}tH') \cdot \exp(-\mathrm{i}tH).$$

This is a well-defined one-parameter family of unitary multipliers because $H' - H$ is bounded. The computation

$$
\begin{aligned}
u_{t+s} &= \exp(\mathrm{i}(t+s)H')\exp(-\mathrm{i}(t+s)H) \\
&= \exp(\mathrm{i}tH')\exp(\mathrm{i}sH')\exp(-\mathrm{i}sH)\exp(-\mathrm{i}tH) \\
&= \exp(\mathrm{i}tH')\exp(-\mathrm{i}tH)\exp(\mathrm{i}tH)\exp(\mathrm{i}sH')\exp(-\mathrm{i}sH)\exp(-\mathrm{i}tH) \\
&= u_t\, \alpha_t(u_s)
\end{aligned}
$$

shows that α'_t is a cocycle. The relation $\alpha'_t = (\mathrm{Ad}\, u_t) \circ \alpha_t$ is immediate from the definitions.   □

Proof of Theorem 10.12 from Lemma 10.16. If ϕ is to be natural and compatible with suspension, it is enough to define it on classes of projections $e \in A$. Since we can perturb a projection to a smooth projection, and close projections are equivalent in K_0, we may assume that e is smooth. Applying Lemmas 10.16 and 10.13, we get an action β on $\mathbb{M}_2(A)$ with α in one corner and α' in the other corner, where α' fixes e. The inclusions $A \hookrightarrow \mathbb{M}_2(A)$ into the two corners are both isomorphisms on K-theory, and are equivariant for α and α', respectively. Hence we can reduce to the case where e is fixed. Then $1 \mapsto e$ is an equivariant map $\mathbb{C} \to A$, so that $\phi([e])$ is defined by naturality from the trivial case $A = \mathbb{C}$, $A \rtimes \mathbb{R} \cong C_0(\mathbb{R})$, where there is an obvious isomorphism $K_0(\mathbb{C}) \to K_1(C_0(\mathbb{R}))$. This yields a natural transformation $\phi_\alpha \colon K_*(A) \to K_{*+1}(A \rtimes_\alpha \mathbb{R})$. Now consider the composite

$$\phi_{\widehat{\alpha}} \circ \phi_\alpha \colon K_*(A) \to K_{*+2}\big((A \rtimes_\alpha \mathbb{R}) \rtimes_{\widehat{\alpha}} \mathbb{R}\big). \tag{10.17}$$

By Bott periodicity and Takai Duality (Theorem 10.10), the right-hand side in (10.17) may be identified with $K_*(A \widehat{\otimes}_{C^*} \mathcal{K}) \cong K_*(A)$, and we need to show that this map $\phi_{\widehat{\alpha}} \circ \phi_\alpha$ is the identity on a class $[e] \in K_0(A)$. But we have already reduced to the case where e is a self-adjoint projection in A fixed by α. In this case, everything comes by naturality from the case $A = \mathbb{C}$ (since $1 \mapsto e$ is an equivariant map $\mathbb{C} \to A$), where ϕ is an isomorphism and $\phi_{\widehat{\alpha}} \circ \phi_\alpha$ is the identity by construction. Hence ϕ is always an isomorphism by naturality. $\qquad\square$

10.2.2 Another proof

We give another proof based on the Pimsner–Voiculescu sequence. This is based on ideas from a different proof by Rieffel [105]. An advantage of this proof is that it might work for local Banach algebras. Start by defining an action of $\mathbb{R}$ on $C_0([0,1), A)$ by

$$(\widetilde{\alpha}_t f)(s) = \alpha_{ts}\big(f(s)\big).$$

The crossed product by $\widetilde{\alpha}$ of the ideal $C_0\big((0,1), A\big) = SA$ is isomorphic to $S(A \rtimes_\alpha \mathbb{R})$, since $C_c\big((0,1), A\big)$ is dense in $C_0\big((0,1), A\big)$ and $C_c\big((0,1), A\big) \rtimes_{\widetilde{\alpha}} \mathbb{R}$ is the completion of $C_c(\mathbb{R} \times (0,1), A)$ under the convolution product

$$(f \star g)(s, u) = \int_{\mathbb{R}} f(s, u)\alpha_{us}\big(g(t - s, u)\big)\, \mathrm{d}t.$$

But we have a linear automorphism Φ of $C_c(\mathbb{R} \times (0,1), A)$ given by $\Phi(f)(s, u) = f(us, u)$, which carries this multiplication over to the multiplication for $S(A \rtimes_\alpha \mathbb{R})$. The quotient by the ideal $C_0\big((0,1), A\big) = SA$ is isomorphic to $A \widehat{\otimes}_{C^*} C^*(\mathbb{R}) \cong SA$ by evaluation at 0. Thus we get a C^*-algebra extension

$$S(A \rtimes_\alpha \mathbb{R}) \rightarrowtail C_0([0,1), A) \rtimes_{\widetilde{\alpha}} \mathbb{R} \twoheadrightarrow SA. \tag{10.18}$$

The desired isomorphism ϕ is defined as the index map for the corresponding K-theory exact sequence. Since $C_0([0,1), A)$ is contractible, its invertibility follows if $K_*(B) = 0$ implies $K_*(B \rtimes \mathbb{R}) = 0$: take $B = C_0([0,1), A)$ with the action $\widetilde{\alpha}$. Conversely, it is clear that this implication follows from Connes' Thom Isomorphism Theorem.

Hence Connes' Thom Isomorphism Theorem is equivalent to the statement that $K_*(B) = 0$ implies $K_*(B \rtimes_\beta \mathbb{R}) = 0$ for all C^*-algebras B with an action β of $\mathbb{R}$. Since $\mathbb{R}$ is torsion-free, this statement is equivalent to the Baum–Connes property as formulated in §5.3.

In order to use the Pimsner–Voiculescu exact sequence, we now want to relate crossed products by $\mathbb{R}$ to crossed products by $\mathbb{Z}$. This uses the *Packer–Raeburn trick*:

Theorem 10.19 (Packer–Raeburn [95]). *Let β be an action of a locally compact group G on a C^*-algebra B, and let N be a closed normal subgroup of G. Then after stabilising, $B \rtimes_\beta G$ is an iterated crossed product first by N and then by G/N, that is,*

$$(B \rtimes_\beta G) \,\widehat{\otimes}_{C^*}\, \mathcal{K} \cong \big((B \rtimes_{\beta|_N} N) \,\widehat{\otimes}_{C^*}\, \mathcal{K}\big) \rtimes (G/N),$$

for a suitable action of G/N on $(B \rtimes_{\beta|_N} N) \,\widehat{\otimes}_{C^*}\, \mathcal{K}$.

Proof. We only sketch how to prove this in the special case where G is Abelian. We may restrict the dual action $\widehat{\beta}$ of $\widehat{G}$ on $B \rtimes_\beta G$ to the subgroup $N^\perp \subseteq \widehat{G}$. A generalisation of Takai Duality yields a natural isomorphism

$$(B \rtimes_\beta G) \rtimes_{\widehat{\beta}|_{N^\perp}} N^\perp \cong (B \rtimes_{\beta|_N} N) \,\widehat{\otimes}_{C^*}\, \mathcal{K}(L^2 N).$$

The double crossed product $(B \rtimes_\beta G) \rtimes_{\widehat{\beta}|_{N^\perp}} N^\perp$ carries a dual action γ of $\widehat{N^\perp} \cong G/N$. Takai Duality yields

$$(B \rtimes_\beta G) \rtimes_{\widehat{\beta}|_{N^\perp}} N^\perp \rtimes_\gamma G/N \cong (B \rtimes_\beta G) \,\widehat{\otimes}_{C^*}\, \mathcal{K}(L^2 N^\perp),$$

and the theorem follows. An extension of this argument to non-Abelian groups replaces actions of the dual group $\widehat{G}$ by *coactions* of G. $\qquad\square$

Example 10.20. Suppose $G = \mathbb{R}$, $N = \mathbb{Z}$, $G/N \cong \mathbb{T}$, and $B = \mathbb{C}$. Then $B \rtimes G = C^*(\mathbb{R}) \cong C_0(\mathbb{R})$, while $B \rtimes N = C^*(\mathbb{Z}) \cong C(\mathbb{T})$. For the trivial action of G/N, we get $C(\mathbb{T}) \rtimes (G/N) \cong C(\mathbb{T}) \,\widehat{\otimes}_{C^*}\, C_0(\widehat{\mathbb{T}}) \cong C(\mathbb{T}) \,\widehat{\otimes}_{C^*}\, C_0(\mathbb{Z})$, which is certainly not isomorphic to $C_0(\mathbb{R})$. But there is a non-trivial action of $\mathbb{T}$ on $A = C(\mathbb{T}) \,\widehat{\otimes}_{C^*}\, \mathcal{K}(L^2(\mathbb{T})) = C(\mathbb{T}, \mathcal{K}(L^2\mathbb{T}))$, which fixes the dual space $\mathbb{T}$ and is *locally* inner but not *globally* inner. Namely, we can think of the automorphisms of A that fix the dual space $\mathbb{T}$ as $\mathrm{Aut}_\mathbb{T}(A) = C(\mathbb{T}, PU(L^2\mathbb{T}))$, which is the free loop space $\Lambda K(\mathbb{Z}, 2)$ of $PU \simeq K(\mathbb{Z}, 2)$. Thus $\pi_1(\mathrm{Aut}_\mathbb{T}(A)) \cong \pi_1(\Lambda K(\mathbb{Z}, 2)) \cong \mathbb{Z}$, and we choose the homomorphism $\mathbb{T} \to \mathrm{Aut}_\mathbb{T}(A)$ so that it induces an isomorphism on π_1. Calculation of the crossed product shows that $A \rtimes \mathbb{T} \cong C_0(\mathbb{R}) \,\widehat{\otimes}_{C^*}\, \mathcal{K}$.

Now we can complete the second proof of Theorem 10.12. Recall that it remains to show that $\mathrm{K}_*(B) = 0$ implies $\mathrm{K}_*(B \rtimes \mathbb{R}) = 0$ for any action of $\mathbb{R}$ on a C^*-algebra B. Let $D := (B \,\widehat{\otimes}_{C^*}\, \mathcal{K}) \rtimes \mathbb{Z}$. The Packer–Raeburn trick yields

$$(B \,\widehat{\otimes}_{C^*}\, \mathcal{K}) \rtimes \mathbb{R} \cong D \rtimes_\beta (\mathbb{R}/\mathbb{Z}).$$

By the Pimsner–Voiculescu exact sequence (Theorem 5.9), $\mathrm{K}_*(B) = 0$ implies $\mathrm{K}_*(D) = 0$. We must show that this implies $\mathrm{K}_*(D \rtimes_\beta (\mathbb{R}/\mathbb{Z})) = 0$. We use Takai Duality (Theorem 10.10), possibly with additional stabilisations, to write

$$D \cong (D \rtimes_\beta (\mathbb{R}/\mathbb{Z})) \rtimes_{\widehat{\beta}} \mathbb{Z}.$$

The Pimsner–Voiculescu sequence and $K_*(D) = 0$ yield that $1 - (\widehat{\beta})_*$ is an isomorphism on $K_*(D \times_\beta (\mathbb{R}/\mathbb{Z}))$. But $\widehat{\beta}$ is the restriction of the $\widehat{\mathbb{R}}$-action $\widehat{\alpha}$. Since $\widehat{\mathbb{R}} \cong \mathbb{R}$ is contractible, it acts trivially on K-theory. Thus $1 - (\widehat{\beta})_*$ is both 0 and bijective. This forces $K_*(D \times_\beta (\mathbb{R}/\mathbb{Z})) = 0$ and hence $K_*(B \rtimes \mathbb{R}) = 0$ as desired. This finishes the second proof of Connes' theorem.

Exercise 10.21. Deduce from Connes' Thom isomorphism theorem that for a connected, simply connected solvable Lie group G of dimension n, $K_*(C^*(G))$ depends only on n mod 2.

Hint: G has a closed connected normal subgroup of codimension 1.

Exercise 10.22. Let $\mathbb{R}$ act on $\mathbb{T}^2 = \mathbb{R}^2/\mathbb{Z}^2$ by flow along lines of slope θ:

$$\alpha_t(x, y) = (x + t, y + \theta t) \quad \text{mod } \mathbb{Z} \times \mathbb{Z}.$$

Compute the K-theory of the crossed product $\mathbb{T}^2 \rtimes_\alpha \mathbb{R}$ (as a group). It is harder to find specific generators for $K_*(\mathbb{T}^2 \rtimes_\alpha \mathbb{R})$.

This is an example of an *induced action*. Thus the K-theory can also be computed by the Pimsner–Voiculescu sequence for the action of $\mathbb{Z}$ on $\mathbb{T}$ by rotation by $2\pi\theta$. We have considered this equivalent situation in Example 5.12.

Now we abstract out certain features of Exercise 10.22.

Exercise 10.23. Let H be a closed subgroup of a locally compact group G, and let α be an action of H on a C^*-algebra A by $*$-automorphisms. Define $\mathrm{Ind}_H^G(A, \alpha)$ to be the pair consisting of the C^*-algebra

$$\mathrm{Ind}_H^G A := \{f \in C(G, A) \mid f(gh) = \alpha_h^{-1}(f(g)) \text{ for all } h \in H, g \in G,$$
$$\text{and } \|f(g)\| \to 0 \text{ as } gH \to \infty \text{ in } G/H\} \quad (10.24)$$

and the action $\mathrm{Ind}\,\alpha$ on this algebra of A-valued functions by left translation: $(\mathrm{Ind}\,\alpha)_g f(s) = f(g^{-1}s)$. Note that the condition of equation (10.24) is preserved since left and right translations commute.

(a) Prove *Green's Imprimitivity Theorem*

$$(\mathrm{Ind}_H^G A) \rtimes_{\mathrm{Ind}\,\alpha} G \cong (A \rtimes_\alpha H) \widehat{\otimes}_{C^*} \mathcal{K}(L^2(G/H)), \quad (10.25)$$

or at least that these two algebras are Morita equivalent (see [102, Lemma 3.1]). (The proof uses some of the same ideas as the proof of Theorem 10.10.)

(b) Show that (10.25) implies $K_*(A \rtimes_\alpha \mathbb{Z}) \cong K_*((\mathrm{Ind}_\mathbb{Z}^\mathbb{R} A) \rtimes_{\mathrm{Ind}\,\alpha} \mathbb{R})$. Use this to prove that the C^*-algebraic version of the Pimsner–Voiculescu sequence (Theorem 5.9) follows from Connes' Thom Isomorphism Theorem.

Thus our first proof of Theorem 10.12 yields a new proof of the Pimsner–Voiculescu sequence.

(c) Show on the other hand, using Connes' Theorem and the Pimsner–Voiculescu sequence, but not using (10.25), that $K_*(A \rtimes_\alpha \mathbb{Z}) \cong K_*((\mathrm{Ind}_\mathbb{Z}^\mathbb{R} A) \rtimes_{\mathrm{Ind}\,\alpha} \mathbb{R})$. This can be viewed as a K-theoretic version of (10.25).

Chapter 11

Applications to physics

11.1 K-theory in physics

K-theory, including twisted K-theory, is starting to appear in the physics literature quite frequently. Good first places to look are [49, 91, 128]. Examples of more technical (but also more detailed) references are [16, 47, 48, 80, 89, 118, 127].

The idea, to quote Witten [127], is that "D-brane charge takes values in the K-theory of space-time." In string theory, a *D-brane* is a submanifold of space-time on which strings can begin and end. The "D" stands for "Dirichlet" and has to do with the boundary conditions on "open" strings. The twisting of K-theory [16, 47, 69] comes in because of a background field, called the *H-flux*, given by a 3-dimensional cohomology class.

To motivate these statements, it is useful to think about some analogous statements in more familiar areas of physics. In classical physics, electrical charge can vary continuously and takes real values. This does not, however, agree with experiment: physically observed charges (as in the Millikan oil-drop experiment) are always integral multiples of the charge of the electron. We can explain this by hypothesising that electrical charge takes values in an infinite cyclic group, of which the charge of the electron is a generator. However, even this may be incorrect because quarks presumably have charge in a larger cyclic group, with generator $1/3$ of the electron charge; it is just because of quark confinement that these fractional charges cannot be observed in practice. Thus the notion of "charges" in some Abelian group is well established in physics.

The idea that certain "charges" should live in topological invariants of space-time also has a long history. Dirac's famous theory of magnetic monopoles hypothesises that magnetic monopoles should correspond to non-trivial line bundles over space-time. But two line bundles with opposite Chern classes can cancel each other out, so that magnetic monopoles should have charges that live in the Grothendieck group of line bundles. This group, called the *Picard group*, is known to be $H^2(X, \mathbb{Z})$.

From the Grothendieck group of line bundles to the Grothendieck group of vector bundles, that is, to K-theory, is not such a great leap.

String theory, as indicated before, supposes that space-time is full of submanifolds called D-branes, which are equipped with certain charges. The word "brane" comes from "membrane" and to physicists basically just means "manifold." Branes can split apart or coalesce, but there should be some sort of generalised homology theory (on space-time X) with the D-branes Y as typical cycles.

In fact, each brane Y is to carry a *Chan–Paton bundle* E, and (at least initially) both X and the branes should be Spin^c manifolds: we need spinors in order to have a theory of fermions, and a certain anomaly must cancel. A Spin^c structure on an oriented Riemannian manifold X is defined by a choice of a lifting of the oriented orthonormal frame bundle of X — which is a principal bundle for SO$_n$, $n = \dim X$ — to a principal bundle for

$$\mathrm{Spin}_n^c := \mathrm{Spin}_n \times_{\mathbb{Z}/2} \mathbb{T}.$$

This guarantees the existence of a spinor bundle, and is the minimum geometric structure required in order to have spinors and a Dirac operator. As pointed out by Baum and Douglas [7], a Spin^c (compact) manifold Y, equipped with a complex vector bundle E and mapping into another space X, defines a topological K-homology class on X. Thus we think of D-branes with their Chan–Paton bundles as giving K-homology classes in X, Poincaré dual to K-cohomology classes.

However, up to this point we have not taken into account one additional piece of structure. In string theory, there is a field living on space-time that corresponds to a class δ in $H^3(X, \mathbb{Z})$ called the *H-flux*. Locally, the H-flux is represented by the de Rham class of $d(B)$, where B is the so-called *B-field*, but B is not always globally well-defined, so that the H-flux is not necessarily trivial in cohomology. The condition for anomaly cancellation is not really that Y should be Spin^c, but that it be Spin^c *after twisting*. If, for simplicity, Y is oriented, we can express this in terms of characteristic classes as the vanishing of $w_3(Y) + \iota^*\delta$ in $H^3(Y, \mathbb{Z}/2)$, where w_3 is the third Stiefel–Whitney class, which is the obstruction to the existence of an (untwisted) Spin^c structure, and where $\iota\colon Y \hookrightarrow X$ is the inclusion of the D-brane.

The Dirac operator for fermions should be twisted as well, that is, it should live in the K-homology not of $C_0(Y)$ but of the stable continuous-trace algebra defined by the H-flux. Thus D-brane charges should live in the *twisted* K-cohomology or K-theory of X, with twisting given by δ (or perhaps $-\delta$, depending on sign conventions). This point is explained in more detail in [16, 69].

Exercise 11.1. Let $Y = \mathrm{SU}_3/\mathrm{SO}_3$, where orthogonal matrices are viewed as unitary matrices with real entries.

Check that Y is a simply connected compact 5-manifold with $\pi_2(Y) \cong \mathbb{Z}/2$, using the long exact homotopy sequence of the fibration

$$\mathrm{SO}_3 \cong \mathbb{RP}^3 \to \mathrm{SU}_3 \to Y.$$

Deduce from the Hurewicz Theorem and Poincaré duality that

$$H_j(Y, \mathbb{Z}) \cong \begin{cases} \mathbb{Z}, & j = 0, 5, \\ \mathbb{Z}/2, & j = 2, \\ 0, & \text{otherwise.} \end{cases}$$

From this and the Atiyah–Hirzebruch spectral sequence (see §4.3.1 and Exercise 9.20), compute the K-homology and K-cohomology groups of Y, and show that Poincaré duality fails in K-theory. The reason is that Y is not Spin^c; in fact, it is the simplest example of an oriented manifold that is not.

But there is a non-trivial torsion class $\delta \in H^3(Y, \mathbb{Z})$, and Y becomes Spin^c after twisting.

11.2   T-duality

Another interesting feature of string theory is the notion of T-duality (T stands for "torus"), which postulates an equivalence of theories on two different space-times X and $X^\#$, which are related by the exchange of tori in X by their dual tori in $X^\#$. Here "equivalence" means that physically observable quantities such as the masses of elementary particles should be the same in both theories, even if their field equations look rather different from one another. This duality is really a *metric* duality, in that small circles in one space-time are replaced by large circles in the other. But following [15], we consider only the *topological* aspects of this duality, which still captures an important part of the theory.

Let us try to make this precise in the case where the tori involved are 1-dimensional. The duality in this case should exchange Type IIA and Type IIB theories (for those who know what this means — roughly speaking, in type A, symplectic geometry is paramount, whereas in type B, complex geometry is dominant). For our purposes, the one thing we need to know about this is that charges that live in K_0 for one theory should live in K_1 for the other, and vice versa.

We consider two principal $\mathbb{T}$-bundles X and $X^\#$ over a common base Z:

$$\text{(11.2)}$$

To simplify the discussion and to avoid some pathologies, the following technical assumptions will be in force for the rest of the chapter, without any further special mention. (These assumptions are definitely satisfied in all cases of physical interest.) Namely, X, $X^\#$, and Z are all assumed *locally compact, second countable (that is, having a countable base for the topology), and of the homotopy type of a finite CW-complex.* Each of X and $X^\#$ is supposed to be equipped with an

H-flux, with associated cohomology classes δ and $\delta^\#$ in $H^3(X)$ and $H^3(X^\#)$ and continuous-trace algebras $CT(X, \delta)$ and $CT(X^\#, \delta^\#)$, respectively.

The circle group $\mathbb{T}$ acts freely on X and $X^\#$, but not necessarily on $CT(X, \delta)$ and $CT(X^\#, \delta^\#)$. In fact, given an action of a group G on a space X and a class $\delta \in H^3(X)$, the action lifts to an action on $CT(X, \delta)$ if and only if

(a) G fixes δ in H^3, and

(b) the G-action on X lifts to an action on the principal PU-bundle associated to δ.

In our situation, (a) is obvious because the group involved is connected and H^3 is homotopy invariant, but (b) is unclear.

Lemma 11.3 (Raeburn–Williams–Rosenberg [102, 103]). *The $\mathbb{T}$-action on X lifts to an action on the principal bundle associated to δ if and only if $\delta \in p^*\big(H^3(Z)\big)$. If we view $\mathbb{T}$ as $\mathbb{R}/\mathbb{Z}$, the action always lifts to $\mathbb{R}$.*

Proof. We give two proofs of the first assertion, the first one purely topological. If the principal PU-bundle E over X associated to δ admits a lifting of the $\mathbb{T}$-action on X, then we get a free action of $\mathbb{T}$ on E that commutes with the free action of PU. Dividing out by this action of $\mathbb{T}$, we get a principal PU-bundle over $X/\mathbb{T} = Z$, say with characteristic class $\eta \in H^3(Z)$. Then $\delta = p^*(\eta)$ by construction, finishing the first proof.

Now to the second proof. Since $\mathbb{T}$ acts transitively on fibres of p, if there were an action α of $\mathbb{T}$ on $CT(X, \delta)$ compatible with the given action of $\mathbb{T}$ on X, then $CT(X, \delta) \rtimes_\alpha \mathbb{T}$ would be a continuous-trace algebra over Z, say with Dixmier–Douady class $c \in H^3(Z)$. Takai Duality then yields

$$CT(X, \delta) \cong CT(Z, c) \rtimes_{\hat\alpha} \mathbb{Z} \cong p^* CT(Z, c).$$

For the second assertion, we assume that X and Z are manifolds and everything is smooth. (This is no loss of generality.) Then we choose a connection on the bundle E and use it to lift the generator of the torus action on X to a horizontal vector field on E. This vector field generates an $\mathbb{R}$-action on E that lifts the action of $\mathbb{R}/\mathbb{Z}$ on X. $\square$

Now we come back to T-duality. If $X \xrightarrow{p} Z$ and $X^\# \xrightarrow{p^\#} Z$ are T-dual, then

(a) the fibres of $p^\#$ should be dual to the fibres of p;

(b) there should be a well-defined procedure for creating $(X^\#, \delta^\#)$ from (X, δ);

(c) applying this process twice should get us back where we started;

(d) there should be a natural isomorphism of twisted K-theories

$$K^*(X, \delta) \cong K^{*+1}(X^\#, \delta^\#).$$

The last condition is forced by the equivalence of the IIA string theory on X and the IIB theory on $X^{\#}$, since D-brane charges are supposed to live in these twisted K-groups. The following theorem achieves all of these conditions:

Theorem 11.4 (Raeburn–Rosenberg [102]). *Lift the $\mathbb{T}$-action on X to an $\mathbb{R}$-action α on $CT(X, \delta)$. All such choices are exterior equivalent. Then*

$$CT(X, \delta) \rtimes_\alpha \mathbb{R} \cong CT(X^{\#}, \delta^{\#}),$$
$$K^*(X, \delta) \cong K^{*+1}(X^{\#}, \delta^{\#}).$$

Here $X^{\#} \xrightarrow{p^{\#}} Z$ is a principal $\mathbb{T}$-bundle over Z whose fibres are naturally dual to the fibres of p. Doing this twice gets us back to (X, δ).

We can compute $p^{\#}$ and $\delta^{\#}$ as follows. Recall that a principal $\mathbb{T}$-bundle over Z is determined by a characteristic class $[p] \in H^2(Z)$, and that for any circle bundle, we have a Gysin sequence

$$\cdots \to H^1(Z) \xrightarrow{\cup[p]} H^3(Z) \xrightarrow{p^*} H^3(X) \xrightarrow{p_!} H^2(Z) \to \cdots.$$

Then

$$p_!(\delta) = [p^{\#}], \qquad (p^{\#})_!(\delta^{\#}) = [p].$$

Furthermore, the diagram (11.2) can be completed to a commuting diagram of principal $\mathbb{T}$-bundles

$$
\begin{array}{ccc}
 & Y & \\
{\scriptstyle p_1}\swarrow & & \searrow{\scriptstyle p_1^{\#}} \\
X & & X^{\#} \\
{\scriptstyle p}\searrow & & \swarrow{\scriptstyle p^{\#}} \\
 & Z. &
\end{array}
\qquad (11.5)
$$

We have $[p_1] = p^*([p^{\#}])$, $[p_1^{\#}] = (p^{\#})^*([p])$ and $p_1^*(\delta) = (p_1^{\#})^*(\delta^{\#})$.

Sketch of the proof. We use Lemma 11.3 to lift the $\mathbb{T}$-action on X to an $\mathbb{R}$-action α on $CT(X, \delta)$. To show that the lifting is unique up to exterior equivalence, consider two such liftings, α and α', and look at $t \mapsto \alpha_t \alpha'_{-t}$. This is a continuous 1-cocycle from $\mathbb{R}$ to $\operatorname{Aut} CT(X, \delta)$. Its image lies in the spectrum-fixing automorphisms since the actions of α and α' on X cancel out. Now the identity component of $\operatorname{Aut}_X CT(X, \delta)$ is the projective unitary group of the multiplier algebra, that is, the unitary group divided out by the centre of the unitary group, $C(X, \mathbb{T})$. Thus we have a lifting problem: we want to lift a 1-cocycle with values in the projective unitary group to a 1-cocycle with values in the unitary group. The obstruction to such a lifting lies in $H^2(\mathbb{R}, C(X, \mathbb{T}))$ for an appropriate group cohomology theory. This is not quite Eilenberg–Mac Lane group cohomology since we have to take the topology of the group and the module into account. The appropriate theory, sometimes called "group cohomology with Borel cochains," was defined and studied by Calvin Moore in [90]. The relevant cohomology group turns out to vanish because the topological group $\mathbb{R}$ has homological dimension 1.

Next, we observe that $CT(X,\delta) \rtimes_\alpha \mathbb{R}$ must be a continuous-trace algebra whose spectrum is a circle bundle over Z whose fibres are in some natural sense dual to the circle fibres of $p\colon X \to Z$. To prove this, note that the statement is local, so we may cut down to a small $\mathbb{T}$-invariant open set in X trivialising δ. Then the situation becomes that of $X = \mathbb{S}^1 \times Z$, with p projecting onto the second factor and with $\mathbb{R}$ acting transitively on the first factor, with Z acting trivially and with $\mathbb{R}/\mathbb{Z}$ acting simply transitively. Hence $CT(X,\delta) = C(\mathbb{S}^1) \widehat{\otimes}_{C^*} C(Z,\mathcal{K})$, with α acting only on the first factor, so that

$$CT(X,\delta) \rtimes_\alpha \mathbb{R} \cong \big(C(\mathbb{R}/\mathbb{Z}) \rtimes \mathbb{R}\big) \widehat{\otimes}_{C^*} C_0(Z,\mathcal{K})$$
$$\cong \big(C^*(\mathbb{Z}) \widehat{\otimes}_{C^*} \mathcal{K}\big) \widehat{\otimes}_{C^*} C_0(Z,\mathcal{K}) \cong C_0(\widehat{\mathbb{Z}} \times Z, \mathcal{K}).$$

As required, this is a stable continuous-trace algebra over a space $X^\#$ which is a principal $\mathbb{T}$-bundle over Z with fibres dual to the fibres of p.

Connes' Thom Isomorphism Theorem 10.12, gives the required isomorphism of twisted K-theories. Furthermore, Takai Duality (Theorem 10.10) shows that X and $X^\#$ play symmetrical roles: if we repeat the T-duality process, we get back $CT(X,\delta)$.

Next, we explain the diagram (11.5). The action α of $\mathbb{R}$ on $CT(X,\delta)$ restricts to the trivial action of Z on the dual space X. Hence the crossed product $CT(X,\delta) \rtimes Z$ is a continuous-trace algebra whose spectrum Y is a principal $(\widehat{\mathbb{Z}} \cong \mathbb{T})$-bundle $p_1\colon Y \to X$ over X; its Dixmier–Douady invariant is $p_1^*(\delta)$. Similarly, $CT(X^\#,\delta^\#) \rtimes_{\widehat{\alpha}|_Z} Z$ is a continuous-trace algebra whose spectrum is a principal $\mathbb{T}$-bundle $p_1^\#$ over $X^\#$, and whose Dixmier–Douady invariant is $(p_1^\#)^*(\delta^\#)$. We claim that these two crossed product algebras are isomorphic because of Takai Duality. Using the Packer–Raeburn trick (Theorem 10.19) to split up the crossed products, we get

$$CT(X^\#,\delta^\#) \rtimes_{\widehat{\alpha}|_Z} Z \cong \big(CT(X,\delta) \rtimes_\alpha \mathbb{R}\big) \rtimes_{\widehat{\alpha}|_Z} Z$$
$$\cong \Big(\big(CT(X,\delta) \rtimes_{\alpha|_Z} Z\big) \rtimes_\beta \mathbb{T}\Big) \rtimes_{\widehat{\alpha}|_Z} Z$$
$$\cong CT\big(Y, p_1^*(\delta)\big) \rtimes_\beta \mathbb{T} \rtimes_{\widehat{\beta}} Z$$
$$\cong CT\big(Y, p_1^*(\delta)\big)$$

because $\widehat{\alpha}_{\mathbb{Z}}$ is dual to the action β of $\mathbb{R}/\mathbb{Z}$. Hence the total spaces of p_1 and $p_1^\#$ agree, $[p_1] = p^*([p^\#])$, $[p_1^\#] = (p^\#)^*([p])$, and $p_1^*(\delta) = (p_1^\#)^*(\delta^\#)$.

The characteristic class formula is proved by checking certain examples and using functoriality. We will use the Gysin sequence for a circle bundle, which may be found in any standard algebraic topology text, for example, [116, §5.7 and §9.5].

To start with, suppose $\delta = p^*(\eta)$, $\eta \in H^3(Z)$, is in the image of p^*. By the Gysin sequence, this implies $p_!(\delta) = 0$. By Lemma 11.3, there is an action of $\mathbb{T}$

on $CT(X, \delta)$ compatible with the action on X. Then we can choose α to factor through the quotient map $\mathbb{R} \rightarrow \mathbb{T}$, making α trivial on $\mathbb{Z} = \ker(\mathbb{R} \rightarrow \mathbb{T})$. Then $CT(X, \delta) \rtimes_{\alpha|_\mathbb{Z}} \mathbb{Z} \cong CT(X, \delta) \hat{\otimes}_{C^*} C(\mathbb{S}^1)$. The Packer–Raeburn trick yields

$$CT(X^\#, \delta^\#) \cong CT(X \times \mathbb{S}^1, \delta \times 1) \rtimes \mathbb{T},$$

because both sides are stable, with $\mathbb{T}$ acting freely on X with quotient Z and trivially on $\mathbb{S}^1$, so that $X^\# = Z \times \mathbb{S}^1$ and $p^\#$ is a trivial bundle. This confirms that $[p^\#] = p_!(\delta)$ in this case. Furthermore, we see in this case that in the diagram (11.5), $Y = X \times \mathbb{S}^1$, p_1 is a trivial bundle, and $p_1^\#$ is $p \times 1$. Hence $p_1^*(\delta) = \delta \times 1 = (p \times 1)^*(\delta^\#)$, and $\delta^\# = \eta \times 1 + x \times a$ for some $x \in H^2(Z)$, where $a \in H^1(\mathbb{S}^1)$ is the generator of $H^1(\mathbb{S}^1)$. Since $p^*(x)$ must vanish, to ensure that $(p \times 1)^*(\delta^\#) = \delta \times 1$, $x = (p^\#)_!(\delta^\#)$ is a multiple of $[p]$ by the Gysin sequence. In fact, x turns out to be precisely equal to $[p]$. At least if $\delta = 0$, $Z = \mathbb{S}^2$, $X = \mathbb{S}^3$, and p is the Hopf fibration, this is easy to see because we need to have $K^{*+1}_{\delta^\#}(\mathbb{S}^2 \times \mathbb{S}^1) \cong K^*(\mathbb{S}^3) \cong \mathbb{Z}$ for both $* = 0$ and $* = 1$, which requires $\delta^\#$ to be a generator of $H^3(\mathbb{S}^2 \times \mathbb{S}^1) \cong \mathbb{Z}$ (see Exercise 9.20). If $\delta^\# = 0$, the twisted K-theory is too big, and if it is not primitive, then the twisted K-theory has torsion.

Now suppose that p is trivial (so that $X = \mathbb{S}^1 \times Z$) and $\delta = a \times b$, where a is the generator of $H^1(\mathbb{S}^1)$ and $b \in H^2(Z)$, so that $p_!(\delta) = b$. It is known that there is an action θ of $\mathbb{Z}$ on $C_0(Z, \mathcal{K})$ with $C_0(Z, \mathcal{K}) \rtimes_\theta \mathbb{Z}$ having spectrum T, where $T \rightarrow Z$ is the principal $\mathbb{T}$-bundle with characteristic class b. (See Exercise 11.8.) It turns out that $\mathrm{Ind}_\mathbb{Z}^\mathbb{R} C_0(Z, \mathcal{K})$ is isomorphic to $CT(X, \delta)$. Thus we can assume $\alpha = \mathrm{Ind}_\mathbb{Z}^\mathbb{R} \theta$, so that

$$CT(X^\#, \delta^\#) \cong \left(\mathrm{Ind}_\mathbb{Z}^\mathbb{R} C_0(Z, \mathcal{K}) \right) \rtimes_{\mathrm{Ind}\, \theta} \mathbb{R} \simeq_{\mathrm{Morita}} C_0(Z, \mathcal{K}) \rtimes_\theta \mathbb{Z},$$

which has dual space T. (See Exercise 10.23.) So $[p^\#] = b = p_!(\delta)$, and in (11.5), $Y = \mathbb{S}^1 \times T$. We have $p_1 = 1 \times p^\#$ and $p_1^*(\delta) = a \times (p^\#)^*(b) = (p_1^\#)^*(\delta^\#)$. But $(p^\#)^*(b) = (p^\#)^*([p^\#])$ vanishes by the Gysin sequence, so that $(p_1^\#)^*(\delta^\#) = 0$. Since $p_1^\#$ is a trivial bundle, this implies $\delta^\# = 0$, and $(p^\#)_!(\delta^\#) = [p]$, as claimed. The general cases are reduced to these. $\qquad\square$

As a result, the use of crossed products of continuous-trace algebras, twisted K-theory, and the Connes Thom Isomorphism enables us to put on a firm mathematical basis a phenomenon suggested empirically by physicists.

Exercise 11.6. (Compare Lemma 11.3.) Suppose a compact group T acts freely on a (reasonably nice) space X, with the quotient map $X \rightarrow Z$ a principal T-bundle, and suppose $E \xrightarrow{p} X$ is a principal G-bundle over X, for G some other group (in our applications PU). Show that the T-action on X lifts to an action on E by bundle automorphisms if and only if p is pulled back from a G-bundle over Z.

Exercise 11.7. With notation as in the last exercise, verify that

$$\mathrm{Ind}_\mathbb{Z}^\mathbb{R} C_0(Z, \mathcal{K}) \cong CT(\mathbb{S}^1 \times Z, \delta),$$

where $\delta = a \times [p]$, $a \in H^1(\mathbb{S}^1)$ is a generator, and $[p] \in H^2(Z)$ is the characteristic class of the $\mathbb{T}$-bundle $p\colon T \to Z$.

Exercise 11.8. Let $p\colon T \to Z$ be a principal $\mathbb{T}$-bundle, with T and Z locally compact. Let $\mathbb{T}$ act on $C_0(T)$ in the obvious way. Show that $C_0(T) \rtimes \mathbb{T} \cong C_0(X, \mathcal{K})$ and that the dual action θ of $\mathbb{Z}$ on $C_0(Z, \mathcal{K})$ satisfies $C_0(Z, \mathcal{K}) \rtimes_\theta \mathbb{Z} \cong C_0(T, \mathcal{K})$.

Chapter 12

Some connections with index theory

Index computations provided one of the main motivations for the development of K-theory. Therefore, we briefly discuss here some aspects of index theory that are related to bivariant K-theory.

The index problems most relevant in topology come from *elliptic differential operators*. The most remarkable fact about these operators is the Atiyah–Singer Index Theorem, which provides a topological formula for their indices. This topological formula is local, that is, it can be expressed as an integral of certain differential forms related to the index problem. The goal of this chapter is to indicate how the Atiyah–Singer Index Theorem fits into our general framework. We are mainly interested in variants of this theorem due to Kasparov and Baum–Douglas–Taylor, which deal with certain bivariant K-theory classes related to the index problem.

It is useful to replace differential operators by *pseudo-differential operators*. We briefly sketch in §12.1.1 how these are constructed. Let $\Psi(M)$ be the C^*-algebra of pseudo-differential operators of order 0 on a closed Riemannian manifold M. It is part of a cpc-split extension of C^*-algebras

$$\mathbf{E}_\Psi \colon \mathcal{K}(L^2 M) \rightarrowtail \Psi(M) \xrightarrow{\Sigma} C(S^* M), \tag{12.1}$$

where $S^* M$ is the cosphere bundle on M; the projection $\Sigma \colon \Psi(M) \to C(S^* M)$ is called the *symbol map*.

The index of elliptic pseudo-differential operators is closely related to the index map for this extension. But we must consider elliptic differential operators between non-trivial, possibly different vector bundles on M. The symbol of such a pseudo-differential operator is an element $\sigma(P) \in \mathrm{K}^0(T^* M)$. Hence the *analytic index map* is a map $\mathrm{ind} \colon \mathrm{K}^0(T^* M) \to \mathbb{Z}$.

An elliptic pseudo-differential operator P on M determines a Kasparov module over $C(M)$, which defines a class $[P]$ in $\mathrm{KK}_0(C(M), \mathbb{C})$. Actually, this class of examples was one of the main motivations for Kasparov's definition of KK. The class $[P]$ refines the numerical index $\operatorname{ind} P \in \mathbb{Z}$ because $\operatorname{ind} P$ is the Kasparov product $[u]\#[P]$, where $[u] \in \operatorname{KK}_0(\mathbb{C}, C(M))$ is the class of the unit map $\mathbb{C} \to C(M)$. The *Kasparov Index Theorem* [70] is a refinement of the Atiyah–Singer Index Theorem that describes $[P] \in \operatorname{KK}_0(C(M), \mathbb{C})$ in terms of the symbol $[\sigma(P)] \in \operatorname{K}^0(T^*M)$.

The extension in (12.1) also has a class $[\mathbf{E}_\Psi]$ in $\operatorname{KK}_1\big(C(S^*M), \mathcal{K}(L^2M)\big) \cong \operatorname{KK}_1(C(S^*M), \mathbb{C})$. The *Baum–Douglas–Taylor Index Theorem* computes this class [8]. Since the index map for the extension (12.1) is the Kasparov product with $[\mathbf{E}_\Psi]$, the Baum–Douglas–Taylor Index Theorem implies the Atiyah–Singer Index Theorem, at least for index problems coming from $\operatorname{K}^1(S^*M)$. We will see that it also implies the Kasparov Index Theorem in this special case.

The main ingredient in these index formulas is the class in $\operatorname{KK}_0(C_0(T^*M), \mathbb{C})$ associated to the *Dolbeault operator* on T^*M. The Atiyah–Singer Index Theorem also involves the relationship between the Dolbeault operator and the Thom Isomorphism. Furthermore, the proofs require explicit formulas for the boundary map in KK on special Kasparov modules.

12.1 Pseudo-differential operators

We assume that the reader is already somewhat familiar with pseudo-differential operators. We briefly sketch how to define them and how they give rise to extensions of C^*-algebras and bornological algebras. We formulate the three index problems that are addressed by the Atiyah–Singer, Kasparov, and Baum–Douglas–Taylor Index Theorem, respectively. Finally, we introduce a class in $\operatorname{KK}_0(C_0(X), \mathbb{C})$ for a complex manifold associated to the Dolbeault operator that plays a crucial role in all three index theorems.

12.1.1 Definition of pseudo-differential operators

Let M be a compact m-dimensional Riemannian manifold. The *Laplace operator* is a certain homogeneous differential operator of order 2 on M. For instance, the Laplace operator on the m-torus $\mathbb{T}^m = (\mathbb{R}/\mathbb{Z})^m$ is

$$\Delta := -\sum_{j=1}^m \frac{\partial^2}{\partial x_j^2}.$$

This is an unbounded operator on L^2M with compact resolvent. Even if we are only interested in differential operators originally, it is very useful to adjoin operators like $(\Delta + \lambda)^{-1}$ or $(1 + \Delta^2)^{-1/2}$, which are not differential operators any more. This leads to the algebra of pseudo-differential operators. The definition of pseudo-differential operators on a general manifold is reduced to the case of open subsets of $\mathbb{R}^n$ using a covering by charts and a subordinate partition of unity. We only consider the case of pseudo-differential operators on $\mathbb{R}^n$. Our discussion is very sketchy. A more thorough account can be found in several textbooks or in [64].

The multiplication operators and differentiation operators

$$q_i f(x) := x_i \cdot f(x), \qquad p_i f(x) := \frac{\partial}{\mathrm{i}\partial x_i} f(x) = -\mathrm{i}\frac{\partial f(x)}{\partial x_i}$$

for $i = 1, \ldots, n$ generate the algebra of *differential operators with polynomial coefficients* on $\mathbb{R}^n$. This algebra is a higher-dimensional analogue W_n of the *Weyl algebra* (see §7.5) because of the commutation relations

$$[\mathrm{i} \cdot p_j, q_k] = \delta_{jk} 1, \qquad [q_j, q_k] = 0, \qquad [\mathrm{i} \cdot p_j, \mathrm{i} \cdot p_k] = 0.$$

Any differential operator with polynomial coefficients can be written (using the summation convention) in the form $C_{\alpha,\beta} q^\alpha p^\beta$ for multi-indices α, β. This prescription yields a vector space isomorphism

$$\mathbb{C}[q_1, \ldots, q_n, p_1, \ldots, p_n] \xrightarrow{\cong} W_n, \qquad f \mapsto \mathrm{Op}(f).$$

For example, the polynomial $p_1^2 + \cdots + p_n^2$ corresponds to the *Laplace operator* on $\mathbb{R}^n$. The idea of pseudo-differential operators is to extend the map Op to functions other than polynomials. This requires a formula for $\mathrm{Op}(f)$ in terms of f.

The *Fourier transform* yields a bornological isomorphism

$$\mathfrak{F} \colon \mathscr{S}(\mathbb{R}^n) \to \mathscr{S}(\mathbb{R}^n), \qquad \mathfrak{F}f(\xi) := \int_{\mathbb{R}^n} f(x) \exp(\mathrm{i}x\xi) \, \mathrm{d}x.$$

We denote the inverse Fourier transform by $\mathfrak{F}^*$.

Using that the Fourier transform turns differentiation into multiplication operators, that is, $q_i \circ \mathfrak{F}^* = \mathfrak{F}^* \circ p_i$, we get

$$(\mathrm{Op}(f)h)(x) = \frac{1}{(2\pi)^n} \int_{\mathbb{R}^n} \int_{\mathbb{R}^n} f(x, \xi) \cdot h(y) \cdot \exp(\mathrm{i}(x-y) \cdot \xi) \, \mathrm{d}y \, \mathrm{d}\xi$$

$$= \frac{1}{(2\pi)^n} \int_{\mathbb{R}^n} h(y) \left(\int_{\mathbb{R}^n} f(x, \xi) \cdot \exp(\mathrm{i}(x-y) \cdot \xi) \, \mathrm{d}\xi \right) \mathrm{d}y$$

for all $h \in \mathscr{S}(\mathbb{R}^2)$ and all $f \in \mathbb{C}[q_1, \ldots, q_n, p_1, \ldots, p_n]$; it suffices to check this formula for monomials $q^\alpha p^\beta$. The basic idea of pseudo-differential operators is to take this as a definition of $\mathrm{Op}(f)$ for other classes of functions $f \colon \mathbb{R}^{2n} \to \mathbb{C}$. We do not specify which functions we allow here. The main issue is to control f and its derivatives for $\xi \to \infty$.

Example 12.2. Since $\mathfrak{F}^* p_i \mathfrak{F} = q_i$, the operator $1 + \Delta$ on $\mathscr{S}(\mathbb{R}^n)$ is equivalent to the operator of multiplication by $1 + q_1^2 + \cdots + q_n^2$ and hence invertible. Its inverse is no longer a differential operator: it is the pseudo-differential operator $\mathrm{Op}\big((1 + p_1^2 + \cdots + p_n^2)^{-1}\big)$.

Example 12.3. The operators $\mathrm{Op}(f)$ for $f \in \mathscr{S}(\mathbb{R}^{2n})$ are precisely the smoothing operators on $\mathbb{R}^n$ with integral kernel in $\mathscr{S}(\mathbb{R}^{2n})$.

Example 12.4. If f only depends on the variables $q_1, \ldots, q_n$, then $\mathrm{Op}(f)$ is the pointwise multiplication operator $h \mapsto f \cdot h$. Hence Op restricts to the usual $*$-homomorphism $C_0(\mathbb{R}^n) \to \mathcal{L}(L^2\mathbb{R}^n)$.

When we pass from $\mathbb{R}^n$ to a smooth manifold M, then we associate operators on $L^2(M)$ to suitable functions on the cotangent bundle T^*M. Although the resulting map is not canonical, depending on charts and a partition of unity, the range of this map, that is, the resulting algebra of pseudo-differential operators is canonical. Of course, the important aspects of the theory are those that are independent of auxiliary choices.

We let $\Psi^\infty(M)$ be the algebra of classical pseudo-differential operators with compact support and smooth symbols; this is the smallest useful algebra of pseudo-differential operators. It comes with a canonical filtration by the *order* of pseudo-differential operators. Let $\Psi^\infty(M)_k \subseteq \Psi^\infty(M)$ be the subalgebra of pseudo-differential operators of order (at most) k. Since the order is submultiplicative, $\Psi^\infty(M)_0$ is a subalgebra and $\Psi^\infty(M)_{-1}$ is a closed ideal in $\Psi^\infty(M)_0$. Our definition of $\Psi^\infty(M)$ excludes operators of fractional order like $(1 + \Delta)^{1/4}$, so that any operator of order < 0 already has order -1. Due to this convention, the *symbol map* provides canonical isomorphisms

$$\Psi^\infty(M)_k / \Psi^\infty(M)_{k-1} \cong C_c^\infty(S^*M)$$

for all $k \in \mathbb{Z}$, where S^*M is the cosphere bundle. For a differential operator of order k, the symbol map picks out the leading terms that involve exactly k derivatives and then, in local coordinates, replaces $\partial/\partial x_i$ by $i\xi_i$. Putting these symbol maps together, we get isomorphisms

$$\Psi^\infty(M)_0 / \Psi^\infty(M)_{-\infty} \cong \prod_{j=0}^{-\infty} C_c^\infty(S^*M), \qquad \Psi^\infty(M) / \Psi^\infty(M)_0 \cong \bigoplus_{j=1}^{\infty} C_c^\infty(S^*M).$$

These are bornological isomorphisms if we equip $\Psi^\infty(M)$ with the standard bornology (this is the von Neumann bornology associated to the standard topology). Although the symbol map is canonical, the isomorphisms above are not canonical. The ideal $\Psi^\infty(M)_{-\infty}$ is canonically isomorphic to the algebra $C_c^\infty(M \times M)$ of compactly supported smoothing operators on M.

Pseudo-differential operators act on L^2M by closed unbounded operators. Those of order 0 act by bounded operators, those of order -1 act by compact operators. Even more, we have $\Psi^\infty(M)_{-1} \subseteq \mathscr{L}^p(L^2M)$ for all $p > \dim M$. It should also be possible to compare $\Psi^\infty(M)_{-1}$ with the algebra $\mathcal{CK}^r$ introduced in Definition 3.19, but we have not yet checked the estimates for this.

Now we fix $p > \dim M$ and replace $\Psi^\infty(M)_0$ by

$$\Psi'(M) := \Psi^\infty(M)_0 + \mathscr{L}^p(L^2M) \cong C_c^\infty(S^*M) \oplus \mathscr{L}^p(L^2M).$$

By construction, we get a semi-split extension of bornological algebras

$$\mathbf{E}_{\Psi}^{\infty} = \left(\mathscr{L}^p(L^2 M) \rightarrowtail \Psi'(M) \twoheadrightarrow C_c^{\infty}(S^* M)\right).$$

We may also pass to the C^*-completion $\Psi(M)$ of $\Psi'(M)$; it fits in an extension

$$\mathbf{E}_{\Psi} = \left(\mathcal{K}(L^2 M) \rightarrowtail \Psi(M) \twoheadrightarrow C_0(S^* M)\right).$$

This extension is cpc-split because $C_0(S^* M)$ is nuclear.

Remark 12.5. Recall that $\Psi^{\infty}(M)_{-\infty}$ is isomorphic to the algebra of smoothing operators $C_c^{\infty}(M \times M)$, which is isomorphic to $\mathcal{K}_{\mathscr{S}}$ if M is compact. But the extension of bornological algebras $C_c^{\infty}(M \times M) \rightarrowtail \Psi^{\infty}(M)_0 \twoheadrightarrow \Psi^{\infty}(M)_0 / \Psi^{\infty}(M)_{-\infty}$ does *not* admit a bounded linear section. Therefore, we cannot use this extension.

12.1.2 Index problems from pseudo-differential operators

In order to get interesting index problems, we must allow pseudo-differential operators acting on sections of vector bundles. The algebra of pseudo-differential operators on a trivial vector bundle $M \times \mathbb{R}^n$ is $\mathbb{M}_n\left(\Psi^{\infty}(M)\right)$. Let $E_{\pm}$ be two vector bundles over M; by the smooth version of Swan's Theorem 1.22, we have $\Gamma^{\infty}(E_{\pm}) = C_c^{\infty}(M)^n \cdot e_{\pm}$ for suitable $e_{\pm} \in \operatorname{Idem} \mathbb{M}_n\left(C_c^{\infty}(M)\right)$. The space of pseudo-differential operators $E_+ \to E_-$ is now defined to be $e_+ \mathbb{M}_n\left(\Psi^{\infty}(M)\right) e_-$. The symbol of such an operator belongs to

$$e_+ \mathbb{M}_n\left(C_c^{\infty}(S^* M)\right) e_- \cong \operatorname{Hom}_{C_c^{\infty}(S^* M)}(C_c^{\infty}(S^* M)^n \cdot e_+, C_c^{\infty}(S^* M)^n \cdot e_-)$$
$$\cong \operatorname{Hom}_{C_c^{\infty}(S^* M)}\left(\Gamma^{\infty}(\pi^* E_+), \Gamma^{\infty}(\pi^* E_-)\right) \cong \operatorname{Hom}(\pi^* E_+, \pi^* E_-),$$

where $\operatorname{Hom}_{C_c^{\infty}(S^* M)}$ denotes module homomorphisms and $\operatorname{Hom}(\pi^* E_+, \pi^* E_-)$ denotes smooth vector bundle morphisms; we use the projection $\pi \colon S^* M \to M$ to pull back $E_{\pm}$ to vector bundles on $S^* M$ and to embed $C_c^{\infty}(M) \to C_c^{\infty}(S^* M)$.

   If M is compact, then $\Psi(M)$ is a unital C^*-algebra, and vice versa. In the non-compact case, we adjoin multiplication operators with arbitrary support and enlarge $\Psi(M)$ to $\Psi(M) + \pi^* C_b^{\infty}(M)$. Similarly, for pseudo-differential operators between vector bundles, we adjoin the space of bounded smooth sections of $\operatorname{Hom}(E_+, E_-)$.

Definition 12.6. A pseudo-differential operator $\Gamma^{\infty}(E_+) \to \Gamma^{\infty}(E_-)$ is called *elliptic* if its symbol in $\operatorname{Hom}(\pi^* E_+, \pi^* E_-)$ is a vector bundle isomorphism.

   Assume first that M is compact. Let $P \colon \Gamma^{\infty}(E_+) \to \Gamma^{\infty}(E_-)$ be an elliptic pseudo-differential operator. Then P is a Fredholm operator as such; if the order of P is equal to 0, then the associated bounded operator on $L^2(M)$ is Fredholm as well and has the same index. This index is called the *analytic index* of P and denoted by $\operatorname{ind} P$. This is the index that is computed by the *Atiyah–Singer Index Theorem*.

Even if M is not compact, an elliptic pseudo-differential operator gives rise to a class in $\mathrm{KK}_0(C_0(M),\mathbb{C})$. Let $L^2(E_\pm)$ be the spaces of L^2-sections of $E_\pm$ and let $C_0(M)$ act on $L^2(E_\pm)$ by pointwise multiplication. We assume that P is of order 0 and that its symbol has unitary values. Then P defines a bounded operator $P\colon L^2(E_+) \to L^2(E_-)$. The above representations of $C_0(M)$ together with

$$F := \begin{pmatrix} 0 & P^* \\ P & 0 \end{pmatrix},$$

define a Kasparov $(C_0(M),\mathbb{C})$-module, that is, F is odd and self-adjoint and $[F,h]$ and $(1-F^2)h$ are compact for all $h \in C_0(M)$. The resulting class in $\mathrm{KK}_0(C_0(M),\mathbb{C})$ is denoted by $[P]$. *Kasparov's Index Theorem* provides a topological formula for this class $[P]$.

To turn this into an abstract even Kasparov module, we embed $L^2(E_\pm)$ in an auxiliary Hilbert space $\mathcal{H}$ and let $\alpha_\pm\colon C_0(M) \to \mathcal{L}(\mathcal{H})$ be the representation on $L^2(E_\pm)$ extended by 0 on the orthogonal complement; we let $F \in \mathcal{L}(\mathcal{H})$ be an invertible operator whose restriction to $L^2(E_+) \subseteq \mathcal{H}$ is a compact perturbation of $P\colon L^2(E_+) \to L^2(E_-) \subseteq \mathcal{H}$ (we can find such an operator if $\mathcal{H}$ is sufficiently big). Then (α_+,α_-,P) is an abstract even Kasparov module that realises the class $[P]$ in $\mathrm{KK}_0(C_0(M),\mathbb{C})$.

Suppose again that M is compact, so that $C_0(M) = C(M)$ is unital. Let $u \in \mathrm{KK}_0(\mathbb{C},C(M))$ be the class of the unit map $\mathbb{C} \to C(M)$. We claim that

$$\mathrm{ind}\, P = u\#[P]. \tag{12.7}$$

Since u is a $*$-homomorphism, this Kasparov product is easy to compute. It is the element of $\mathrm{K}_0(\mathcal{K}(\mathcal{H})) \cong \mathbb{Z}$ represented by the pair of idempotents (e_+, Fe_-F^{-1}), where $e_\pm \in \mathcal{L}(\mathcal{H})$ are the orthogonal projections onto $L^2(E_\pm)$. It is easy to see that this is equivalent to the pair $(\ker P, \mathrm{coker}\, P)$ and hence is mapped to $\mathrm{ind}\, P \in \mathbb{Z}$. As a result, $[P]$ is a finer invariant than $\mathrm{ind}\, P$.

The extension of pseudo-differential operators $\mathbf{E}_\Psi$ determines a class $[\mathbf{E}_\Psi]$ in $\mathrm{KK}_{-1}(C_0(S^*M),\mathbb{C})$. This class is computed by the *Baum–Douglas–Taylor Index Theorem*. In the following, we will formulate these index theorems and discuss how they are related.

12.1.3 The Dolbeault operator

Let X be a complex manifold (or a manifold with an almost complex structure). Then TX inherits a complex structure. The bundle $\Lambda^* T^*X$ decomposes canonically into subbundles $\Lambda^{p,q} T^*X$, where a form of type (p,q) is locally a linear combination of forms $f\, \mathrm{d}z_{i_1} \wedge \cdots \wedge \mathrm{d}z_{i_p} \wedge \mathrm{d}\bar{z}_{j_1} \wedge \cdots \wedge \mathrm{d}\bar{z}_{j_q}$ with respect to local complex coordinates $z_i, \bar{z}_i$.

The *Dolbeault operator*

$$\bar\partial\colon \Gamma^\infty(\Lambda^{0,p}) \to \Gamma^\infty(\Lambda^{0,p+1})$$

is the order 1 differential operator given in local coordinates by

$$\bar{\partial} f := \sum \frac{\partial f}{\partial \bar{z}_i} \, d\bar{z}_i.$$

Notice that $\bar{\partial}^2 = 0$, that is, we have a chain complex; its homology is the sheaf cohomology of the sheaf of holomorphic functions on X. This is an example of an *elliptic chain complex*. To get an elliptic differential operator, we take

$$\bar{D} := \begin{pmatrix} 0 & \bar{\partial} \\ \bar{\partial}^* & 0 \end{pmatrix},$$

acting on sections of the $\mathbb{Z}/2$-graded vector bundle

$$\Lambda^{0,*} T^* X := \Lambda^{0,\mathrm{even}} T^* X \oplus \Lambda^{0,\mathrm{odd}} T^* X.$$

Exercise 12.8. Compute the symbol of $\bar{D}$ and check that it is elliptic.

Since $\bar{D}$ is a differential operator of order 1, the commutator $[D, f]$ with a function $f \in C_c^\infty(X)$ is a differential operator of order 0, that is, a multiplication operator, and therefore bounded. The operator $\bar{D}$ does not have compact resolvent because X usually is not compact. But if $f \in C_c^\infty(X)$, then $f \cdot (1 + \bar{D}^2)^{-1/2}$ is compact. Hence we get a spectral triple $(C_c^\infty(X), \mathcal{H}, \bar{D})$ where $\mathcal{H}$ is the $\mathbb{Z}/2$-graded Hilbert space of L^2-sections of $\Lambda^{0,*} T^* X$ and $C_c^\infty(X)$ acts by pointwise multiplication. This spectral triple is p-summable for $p > \dim_{\mathbb{R}} X$.

We get a p-summable even Fredholm module as in (8.38) by taking

$$F_{\bar{\partial}} := \frac{\bar{D}}{\sqrt{1 + \bar{D}^2}}.$$

This defines an element $[\bar{\partial}_X]$ in $\mathrm{KK}_0(C_0(X), \mathbb{C})$. We shall see that this element plays a crucial role for index theory.

Similarly, we can define an element $[\bar{\partial}_X]$ in $\mathrm{kk}_0^{\mathscr{S}}(C_c^\infty(X), \mathbb{C})$. But a thorough treatment leads to technical complications, which we want to avoid. Therefore, we only consider the Dolbeault operator in the setting of Kasparov theory for C^*-algebras.

Observe that $\Lambda^{0,0} T^* X$ is the trivial vector bundle, whose sections are the scalar-valued functions. If $f \colon X \to \mathbb{C}$ is a smooth scalar-valued function, then $\bar{\partial}_X(f) = 0$ means that f satisfies the Cauchy–Riemann differential equations and therefore is holomorphic. Hence the kernel of $\bar{D}$ is the *Bergman space* $H^2(X)$ of holomorphic functions in $L^2(X)$.

If X is a strictly pseudo-convex domain in a complex manifold, then the chain complex

$$\Gamma^\infty(\Lambda^{0,0}) \xrightarrow{\bar{\partial}} \Gamma^\infty(\Lambda^{0,1}) \xrightarrow{\bar{\partial}} \Gamma^\infty(\Lambda^{0,2}) \xrightarrow{\bar{\partial}} \Gamma^\infty(\Lambda^{0,3}) \xrightarrow{\bar{\partial}} \Gamma^\infty(\Lambda^{0,4}) \xrightarrow{\bar{\partial}} \cdots$$

is exact except in degree 0, and 0 is an isolated point in the spectrum of $\bar{D}$. A proof of this fact can be found in [56]. We shall be mainly interested in this case.

If 0 is an isolated point in the spectrum of $\bar{D}$, then the sign function is continuous on the spectrum of $\bar{D}$, and we may replace the bounded operator $F_{\bar{\partial}}$ by the partial isometry $\text{sign}(\bar{D})$. This defines the same cycle in $\text{KK}_0(C_0(X), \mathbb{C})$.

12.2   The index theorem of Baum, Douglas, and Taylor

Let M be a smooth manifold. We want to compute the class of the extension $[\mathbf{E}_\Psi]$ in $\text{KK}_{-1}(C_0(S^*M), \mathbb{C})$. The index theorem of Baum, Douglas, and Taylor [8] asserts that

$$[\mathbf{E}_\Psi] = [\mathbf{E}_{B^*M}] \# [\bar{\partial}_{T^*M}], \tag{12.9}$$

where $[\mathbf{E}_{B^*M}] \in \text{KK}_{-1}(C_0(S^*M), C_0(T^*M))$ is the class of the extension

$$\mathbf{E}_{B^*M} = (C_0(T^*M) \rightarrowtail C_0(B^*M) \twoheadrightarrow C_0(S^*M)) \tag{12.10}$$

and $[\bar{\partial}_{T^*M}] \in \text{KK}_0(C_0(T^*M), \mathbb{C})$ is determined by the Dolbeault operator on T^*M (with a suitable almost complex structure).

The proof proceeds in three steps. First, we use a theorem of Louis Boutet de Monvel [12, 13] that identifies $\mathbf{E}_\Psi$ with another extension: the Toeplitz extension of S^*M with respect to a suitable complex structure on T^*M. Then we prove an abstract theorem that computes the boundary map $\text{KK}_0(I, K) \to \text{KK}_{-1}(Q, K)$ for a cpc-split extension $I \rightarrowtail E \twoheadrightarrow Q$ on special elements of $\text{KK}_0(I, K)$. Finally, we apply this general result to the class $[\bar{\partial}_{T^*M}]$.

Along the way, we very briefly define the Toeplitz extension for suitable almost complex manifolds with boundary; this construction yields the usual Toeplitz extension in the case of the unit disk. The same method that we use to prove (12.9) yields an analogous result for the Toeplitz extension

$$\mathbf{E}_T = (\mathcal{K} \rightarrowtail T(X) \twoheadrightarrow C_0(\partial X))$$

of strictly pseudo-convex domains in complex manifolds. Such a domain X is required to have a smooth boundary ∂X, so that we get a manifold with boundary $\overline{X} = X \cup \partial X$. Letting

$$\mathbf{E}_X = (C_0(X) \rightarrowtail C_0(\overline{X}) \twoheadrightarrow C_0(\partial X)),$$

we have

$$[\mathbf{E}_T] = [\mathbf{E}_X] \# [\bar{\partial}_X].$$

12.2.1   Toeplitz operators

Toeplitz operators on the circle play a crucial role in our proof of Bott periodicity in Chapter 4. Now we generalise this notion and study Toeplitz operators on strictly pseudo-convex domains in complex manifolds. These are bounded domains with a smooth boundary that satisfies a certain condition [56].

In the case of the unit disk $\mathbb{D}$, we have considered Toeplitz operators on the Hardy space, which is a subspace of $L^2(\partial\mathbb{D}) = L^2(\mathbb{S}^1)$. We could also use the Bergman space $H^2(\mathbb{D}) \subseteq L^2(\mathbb{D})$ instead: both yield equivalent algebras of Toeplitz operators. In general, we also have two (or even more) parallel theories, depending on whether we use the analogue of the Hardy or the Bergman space. Following [56], we shall favour the Bergman space.

Let X be a bounded domain in a complex manifold whose boundary ∂X is strictly pseudo-convex, and let $\overline{X} := X \cup \partial X$; this is a compact manifold with boundary. Let $H^2(X) \subseteq L^2(X)$ be the Bergman space considered already in §12.1.3, and let $P\colon L^2(X) \to H^2(X)$ be the orthogonal projection. (Since X is not compact, the choice of Riemannian metric on X really matters for $L^2(X)$; the right choice is explained in [56].) If $f \in C(\overline{X})$, then we get an operator $T_f := PM_fP \in \mathcal{L}\big(H^2(X)\big)$, where $M_f \in \mathcal{L}(L^2(X))$ is the pointwise multiplication operator. Such operators are called Toeplitz operators. It is shown in [56] that $[P, M_f]$ is compact for all $f \in C(\overline{X})$ and that T_f is compact if $f|_{\partial X} = 0$. Hence $f \mapsto T_f \bmod \mathcal{K}$ defines a $*$-homomorphism $C(\partial X) \to \mathcal{L}/\mathcal{K}$. Thus

$$\mathcal{T}(X) := \mathcal{K}\big(H^2(X)\big) + T\big(C(X)\big) \subseteq \mathcal{L}\big(H^2(X)\big)$$

is a C^*-subalgebra and we get an extension of C^*-algebras

$$\mathbf{E}_{\mathcal{T}} = \big(\mathcal{K}\big(H^2(X)\big) \rightarrowtail \mathcal{T}(X) \twoheadrightarrow C(\partial X)\big).$$

This extension is cpc-split with section $f \mapsto T_{s(f)}$, where $s\colon C(\partial X) \to C(X)$ is some completely positive section for the extension

$$\mathbf{E}_X = \big(C_0(X) \rightarrowtail C_0(\overline{X}) \twoheadrightarrow C_0(\partial X)\big),$$

We are mainly interested in the case where $X = B^*M$ for a closed manifold M. In order to view B^*M as a bounded complex domain, we first equip M with a real-analytic structure. This furnishes us with a complexification $M \to M_{\mathbb{C}}$. It is easy to identify the normal bundle of this embedding with $TM \cong T^*M$. Hence the tubular neighbourhood theorem provides an embedding $T^*M \to M_{\mathbb{C}}$, which yields the desired complex structure on the closed disk bundle $B^*M \subseteq T^*M$. If we shrink this disk bundle sufficiently, we can ensure that its boundary S^*M becomes strictly pseudo-convex, so that the $\bar\partial_X$-complex is exact except in degree 0. Hence we get a Toeplitz extension for $C(S^*M)$. (Here the metric and complex structure on T^*M are the one for which the boundary S^*M becomes strictly pseudo-convex, not the original one from the tubular neighbourhood theorem.) The following theorem is due to Louis Boutet de Monvel (see [12, §3] and [13]).

Theorem 12.11. *The Toeplitz extension $\mathbf{E}_{\mathcal{T}}$ and the pseudo-differential operator extension $\mathbf{E}_\Psi$ of $C(S^*M)$ are unitarily equivalent, that is, there is a commuting* *diagram*

$$\begin{array}{ccccc}
\mathcal{K}(L^2 M) & \rightarrowtail & \Psi(M) & \twoheadrightarrow & C(S^* M) \\
\cong \downarrow \mathrm{Ad}_U & & \cong \downarrow \mathrm{Ad}_U & & \| \\
\mathcal{K}\big(H^2(T^* M)\big) & \rightarrowtail & \mathcal{T}(S^* M) & \twoheadrightarrow & C(S^* M)
\end{array}$$

for some unitary operator $U\colon L^2 M \overset{\cong}{\to} H^2(T^* M)$. As a consequence, both extensions define the same class in $\mathrm{KK}_1(C(S^* M), \mathbb{C})$.

There are smooth analogues of the Toeplitz extension and Theorem 12.11 as well. It seems likely that there is a similar result for non-compact M, but the analysis of the Toeplitz operators becomes more difficult.

Actually, Boutet de Monvel's Theorem involves the Hardy space realisation of Toeplitz operators. We omit the proof that this Toeplitz extension is equivalent to the one on Bergman space.

12.2.2   A formula for the boundary map

Let $I \overset{i}{\rightarrowtail} E \overset{p}{\twoheadrightarrow} Q$ be an extension of bornological algebras with a bounded linear section $s\colon Q \to E$. Let K be another bornological algebra. The long exact sequence for $\mathrm{kk}^?$ provides a boundary map

$$\partial\colon \mathrm{kk}_0^?(I, K) \to \mathrm{kk}_{-1}^?(Q, K). \tag{12.12}$$

Our goal is to compute this map explicitly on elements of $\mathrm{kk}_0^?(I, K)$ that are represented by quasi-homomorphisms with some additional properties.

Before we go into this, we claim that this boundary map is closely related to the index map of the extension. Since the index map defined in §1.3.2 involves algebraic K-theory, we restrict attention to local Banach algebras, where we know that algebraic and topological K-theory agree by Proposition 7.36.

Lemma 12.13. *Let $I \rightarrowtail E \twoheadrightarrow Q$ be an extension of local Banach algebras and let $\mathbf{E} \in \Sigma\mathrm{Ho}_1(Q, I)$ be the class of its classifying map. Then the following diagram commutes:*

$$\begin{array}{ccc}
\mathrm{K}_1(Q) & \overset{\mathrm{ind}}{\longrightarrow} & \mathrm{K}_0(I) \\
\uparrow \cong & & \uparrow \cong \\
\mathrm{kk}_1(\mathbb{C}, Q) & \overset{\_\#\mathbf{E}}{\longrightarrow} & \mathrm{kk}_0(\mathbb{C}, I) \, .
\end{array}$$

The vertical isomorphisms are constructed as in §7.4. A similar result holds for C^*-algebras.

Proof. Denote the quotient map $E \to Q$ by p. A diagram chase using the naturality of the index map applied to the morphisms of extensions

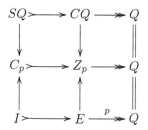

shows that the index map for our original extension **E** agrees with the composite map

$$\mathrm{K}_1(Q) \xrightarrow{\cong} \mathrm{K}_0(SQ) \to \mathrm{K}_0(C_p) \xrightarrow{\cong} \mathrm{K}_0(I),$$

where the first isomorphism is the index map for the cone extension (see Theorem 2.31). The same ingredients give rise to an isomorphism $\Sigma\mathrm{Ho}_{-1}(Q,I) \cong \Sigma\mathrm{Ho}_0(SQ, C_p)$, which maps $[\mathbf{E}]$ to the class of the canonical embedding $SQ \to C_p$. These two computations imply the assertion. □

Hence a computation of the boundary map in (12.12) implies statements about the index map and can therefore be viewed as a kind of index theorem.

Now we explain the special case that we want to treat. Let L be a unital bornological algebra that contains K as a (generalised) ideal and let $\varphi_\pm\colon I \rightrightarrows L \triangleright K$ be a quasi-homomorphism. We assume that the maps $\varphi_\pm$ extend to bounded homomorphisms $\varphi_\pm\colon E \rightrightarrows L$. Furthermore, we assume that there is a bounded linear map $\tau\colon E \to L$ with $\tau(x)\varphi_-(y) = 0 = \varphi_-(y)\tau(x)$ for all $x, y \in E$ such that $\varphi_+ - \varphi_- - \tau$ restricts to a bounded map $\tau'\colon E \to K$. We will compute the boundary map under these assumptions.

We prepare with some computations. Computing in L/K, we have $\tau \equiv \varphi_+ - \varphi_-$, so that our condition means that

$$\varphi_+(x) \cdot \varphi_-(y) \equiv \varphi_-(xy) \equiv \varphi_-(x) \cdot \varphi_+(y) \bmod K.$$

This implies that $\varphi_+ - \varphi_- \equiv \tau$ induces an algebra homomorphism $E \to L/K$. Even more, since $\varphi_+ - \varphi_-$ maps I to K, it descends to an algebra homomorphism $Q \to L/K$. Let

$$\psi := \tau \circ s\colon Q \to L,$$

then it is easy to check that $\omega_\psi(x, y) := \psi(x)\psi(y) - \psi(xy)$ defines a bounded bilinear map $Q \times Q \to K$. Hence we have a singular morphism-extension

$$K \rightarrowtail L \xrightarrow{q} L/K$$

$$\psi \nwarrow \quad \uparrow$$

$$Q,$$

(12.14)

which defines a class in $\mathrm{kk}_{-1}(Q, K)$ by Lemma 6.26.

Proposition 12.15. *The boundary map ∂ maps $\Sigma\mathrm{Ho}(\varphi_\pm) \in \Sigma\mathrm{Ho}_{-1}(I, K)$ to the class of the singular morphism-extension* (12.14).

Of course, the same result holds in $\mathrm{kk}^?$. This statement and its proof are very close to some of the arguments in [38].

Proof. The section $s\colon Q \to E$ defines a bounded homomorphism $TQ \to E$, which restricts to a classifying map $\gamma_s\colon JQ \to I$ for the extension $I \rightarrowtail E \twoheadrightarrow Q$. We remark for later that any bounded linear map $l\colon Q \to L$ for which $\omega_l(x, y) := l(x)\cdot l(y) - l(x\cdot y)$ is a bounded linear map $Q \to K$ defines a bounded homomorphism $\gamma_l\colon JQ \to K$ in this way.

The boundary map $\Sigma\mathrm{Ho}_0(I, K) \to \Sigma\mathrm{Ho}_{-1}(Q, K)$ is given by composition with $[\gamma_s] \in \Sigma\mathrm{Ho}_{-1}(Q, I)$. By the naturality statement in Proposition 3.3, this is represented by the quasi-homomorphism $(\varphi_+ \circ \gamma_s, \varphi_- \circ \gamma_s)\colon JQ \rightrightarrows L \triangleright K$. Since $\varphi_\pm$ extend to bounded algebra homomorphisms on E, we may form linear maps $\varphi_\pm \circ s$ and get $\varphi_\pm \circ \gamma_s = \gamma_{\varphi_\pm \circ s}$.

We write $f_0 \sim_K f_1$ if two bounded homomorphisms $X \to L$ are smoothly homotopic with a smooth homotopy $F\colon X \to L[0,1]$ that is constant modulo K, that is, $F - \mathrm{const}\, f_0$ is a bounded map $X \to K[0,1]$. If $(\varepsilon_\pm)$ and $(\delta_\pm)$ are quasi-homomorphisms $X \rightrightarrows L \triangleright K$ such that $\varepsilon_+ \sim_K \delta_+$ and $\varepsilon_- \sim_K \delta_-$, then the associated bounded homomorphisms $qX \to K$ are smoothly homotopic as well, so that both yield the same class in $\Sigma\mathrm{Ho}$. We apply this to the maps $\gamma_{\varphi_+ \circ s}$ and $\gamma_{\varphi_- \circ s + \tau \circ s}$; since $\varphi_+ - \varphi_- - \tau$ is a bounded map to K, the linear homotopy between φ_+ and $\varphi_- + \tau$ is constant modulo K and thus defines a smooth homotopy of bounded algebra homomorphisms $JQ \to L[0,1]$ that is constant modulo K. Therefore,

$$\gamma_{\varphi_+ \circ s} \sim_K \gamma_{\varphi_- \circ s + \tau \circ s} = \gamma_{\varphi_- \circ s} + \gamma_{\tau \circ s} = \gamma_{\varphi_- \circ s} + \gamma_\psi,$$

where we use that τ and φ_- are orthogonal to conclude that $\gamma_{\varphi_- \circ s}$ and γ_ψ are orthogonal. Hence

$$\Sigma\mathrm{Ho}(\varphi_\pm) \circ \gamma_s = \Sigma\mathrm{Ho}(\gamma_{\varphi_- \circ s} + \gamma_\psi, \gamma_{\varphi_- \circ s}) = \Sigma\mathrm{Ho}(\gamma_\psi, 0) = \Sigma\mathrm{Ho}(\gamma_\psi)$$

by (3.8) and (3.7). Finally, we observe that $\gamma_\psi\colon Q \to K$ is equal to the classifying map of the singular morphism-extension (12.14) by a remark after Lemma 6.26. $\square$

Replacing bounded homomorphisms by $*$-homomorphisms and bounded linear maps by completely positive contractions, we get a corresponding theorem in the C^*-algebraic case:

Proposition 12.16. *Let $I \rightarrowtail E \twoheadrightarrow Q$ be a cpc-split extension of C^*-algebras, let K be a closed two-sided ideal in a unital C^*-algebra L, and let $\varphi_\pm\colon I \rightrightarrows L \triangleright K$ be a quasi-homomorphism such that $\varphi_\pm$ extend to $*$-homomorphisms $\varphi_\pm\colon E \to L$. Let $\tau\colon E \to L$ be a completely positive contraction such that $\varphi_+ - \varphi_- - \tau$ maps E to K and $\tau(E) \cdot \varphi_-(E) = 0 = \varphi_-(E) \cdot \tau(E)$.*

Then the boundary of $\Sigma \mathrm{Ho}^{C^*}(\varphi_\pm) \in \Sigma \mathrm{Ho}_0^{C^*}(I, K)$ in $\Sigma \mathrm{Ho}_{-1}^{C^*}(Q, K)$ is the class associated to the singular morphism-extension defined as in (12.14).

The proof of Proposition 12.15 carries over literally.

12.2.3 Application to the Dolbeault operator

Now we apply the result of §12.2.2 to the following situation. For the extension $I \rightarrowtail E \twoheadrightarrow Q$, we take

$$\mathbf{E}_X = \left(C_0(X) \rightarrowtail C_0(\overline{X}) \twoheadrightarrow C_0(\partial X) \right)$$

where X is a strictly pseudo-convex domain in a complex manifold. We want to compute the boundary in $\mathrm{KK}_{-1}(C_0(\partial X), \mathbb{C})$ of the class $[\bar{\partial}_X]$ in $\mathrm{KK}_0(C_0(X), \mathbb{C})$. The necessary functional analysis that we omit here can be found mostly in [56].

Recall that 0 is an isolated point in the spectrum of

$$\bar{D} := \begin{pmatrix} 0 & \bar{\partial} \\ \bar{\partial}^* & 0 \end{pmatrix}$$

and that the kernel of $\bar{D}$ is the Bergman space $H^2(X)$. Hence we can form the operator $F := \mathrm{sign}(\bar{D})$. Since it is still odd and self-adjoint, we may write

$$F = \begin{pmatrix} 0 & v \\ v^* & 0 \end{pmatrix}.$$

Since $\ker F = \ker \bar{D}$ is the Bergman subspace and concentrated in the even subspace, v is an isometry ($v^* v = 1$) and $1 - vv^* =: P$ is the projection onto the Bergman subspace.

Now we can define the ingredients of Proposition 12.16. Let

$$\mathcal{H}_+ := \Lambda^{0,\mathrm{even}}(X), \qquad \mathcal{H}_- := \Lambda^{0,\mathrm{odd}}(X),$$

and let $\varphi'_\pm \colon C_0(\overline{X}) \to \mathcal{L}(\mathcal{H}_\pm)$ be the representations by pointwise multiplication. We let

$$\varphi_+ := \varphi'_+, \quad \varphi_- := \mathrm{Ad}_{v,v^*} \circ \varphi'_- \colon C_0(\overline{X}) \to \mathcal{L}(\mathcal{H}_+).$$

Finally, we define $\tau \colon C_0(\overline{X}) \to \mathcal{L}(\mathcal{H}_+)$ by $\tau(f) = P\varphi'_+(f)P$.

We claim that these maps satisfy the requirements of Proposition 12.16 with respect to $K = \mathcal{K}(\mathcal{H}_+)$, $L = \mathcal{L}(\mathcal{H}_+)$. It is clear from the definition that $\varphi_\pm$ are $*$-homomorphisms $C_0(\overline{X}) \to \mathcal{L}(\mathcal{H}_+)$ and that $\tau \colon C_0(\overline{X}) \to \mathcal{L}(\mathcal{H}_+)$ is a completely positive contraction that is orthogonal to φ_-. We have $\varphi_+(f) - \varphi_-(f) \in \mathcal{K}(\mathcal{H}_+)$ if $f \in C_0(X)$ because $[F, \varphi(f)]$ is compact for such f. It is shown in [56] that $[P, \varphi'_+(f)]$ is compact for all $f \in C_0(\overline{X})$. Hence the compactness of $\varphi_+(f) - \varphi_-(f) - \tau(f)$ reduces to the assertion $vv^* \varphi_+(f) \equiv \varphi_-(f)$ for $f \in C_0(\overline{X})$.

Proposition 12.16 shows that

$$[\mathbf{E}_X] \# [\bar{\partial}_X] \in \mathrm{KK}_1(C_0(\partial X), \mathbb{C})$$

is the class of the morphism-extension determined by $\tau \circ s \colon C_0(\partial X) \to \mathcal{L}(\mathcal{H})$, where $s \colon C_0(\partial X) \to C_0(\overline{X})$ is some completely positive section. The resulting extension is exactly the Toeplitz extension of $C(\partial X)$, realised on the Bergman space. We conclude that

$$[\mathbf{E}_T] = [\mathbf{E}_X] \# [\bar{\partial}_X].$$

Finally, we specialise to $X = T^*M$, $\partial X = S^*M$ and use Theorem 12.11 to get the Baum–Douglas–Taylor Index Theorem [8]:

Theorem 12.17. $[\mathbf{E}_\Psi] = [\mathbf{E}_T] = [\mathbf{E}_{B^*M}] \# [\bar{\partial}_{T^*M}].$

12.3   The index theorems of Kasparov and Atiyah–Singer

Recall that any elliptic pseudo-differential operator on M defines a class $[P]$ in $\mathrm{KK}_0(C_0(M), \mathbb{C})$ and has an index $\mathrm{ind}\, P \in \mathbb{Z}$ provided M is compact. We have already observed in §12.1.2 that $\mathrm{ind}\, P = [u] \# [P]$ for the unit map $u \colon \mathbb{C} \to C(M)$. Hence it suffices to compute $[P]$. Nevertheless, we also discuss $\mathrm{ind}\, P$ because this is simpler. We want to compute $[P]$ and $\mathrm{ind}\, P$ from the *symbol* of P, which we have to define first.

Let P be an elliptic pseudo-differential operator between vector bundles $E_\pm$. The ellipticity of P tells us that $\pi^* E_+$ and $\pi^* E_-$ are isomorphic vector bundles on S^*M. For the time being, *we assume that already E_+ and E_- are stably isomorphic.* For a suitable vector bundle $E^\perp$, the direct sums $E_\pm \oplus E^\perp$ are trivial vector bundles. We can lift P to a pseudo-differential operator on $E_\pm \oplus E^\perp$ with the same class in $\mathrm{KK}_0(C_0(M), \mathbb{C})$. Therefore, we may restrict attention to operators between trivial bundles.

Thus we get an elliptic pseudo-differential operator $P \colon C_c^\infty(M)^n \to C_c^\infty(M)^n$ of order 0. Its symbol is an invertible function $S^*M \to \mathbb{M}_n(\mathbb{C})$, that is, an element of $\mathrm{Gl}_n\big(C_c^\infty(S^*M)\big)$. Hence it defines a class $[\Sigma(P)] \in \mathrm{K}_1\big(C_c^\infty(S^*M)\big) \cong \mathrm{K}^1(S^*M)$. Conversely, any element of $\mathrm{K}^1(S^*M)$ is the symbol of an elliptic pseudo-differential operator of the special form we consider.

The index map for the extension $\mathbf{E}_\Psi$ furnishes us with a map

$$\mathrm{ind} \colon \mathrm{K}^1(S^*M) = \mathrm{K}_1\big(C(S^*M)\big) \to \mathrm{K}_0\big(\mathcal{K}(L^2M)\big) \cong \mathbb{Z}.$$

The same computation as for Exercise 1.50 shows that the analytic index of P agrees with $\mathrm{ind}[\Sigma(P)]$. Now we use Lemma 12.13 and the Baum–Douglas–Taylor Index Theorem to conclude that

$$\mathrm{ind}\, P = [\Sigma(P)] \# [\mathbf{E}_\Psi] = [\Sigma(P)] \# [\mathbf{E}_{B^*M}] \# [\bar{\partial}_{T^*M}]. \qquad (12.18)$$

We will explain later what this formula has to do with the Atiyah–Singer Index Theorem.

These computations only apply if the source and target vector bundles of P are stably isomorphic. Now we remove this hypothesis. The symbol of a general index problem consists of the two vector bundles $E_\pm$ over M together with an isomorphism between their pull-backs $\pi^* E_+ \xrightarrow{\cong} \pi^* E_-$. Such data define elements in the *relative* K-theory $\mathrm{K}^{\mathrm{rel}}_*(\pi^*)$ of the map $\pi^* \colon C(M) \to C(S^*M)$ induced by the coordinate projection.

Recall that $\mathrm{K}^{\mathrm{rel}}_*(\pi^*) := \mathrm{K}_*(C_{\pi^*})$. Exercise 2.37 shows that this mapping cone is isomorphic to $C_0(T^*M)$; the mapping cylinder is $C(B^*M)$, where B^*M denotes the closed disk bundle over M. The standard extension $C_f \rightarrowtail Z_f \twoheadrightarrow A$ for a map $f \colon A \to B$ specialises to the extension $\mathbf{E}_{B^*M}$ in (12.10).

As a result, the symbol $\sigma(P)$ of a general elliptic pseudo-differential operator belongs to $\mathrm{K}^0(T^*M)$, not $\mathrm{K}^1(S^*M)$. If it happens that $E_+ \cong E_-$, then $[\sigma(P)] \in \mathrm{K}^0(T^*M)$ is the image of $[\Sigma(P)] \in \mathrm{K}^1(S^*M)$ under the index map for the extension (12.10). By Lemma 12.13, this means that

$$[\sigma(P)] = [\Sigma(P)] \# [\mathbf{E}_{B^*M}], \qquad (12.19)$$

where $[\mathbf{E}_{B^*M}]$ denotes the class in $\mathrm{KK}_1\big(C(S^*M), C_0(T^*M)\big)$ associated to the cpc-split extension (12.10). The isomorphism

$$\mathrm{KK}_1\big(C(S^*M), C_0(T^*M)\big) \cong \mathrm{KK}_0\big(SC(S^*M), C_0(T^*M)\big)$$

maps $[\mathbf{E}_{B^*M}]$ to the class of the $*$-homomorphism $SC_0(S^*M) = C_0(S^*M \times \mathbb{R}) \rightarrowtail C_0(T^*M)$ that we get from an identification $T^*M \setminus M \cong S^*M \times \mathbb{R}$.

We remark that the map $\mathrm{K}^1(S^*M) \to \mathrm{K}^0(M)$ need not be surjective, so that there exist index problems that do not come from $\mathrm{K}^1(S^*M)$. Nevertheless, we will limit proofs to index problems of this special form whenever this simplifies matters considerably.

Equation (12.19) allows us to simplify (12.18):

$$\mathrm{ind}\, P = [\Sigma(P)] \# [\mathbf{E}_{B^*M}] \# [\bar{\partial}_{T^*M}] = [\sigma(P)] \# [\bar{\partial}_{T^*M}]. \qquad (12.20)$$

This formula remains valid even if $\Sigma(P)$ is not defined. We omit the proof.

Now we describe the symbol $\sigma(P)$ explicitly using an abstract even Kasparov module. Write $E_\pm = C(M)^n \cdot e_\pm$ with projections $e_\pm \in \mathbb{M}_n\big(C(M)\big)$. Using Lemma 1.42, we get an invertible element $F \in \mathbb{M}_{2n}\big(C(B^*M)\big)$ such that $F|_{S^*M}$ restricts to the given isomorphism $P \colon \pi^* E_+ \cong \pi^* E_-$ on the range of e_+; hence $F\pi^*(e_+)F^{-1} - \pi^*(e_-) \in C_0(T^*M)$. The triple (e_+, e_-, F) yields the desired abstract Kasparov $\mathbb{C}, \mathbb{M}_{2n}\big(C_0(T^*M)\big)$-module.

We can simplify this if we use non-trivial Hilbert modules over $C_0(T^*M)$. Consider the Hilbert module

$$\mathcal{H} := \Gamma(\pi_T^* E_+) \oplus \Gamma(\pi_T^* E_-)^{\mathrm{op}},$$

where π_T denotes the bundle projection $T^*M \to M$; let $\mathbb{C}$ act on $\mathcal{H}$ by the unit map, and let F be an extension of the symbol $\Sigma(P) \colon S^*M \to \mathrm{Hom}(\pi^* E_+, \pi^* E_-)$ to B^*M, acting on $\mathcal{H}$ by pointwise multiplication. This defines another Kasparov module whose class in $\mathrm{KK}_0\big(\mathbb{C}, C_0(T^*M)\big)$ is $[\sigma(P)]$.

Now we turn to the computation of $[P]$. We must enrich this symbol as follows. First, we use the canonical map

$$\mathrm{KK}_0(A, B) \to \mathrm{KK}_0(C_0(M) \,\widehat{\otimes}_{C^*} A, C_0(M) \,\widehat{\otimes}_{C^*} B)$$

to map $[\Sigma(P)]$ to a class in $\mathrm{KK}_1\big(C_0(M), C_0(M \times S^*M)\big)$. Then we use the $*$-homomorphism $C_0(M \times S^*M) \to C_0(S^*M)$ induced by the map $(\pi, \mathrm{id}) \colon S^*M \to M \times S^*M$ to push this class forward to a class

$$[\![\Sigma(P)]\!] \in \mathrm{KK}_1\big(C_0(M), C_0(S^*M)\big),$$

which we call the *bivariant symbol class*. Similarly, we get a bivariant symbol class $[\![\sigma(P)]\!] \in \mathrm{KK}_0\big(C_0(M), C_0(T^*M)\big)$ if the source and target bundles of P are not isomorphic.

We can describe these bivariant symbols quite explicitly because their construction only involves easy cases of the Kasparov product. In the first step, we simply tensor everything with $C_0(M)$, in the second step, we restrict the Hilbert modules to the subspace $T^*M \subseteq M \times T^*M$ or $S^*M \subseteq M \times S^*M$. The result for $\sigma(P)$ is the triple $(e_+ \cdot \varrho, e_- \cdot \varrho, F)$, where $\varrho \colon C(M) \to \mathcal{L}\big(C_0(T^*M)^n\big)$ lets $f \in C_0(M)$ act by pointwise multiplication with $\pi_T^*(f)$. Since this action is central, $\varrho \cdot e_\pm$ is again a $*$-homomorphism, and the assumptions for a Kasparov module are not affected.

The bivariant symbols determine the usual symbols because we have

$$[\sigma(P)] = [u] \# [\![\sigma(P)]\!], \qquad [\Sigma(P)] = [u] \# [\![\Sigma(P)]\!], \tag{12.21}$$

where $[u] \in \mathrm{KK}_0\big(\mathbb{C}, C(M)\big)$ is the class of the unit map.

Proposition 12.22. *We have* $[P] = [\![\Sigma(P)]\!] \# [\mathbf{E}_\Psi]$.

By a similar construction, a symbol $\Sigma(P) \in \mathrm{KK}^1(T^*M)$ yields an element $[P]$ in $\mathrm{KK}_1(C(M), \mathbb{C})$, which may be non-trivial although $[u] \# [P] \in \mathrm{K}_1(\mathbb{C}) = 0$. Proposition 12.22 also applies in this case; in fact, the proof in the odd case is simpler.

Proof. We consider the following more general situation. We have a cpc-split extension $I \overset{i}{\rightarrowtail} E \overset{p}{\twoheadrightarrow} Q$ of C^*-algebras, a $*$-homomorphism $\varphi \colon A \to \mathcal{M}(E)$, and a unitary element $F \in \mathcal{M}(E)$. We assume that $p \circ F$ commutes with $p \circ \varphi(A)$, that is, $[F, \varphi(a)] \in I$ for all $a \in A$; hence $(\varphi, \mathrm{Ad}_F \circ \varphi) \colon A \rightrightarrows E \rhd I$ is a quasi-homomorphism, which defines a class $[\varphi, F]$ in $\mathrm{KK}_0(A, I)$. We also assume that $(1 - F)\varphi(a) \in E$ for all $a \in A$.

The functional calculus for F provides a $*$-homomorphism $\psi \colon C_0(\mathbb{R}) \to \mathcal{M}(E)$, where we identify $\mathbb{R}$ with $\mathbb{S}^1 \setminus \{1\}$. When we compose with p, we get a tensor product homomorphism $C_0(\mathbb{R}, A) = C_0(\mathbb{R}) \,\widehat{\otimes}_{C^*} A \to Q$, which defines a class $[F \otimes \varphi] \in \mathrm{KK}_1(A, Q)$ (the last assumption above ensures that its range is contained in $Q \subseteq \mathcal{M}(Q)$). We claim that the boundary map for the extension $I \rightarrowtail E \twoheadrightarrow Q$ maps $[F \otimes \varphi]$ to $[\varphi, F]$. This claim yields the assertion.

The following proof is probably not optimal, but we have not yet found a better argument. The first step is the construction of a morphisms of extensions

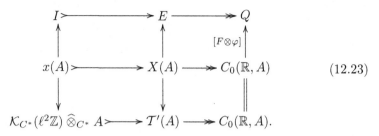

$$(12.23)$$

This will allow us to reduce the general case to the simple special case of the Toeplitz extension in the bottom row.

Let $X(A)$ be the kernel of the natural homomorphism from the free product $C_0(\mathbb{R}) * A$ to the direct sum $C_0(\mathbb{R}) \oplus A$. The coordinate embeddings of $C_0(\mathbb{R})$ and A in $C_0(\mathbb{R})^+ \widehat{\otimes}_{C_*} A^+$ induce a *-homomorphism on $C_0(\mathbb{R}) * A$, whose restriction to $X(A)$ is a map $X(A) \to C_0(\mathbb{R}, A)$. This map is easily seen to be a surjection, so that we get an extension

$$x(A) \rightarrowtail X(A) \twoheadrightarrow C_0(\mathbb{R}, A).$$

This extension is cpc-split, but we omit the proof. By the universal property of the free product, F and φ induce a *-homomorphism $C_0(\mathbb{R}) * A \to E$; its restriction $\tau\colon X(A) \to E$ lifts the homomorphism $C_0(\mathbb{R}, A) \to Q$ and hence restricts to a map $x(A) \to I$. This finishes the construction of the first morphism of extensions in (12.23).

We let operators in $\mathcal{T}^0_{C_*} \subseteq \mathcal{L}(\ell^2 \mathbb{N})$ act by 0 on the orthogonal complement of $\ell^2 \mathbb{N}$ in $\ell^2 \mathbb{Z}$ and enlarge $\mathcal{T}^0_{C_*}$ to

$$\mathcal{T}' := \mathcal{K}_{C_*}(\ell^2 \mathbb{Z}) + \mathcal{T}^0_{C_*} \subseteq \mathcal{L}(\ell^2 \mathbb{Z}).$$

This yields $\mathcal{T}'(A) := \mathcal{T}' \widehat{\otimes}_{C_*} A$. The Toeplitz extension yields an extension as in the last row of (12.23).

The bilateral shift $Uf(n) = f(n-1)$ for all $n \in \mathbb{Z}$ is a unitary operator on $\ell^2(\mathbb{Z})$ that belongs to the multiplier algebra of $\mathcal{T}'$. Hence we also get a corresponding multiplier of $\mathcal{T}'(A)$. We also use the map

$$\varphi'\colon A \to \mathcal{M}(\mathcal{T}'(A)), \qquad a \mapsto P_{\ell^2 \mathbb{N}} \otimes a,$$

where $P_{\ell^2 \mathbb{N}}$ denotes the projection onto $\ell^2 \mathbb{N}$. It is easy to check that $[\varphi'(a), U]$ is compact and $\varphi'(a) \cdot (1 - U) \in \mathcal{T}'(A)$ for all $a \in A$. Copying the construction above, we therefore get the second morphism of extensions in (12.23). This finishes the construction of the diagram (12.23).

Next we claim that the maps $x(A) \to \mathcal{K}_{C_*}(A)$ and $X(A) \to \mathcal{T}'(A)$ in (12.23) are KK-equivalences. It is easy to see that $\mathcal{T}'(A) \sim_{\mathrm{KK}} \mathcal{T}^0_{C_*}(A) \sim_{\mathrm{KK}} 0$. Proposition 8.49 asserts that the map $C_0(\mathbb{R}) * A \to C_0(\mathbb{R}) \oplus A$ is a KK-equivalence. Since we have a cpc-split extension $X(A) \rightarrowtail C_0(\mathbb{R}) * A \twoheadrightarrow C_0(\mathbb{R}) \oplus A$, it follows that $X(A) \sim_{\mathrm{KK}} 0$. Since a morphism of extensions gives rise to a morphism between the extension triangles, the Five Lemma in triangulated categories 6.59 now shows that the map $x(A) \to \mathcal{K}_{C^*}(A)$ is a KK-equivalence as well.

Now we apply the naturality of the boundary map to the two morphisms of extensions in (12.23). It shows that the assertion for a general extension follows from the special case of the Toeplitz extension in the third row. This case is easy and left as an exercise. $\square$

The formulas in Proposition 12.22 and (12.18) are compatible by (12.21).

Combining Theorem 12.17 with Proposition 12.22, we obtain (again assuming that $E_+ \cong E_-$) Kasparov's Index Theorem:

Theorem 12.24 ([72]). *We have* $[P] = [\![\sigma(P)]\!] \# [\bar{\partial}_{T^*M}]$.

12.3.1   The Thom isomorphism and the Dolbeault operator

In order to relate our computation of ind P to the Atiyah–Singer Index Theorem, we first have to discuss the Thom isomorphism for complex vector bundles. This is a generalisation of Bott periodicity.

The Bott Periodicity theorem 7.24 implies that $C_0(\mathbb{R}^2 \times M) = S^2 C_0(M)$ and $C_0(M)$ are KK-equivalent (isomorphic in the category KK) for any locally compact space M. Hence they cannot be distinguished by any split-exact, homotopy invariant, C^*-stable functor on the category of (separable) C^*-algebras. By iteration, we get a KK-equivalence between $C_0(M \times \mathbb{R}^{2n})$ and $C_0(M)$; in particular, $\mathrm{K}^*(M \times \mathbb{R}^{2n}) \cong \mathrm{K}^*(M)$.

Similar assertions hold for $\mathscr{S}(\mathbb{R}^{2n})$ and $\mathbb{C}$ in $\mathrm{kk}^?$. Here we use the identification $\mathscr{S}(\mathbb{R}) \cong \mathbb{C}(0, 1)$, which implies $\mathscr{S}(\mathbb{R}^{2n}) \cong S^{2n}\mathbb{C}$.

Exercise 12.25. Check that the bornological algebras $C_c^\infty(\mathbb{R}^n)$ and $\mathscr{S}(\mathbb{R}^n)$ are smoothly homotopy equivalent.

This exercise allows us to replace $\mathscr{S}(\mathbb{R}^{2n})$ by the smaller and more convenient algebra $C_c^\infty(\mathbb{R}^{2n})$. We may identify $C_c^\infty(M) \hat{\otimes} C_c^\infty(\mathbb{R}^{2n}) \cong C_c^\infty(M \times \mathbb{R}^{2n})$ if M is a smooth manifold. Bott periodicity implies that $C_c^\infty(M \times \mathbb{R}^{2n})$ and $C_c^\infty(M)$ are isomorphic in $\mathrm{kk}^?$.

The above assertions deal with trivial vector bundles $M \times \mathbb{R}^{2n} \to M$ of even dimension. We may ask, more generally, whether these isomorphisms may be combined to a KK-equivalence between $C_0(E)$ and $C_0(M)$ for an even-dimensional vector bundle $E \to M$. We should, however, expect some topological obstruction. The issue is: how equivariant is the Bott periodicity isomorphism $C_0(\mathbb{R}^{2n}) \sim \mathbb{C}$ for the fibres?

For an arbitrary vector bundle, we can always reduce the structure group to the group $\mathrm{O}_{2n}(\mathbb{R})$ of orthogonal matrices. An orientation allows us to further reduce to the special orthogonal group $\mathrm{SO}_{2n}(\mathbb{R})$. As it turns out, this is not yet good enough to get Bott periodicity. We must lift from $\mathrm{SO}_{2n}(\mathbb{R})$ to a certain group called Spin^c, which maps onto SO$_{2n}(\mathbb{R})$. Such a reduction of the structure group is also called a Spin^c *structure* (see also §11.1).

The *Thom Isomorphism Theorem* asserts that K$^*(E) \cong$ K$^*(M)$ for an even-dimensional vector bundle with a Spin^c structure. This result is related to but different from Connes' Thom Isomorphism Theorem 10.12. For simplicity, we shall only consider a special case: that of complex vector bundles, which we may view as even-dimensional real vector bundles. If a real vector bundle has a complex structure, that is, comes from a complex vector bundle, then it also has a Spin^c structure.

To prove the Thom Isomorphism Theorem in Kasparov theory for a complex vector bundle $E \to M$, we need elements $\alpha_E \in$ KK$\big(C_0(E), C_0(M)\big)$ and $\beta_E \in$ KK$\big(C_0(M), C_0(E)\big)$ and must check that the products $\alpha_E \beta_E$ and $\beta_E \alpha_E$ are the identity elements. This is done by Gennadi Kasparov in [71]. We only describe the element α_E here.

We may specialise the Dolbeault element $[\bar{\partial}_X] \in$ KK$_0(C_0(X), \mathbb{C})$ to the case where X is a complex vector space, say, a fibre of our complex vector bundle E. This family of Dolbeault operators along the fibres defines an element $[\alpha_E]$ in KK$_0\big(C_0(E), C_0(M)\big)$. The underlying Hilbert module is the space of C_0-sections of the continuous field of $\mathbb{Z}/2$-graded Hilbert spaces $\big(L^2(\Lambda^{0,*}E_x)\big)_{x \in X}$.

Proposition 12.26 (Kasparov). *The element α_E is invertible, and the Kasparov product with α_E implements the Thom Isomorphism* K$^0(M) \cong$ K$^0(E)$.

The elements α_E have the following crucial multiplicativity property:

Proposition 12.27. *Let X be a complex manifold and E a complex vector bundle over X. Then $[\bar{\partial}_E] = \alpha_E \# [\bar{\partial}_X]$.*

Proof. We first consider the case of a trivial vector bundle, which is particularly simple. In this case, $E = X \times \mathbb{C}^n$ and we have *canonical* isomorphisms $T^*E \cong p^*E \oplus p^*T^*X$ as vector bundles over E and consequently $\Lambda^{0,*}(T^*E) \cong \Lambda^{0,*}(p^*E) \otimes \Lambda^{0,*}(p^*T^*X)$. Hence

$$\mathcal{H} := \mathcal{H}_E \otimes_{C_0(X)} L^2(\Lambda^{0,*}T^*X) \cong L^2(\Lambda^{0,*}T^*E)$$

and we can lift the differential operators $\bar{D}_{\text{fibre}}$ and $\bar{D}_X$ in a canonical way to operators on $\mathcal{H}$. Since they act in different directions, they commute. We let

$$\bar{F}_{\text{fibre}} := \frac{\bar{D}_{\text{fibre}}}{\sqrt{1 + \bar{D}_{\text{fibre}}^2}}, \qquad \bar{F}_X := \frac{\bar{D}_X}{\sqrt{1 + \bar{D}_X^2}}$$

be the associated bounded operators, which lift the operators that we use to define α_E and $[\bar{\partial}_X]$. Notice also that

$$\bar{D}_E = \bar{D}_{\text{fibre}} + \varepsilon \bar{D}_X,$$

where ε is the natural grading operator on the first tensor factor $\mathcal{H}_E$ of $\mathcal{H}$.

Now we use Lemma 8.55 to compute the product in the form

$$F_{\bar{\partial}_{\text{fibre}}} \# F_{\bar{\partial}_X} = M F_{\bar{\partial}_{\text{fibre}}} + \sqrt{1 - M^2}\, \varepsilon F_{\bar{\partial}_X}.$$

There is a canonical choice for M in this case:

$$M := \frac{\sqrt{1 + \bar{D}_{\text{fibre}}^2}}{\sqrt{2 + \bar{D}_{\text{fibre}}^2 + \bar{D}_X^2}}, \qquad \sqrt{1 - M^2} = \frac{\sqrt{1 + \bar{D}_X^2}}{\sqrt{2 + \bar{D}_{\text{fibre}}^2 + \bar{D}_X^2}}.$$

These fractions are well-defined because $\bar{D}_{\text{fibre}}$ and $\bar{D}_X$ commute. They are pseudo-differential operators of order 0, whose symbols we can compute explicitly. This allows us to check that the conditions of Lemma 8.55 are satisfied. We get

$$F_{\bar{\partial}_{\text{fibre}}} \# F_{\bar{\partial}_X} = \frac{\bar{D}_{\text{fibre}}}{\sqrt{2 + \bar{D}_{\text{fibre}}^2 + \bar{D}_X^2}} + \varepsilon \frac{\bar{D}_X}{\sqrt{2 + \bar{D}_{\text{fibre}}^2 + \bar{D}_X^2}} = \frac{\bar{D}_E}{\sqrt{2 + \bar{D}_E^2}}$$

because $\bar{D}_{\text{fibre}}$ and $\varepsilon \bar{D}_X$ anti-commute. Up to a compact perturbation, this agrees with $\bar{F}_{\bar{\partial}_E}$, so that we get $[\bar{\partial}_E]$ as asserted.

So far, we have only treated the case of trivial bundles. The crucial point is that we can reduce the computation in general to the local case, using that vector bundles are locally trivial. We give a few more details about this. Let $(U_i)_{i \in I}$ be an open covering of X such that $E \twoheadrightarrow X$ restricts to a trivial bundle on U_i for all $i \in I$, and let (φ_i) be a subordinate partition of unity on X. Using a trivialisation of E on U_i, we construct operators $\bar{D}_X$, $\bar{D}_{\text{fibre}}$, and M on $L^2(\Lambda^{0,*}(T^*E|_{U_i}))$. If T is one of these operators, then $\varphi_i^{1/2} T \varphi_i^{1/2}$ defines an operator on $L^2(\Lambda^{0,*}(T^*E))$, where we use the standard action of $C_0(X)$ on this Hilbert space. Summing these operators for $i \in I$, we get an operator on $L^2(\Lambda^{0,*}(T^*E))$ that behaves very much like T in the local case. The only issue is that various identities that hold identically for trivial bundles now only hold up to small perturbations because the functions φ_i do not commute with $\bar{D}_{\text{fibre}}$ and $\bar{D}_X$. Nevertheless, it turns out that the conditions that we really need are still satisfied. Furthermore, the resulting Kasparov product agrees with $\bar{F}_{\bar{\partial}_E}$ up to a compact perturbation. $\square$

12.3.2 The Dolbeault element and the topological index map

Finally, it remains to explain why (12.20) implies the Atiyah–Singer Index Theorem [6], which we recall first. Atiyah and Singer define the *topological index map* $\text{ind}_t \colon K^0(T^*M) \to \mathbb{Z}$ in the following way. Choose an embedding of M into $\mathbb{R}^n$ for sufficiently large n and a tubular neighbourhood N of this embedded submanifold in $\mathbb{R}^n$. Then N can be identified with the normal bundle of M in $\mathbb{R}^n$.

We obtain an embedding of $X = T^*M$ into $T^*\mathbb{R}^n \cong \mathbb{C}^n$. Then $E = T^*N$ is a complex vector bundle over X: it is the normal bundle to X in $\mathbb{C}^n$.

The Thom Isomorphism shows that $K^0(T^*M) = K^0(X) \cong K^0(E)$. The inclusion of E into $\mathbb{C}^n$ as an open subset induces an inclusion map $C_0(E) \to C_0(\mathbb{C}^n)$ and hence a map $K^0(E) \to K^0(\mathbb{C}^n) \cong \mathbb{Z}$ (the last isomorphism is Bott periodicity).

The composition

$$K^0(T^*M) = K^0(X) \cong K^0(E) \to K^0(\mathbb{C}^n) \cong \mathbb{Z}$$

is the *topological index map*. The Atiyah–Singer Index Theorem asserts that it is equal to the analytic index map constructed above.

Proposition 12.28. *We have* $\mathrm{ind}_t(x) = x \# [\bar{\partial}_{T^*M}]$ *for all* $x \in K^0(T^*M)$.

Proof. Consider the following diagram:

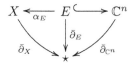

where the entries stand for the K^0-groups of the corresponding spaces and the arrows stand for the maps between them given by product with the indicated bivariant elements α_E, $[\bar{\partial}_X]$, and so on.

In this diagram, the right triangle obviously commutes because the restriction of the Dolbeault element to an open complex submanifold is the Dolbeault element of the submanifold. The left triangle commutes by Proposition 12.27.

The topological index is the composition of the map α_E^{-1}—this is the Thom isomorphism—with the maps $K^0(E) \to K^0(\mathbb{C}^n) \to K^0(\star)$. The commutative diagram shows that this is equal to the product with $[\bar{\partial}_X]$. $\qquad\square$

Combining this with (12.20), we get the Atiyah–Singer formula for $\mathrm{ind}\, P$.

The above computations can be carried over to the setting of smooth functions. This shows that the topological index map sends $x \in K_0\big(C_c^\infty(T^*M)\big)$ to the composition with the corresponding element $[\bar{\partial}_{T^*M}]$ in $\mathrm{kk}_0^{\mathscr{L}}\big(C_c^\infty(T^*M), \mathbb{C}\big)$.

Chapter 13

Localisation of triangulated categories

Let $\mathfrak{T}$ be a triangulated category and let $\mathfrak{N} \subseteq \mathfrak{T}$ be a class of objects. We want to construct a quotient category $\mathfrak{T}/\mathfrak{N}$ in which all objects of $\mathfrak{N}$ become isomorphic to 0, and which should again be triangulated. This process of localisation is the most important construction with triangulated categories. The motivating example of this construction is the passage from the homotopy category of chain complexes over an Abelian category to the derived category. In our context, the most evident example of a localisation is the passage from $\Sigma\mathrm{Ho}$ to $\mathrm{kk}^?$. Less trivial examples are related to Universal Coefficient Theorems and the Baum–Connes conjecture.

In general, the construction of the localisation involves the *Octahedral Axiom*; this was the reason for Jean-Louis Verdier to introduce it in [123]. But in the presence of enough projective objects, we can get away without this axiom, as kindly pointed out to us by Bernhard Keller. Although this special case is good enough for our applications, we discuss the Octahedral Axiom here because it is part of the standard setup of triangulated categories and useful for other purposes.

The following discussion is mostly taken from [87].

We are going to consider classes of objects in $\mathfrak{T}$. Given a class of objects, we get a full subcategory by taking this class of objects with the same morphisms as in $\mathfrak{T}$. Thus classes of objects in $\mathfrak{T}$ are equivalent to full subcategories of $\mathfrak{T}$.

Definition 13.1. A *triangulated subcategory* of $\mathfrak{T}$ is a full subcategory $\mathfrak{N} \subseteq \mathfrak{T}$ that is closed under suspensions and desuspensions and has the exactness property that if $\Sigma B \to C \to A \to B$ is an exact triangle with $A, B \in \mathfrak{N}$, then $C \in \mathfrak{N}$ as well.

In particular, a triangulated subcategory $\mathfrak{N} \subseteq \mathfrak{T}$ is closed under isomorphisms and finite direct sums. When we equip it with the obvious additional structure, then it becomes a triangulated category in its own right, that is, it automatically verifies all the axioms of a triangulated category. This is easy to see: the axioms require certain objects and certain morphisms to exist. These exist in $\mathfrak{T}$; the objects belong to $\mathfrak{N}$ because it is triangulated, the morphisms because it is full.

Definition 13.2. A triangulated subcategory $\mathfrak{N} \subseteq \mathfrak{T}$ is called *thick* if all retracts (direct summands) of objects of $\mathfrak{N}$ belong to $\mathfrak{N}$.

Let $F\colon \mathfrak{T}_1 \to \mathfrak{T}_2$ be an exact functor between two triangulated categories. Let $\ker F \subseteq \mathfrak{T}_1$ be the set of all objects with $F(A) \cong 0$. Clearly, this is always a thick subcategory of $\mathfrak{T}_1$. Conversely, given a thick subcategory, we may ask whether it arises as the kernel of an exact functor.

Definition 13.3. Let $\mathfrak{N}$ be a thick subcategory of a triangulated category $\mathfrak{T}$. The *localisation functor* for $\mathfrak{N} \subseteq \mathfrak{T}$ is an exact functor $\flat\colon \mathfrak{T} \to \mathfrak{T}/\mathfrak{N}$ to a triangulated category $\mathfrak{T}/\mathfrak{N}$ called the *localisation* of $\mathfrak{T}$ at $\mathfrak{N}$, such that $\mathfrak{N} = \ker \flat$ and such that any other exact functor with $\mathfrak{N} \subseteq \ker \flat$ factors uniquely through $\flat$.

Clearly, the localisation functor is unique if it exists. A basic result on triangulated categories asserts that it always exists (see [93]), up to the following set theoretic difficulty: the morphism spaces in $\mathfrak{T}/\mathfrak{N}$ may be classes instead of sets. But this pathology does not arise in the examples that we care about.

The construction of derived categories in homological algebra becomes considerably simpler if there are enough projective or injective objects. This situation can be formalised in the abstract language of triangulated categories:

Definition 13.4 ([87])**.** Let $\mathfrak{T}$ be a triangulated category and let $\mathfrak{P}$ and $\mathfrak{N}$ be subcategories of $\mathfrak{T}$. Suppose that $\mathfrak{P}$ and $\mathfrak{N}$ are closed under isomorphisms (that is, if $A_1 \cong A_2$, then A_1 belongs to one of them if and only if A_2 does) and under suspensions and desuspensions. We call the pair $(\mathfrak{P}, \mathfrak{N})$ *complementary* if $\mathfrak{T}(P, N) = 0$ for all $P \in \mathfrak{P}$, $N \in \mathfrak{N}$ and if for any $A \in \mathfrak{T}$ there is an exact triangle $\Sigma N \to P \to A \to N$ with $P \in \mathfrak{P}$, $N \in \mathfrak{N}$.

Roughly speaking, we require the two subcategories to be orthogonal and to generate the whole category. Notice that it makes little sense to require the existence of an exact triangle of the form $\Sigma P \to N \to A \to P$ instead because the morphism $\Sigma P \to N$ would be forced to vanish, so that $A \cong N \oplus P$ by Lemma 6.61.

We will see below that the subcategories $\mathfrak{P}$ and $\mathfrak{N}$ in a complementary pair are automatically thick.

The situation of Definition 13.4 occurs frequently under different names. For instance, in homotopy theory, the localisation $\mathfrak{T}/\mathfrak{N}$ is often called *smashing* if $\mathfrak{N}$ is part of a complementary pair because this situation has something to do with smash products of topological spaces.

Proposition 13.5. *Let $\mathfrak{T}$ be a triangulated category and let $(\mathfrak{P}, \mathfrak{N})$ be complementary subcategories of $\mathfrak{T}$.*

(1) *We have $N \in \mathfrak{N}$ if and only if $\mathfrak{T}(P, N) = 0$ for all $P \in \mathfrak{P}$, and $P \in \mathfrak{P}$ if and only if $\mathfrak{T}(P, N) = 0$ for all $N \in \mathfrak{N}$. Thus $\mathfrak{P}$ and $\mathfrak{N}$ are thick subcategories and determine each other.*

(2) *The exact triangle* $\Sigma N \to P \to A \to N$ *with* $P \in \mathfrak{P}$ *and* $N \in \mathfrak{N}$ *is unique up to canonical isomorphism and depends functorially on* A. *In particular, its entries define functors* $P \colon \mathfrak{T} \to \mathfrak{P}$ *and* $N \colon \mathfrak{T} \to \mathfrak{N}$.

(3) *The functors* $P, N \colon \mathfrak{T} \to \mathfrak{T}$ *are exact.*

(4) *The localisations* $\mathfrak{T}/\mathfrak{N}$ *and* $\mathfrak{T}/\mathfrak{P}$ *exist.*

(5) *The compositions* $\mathfrak{P} \to \mathfrak{T} \to \mathfrak{T}/\mathfrak{N}$ *and* $\mathfrak{N} \to \mathfrak{T} \to \mathfrak{T}/\mathfrak{P}$ *are equivalences of triangulated categories.*

(6) *The functors* $P, N \colon \mathfrak{T} \to \mathfrak{T}$ *descend to exact functors* $P \colon \mathfrak{T}/\mathfrak{N} \to \mathfrak{P}$ *and* $N \colon \mathfrak{T}/\mathfrak{P} \to \mathfrak{N}$, *respectively, that are inverse (up to isomorphism) to the functors in* (5).

(7) *The functors* $P \colon \mathfrak{T}/\mathfrak{N} \to \mathfrak{T}$ *and* $N \colon \mathfrak{T}/\mathfrak{P} \to \mathfrak{T}$ *are left and right adjoint to the localisation functors* $\mathfrak{T} \to \mathfrak{T}/\mathfrak{N}$ *and* $\mathfrak{T} \to \mathfrak{T}/\mathfrak{P}$, *respectively; that is, we have natural isomorphisms*

$$\mathfrak{T}(P(A), B) \cong \mathfrak{T}/\mathfrak{N}(A, B), \qquad \mathfrak{T}(A, N(B)) \cong \mathfrak{T}/\mathfrak{P}(A, B),$$

for all $A, B \in \mathfrak{T}$.

The following diagram contains the triangulated categories and exact functors found above:

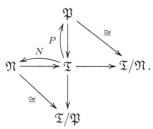

Proof. We can exchange the roles of $\mathfrak{P}$ and $\mathfrak{N}$ by passing to opposite categories. Hence it suffices to prove the various assertions about one of them.

By hypothesis, $N \in \mathfrak{N}$ implies $\mathfrak{T}(P, N) = 0$ for all $P \in \mathfrak{P}$. Conversely, suppose $\mathfrak{T}(P, A) = 0$ for all $P \in \mathfrak{P}$. Let $\Sigma N \to P \to A \to N$ be an exact triangle with $P \in \mathfrak{P}$ and $N \in \mathfrak{N}$. The map $P \to A$ vanishes by hypothesis. Lemma 6.61 implies $N \cong A \oplus \Sigma^{-1}P$. Since $\mathfrak{T}(\Sigma^{-1}P, N) = 0$ by hypothesis, we get $\mathfrak{T}(\Sigma^{-1}P, \Sigma^{-1}P) = 0$, so that $\Sigma^{-1}P = 0$. It follows that the map $A \to N$ is an isomorphism, so that $A \in \mathfrak{N}$ as claimed. Thus $A \in \mathfrak{N}$ if and only if $\mathfrak{T}(P, A) = 0$ for all $P \in \mathfrak{P}$. The latter condition describes a thick subcategory because $\Sigma(\mathfrak{P}) = \mathfrak{P}$. This finishes the proof of (1).

Let $\Sigma N \to P \to A \to N$ and $\Sigma N' \to P' \to A' \to N'$ be exact triangles with $P, P' \in \mathfrak{P}$ and $N, N' \in \mathfrak{N}$, and let $f \in \mathfrak{T}(A, A')$. Since $\mathfrak{T}(P, N') = 0$, the map $P' \to A'$ induces an isomorphism $\mathfrak{T}(P, P') \cong \mathfrak{T}(P, A')$. Hence there is a unique and hence natural way to lift the composite map $P \to A \to A'$ to a map

$P \to P'$. Thus P is unique up to isomorphism and depends functorially on A. Similar remarks apply to N by a dual reasoning. By Axiom 6.53 (TR3), the map $f \colon A \to A'$ and its lifting $P(f) \colon P \to P'$ are part of a morphism of exact triangles

$$
\begin{array}{ccccccc}
\Sigma N & \longrightarrow & P & \longrightarrow & A & \longrightarrow & N \\
\big\downarrow & & \big\downarrow{\scriptstyle P(f)} & & \big\downarrow{\scriptstyle f} & & \big\downarrow \\
\Sigma N' & \longrightarrow & P' & \longrightarrow & A' & \longrightarrow & N'.
\end{array}
$$

The map $N \to N'$ can only be the unique map that lifts f. Therefore, the whole triangle $\Sigma N \to P \to A \to N$ depends functorially on A. This proves (2).

Next we show that P is an exact functor on $\mathfrak{T}$. It clearly commutes with suspensions (up to a canonical isomorphism). Let $\Sigma B \to C \to A \to B$ be an exact triangle. We get an induced map $P(A) \to P(B)$, which we embed in an exact triangle $\Sigma P(B) \to X \to P(A) \to P(B)$. We have $X \in \mathfrak{P}$ because $\mathfrak{P}$ is triangulated. Axiom 6.53 (TR3) shows that the canonical maps $P(A) \to A$ and $P(B) \to B$ are part of a triangle morphism

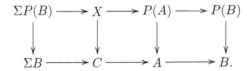

For any $Y \in \mathfrak{P}$, the maps $P(A) \to A$ and $P(B) \to B$ induce isomorphisms on $\mathfrak{T}(Y, \_)$. By the Five Lemma, so does the map $X \to C$. Hence its mapping cone belongs to $\mathfrak{N}$. This yields $X \cong P(C)$; we have already seen that the liftings of the maps $\Sigma B \to C \to A$ to maps $\Sigma P(B) \to P(C) \to P(A)$ are unique. Hence the exact triangle $\Sigma P(B) \to X \to P(A) \to P(B)$ must be the P-image of the triangle $\Sigma B \to C \to A \to B$. Since this triangle is exact by construction, we get the exactness of P; the same argument works for N. This finishes the proof of (3).

Next we construct a candidate $\mathfrak{T}'$ for the localisation $\mathfrak{T}/\mathfrak{N}$. We let $\mathfrak{T}'$ have the same objects as $\mathfrak{T}$ and morphisms $\mathfrak{T}'(A, B) := \mathfrak{T}\big(P(A), P(B)\big)$. The identity map on objects and the map P on morphisms define a canonical functor $\mathfrak{T} \to \mathfrak{T}'$. We define the suspension on $\mathfrak{T}'$ to be the same as for $\mathfrak{T}$. A triangle in $\mathfrak{T}'$ is called exact if it is isomorphic to the image of an exact triangle in $\mathfrak{T}$. We claim that $\mathfrak{T}'$ with this additional structure is a triangulated category and that the functor $\mathfrak{T} \to \mathfrak{T}'$ is the localisation functor at $\mathfrak{N}$.

The uniqueness of the exact triangle $\Sigma N(A) \to P(A) \to A \to N(A)$ yields that the natural map $P(A) \to A$ is an isomorphism for $A \in \mathfrak{P}$. Therefore, the map $\mathfrak{T}(A, B) \to \mathfrak{T}'(A, B)$ is an isomorphism for $A, B \in \mathfrak{P}$. That is, the restriction of the functor $\mathfrak{T} \to \mathfrak{T}'$ to $\mathfrak{P}$ is fully faithful and identifies $\mathfrak{P}$ with a full subcategory of $\mathfrak{T}'$. Moreover, since $P(A) \in \mathfrak{P}$, the map $P^2(A) \to P(A)$ is an isomorphism. This implies that the map $P(A) \to A$ is mapped to an isomorphism in $\mathfrak{T}'$. Thus any object of $\mathfrak{T}'$ is isomorphic to one in the full subcategory $\mathfrak{P}$. Therefore, the category $\mathfrak{T}'$ is equivalent to $\mathfrak{P}$.

We define the functor $P \colon \mathfrak{T}' \to \mathfrak{P}$ to be P on objects and the identity on morphisms. This functor is clearly inverse to the above equivalence $\mathfrak{P} \to \mathfrak{T}'$ and has the property that the composition $\mathfrak{T} \to \mathfrak{T}' \xrightarrow{P} \mathfrak{P} \subseteq \mathfrak{T}$ agrees with $P \colon \mathfrak{T} \to \mathfrak{T}$. Both functors $\mathfrak{P} \to \mathfrak{T}'$ and $\mathfrak{T}' \to \mathfrak{P}$ preserve exactness of triangles because P is an exact functor on $\mathfrak{T}$. These functors commute with suspensions anyway. Since they are equivalences of categories and since $\mathfrak{P}$ is a triangulated category, the category $\mathfrak{T}'$ is triangulated and the equivalence $\mathfrak{P} \cong \mathfrak{T}'$ is compatible with this additional structure.

We have already observed that $\mathfrak{T}(P(A), B) \cong \mathfrak{T}(P(A), P(B))$ for all $A, B \in \mathfrak{T}$. Hence all the remaining assertions follow once we show that $\mathfrak{T}'$ has the universal property of $\mathfrak{T}/\mathfrak{N}$. It is easy to see that $\mathfrak{N}$ is equal to the kernel of $\mathfrak{T} \to \mathfrak{T}'$. If $F \colon \mathfrak{T} \to \mathfrak{T}''$ is an exact functor with kernel $\mathfrak{N}$, then the maps $P(A) \to A$ induce isomorphisms $F\bigl(P(A)\bigr) \to F(A)$ by Lemma 6.61. Therefore, $\mathfrak{T}' \xrightarrow{P} \mathfrak{P} \subseteq \mathfrak{T} \xrightarrow{F} \mathfrak{T}''$ is the required factorisation of F through $\mathfrak{T}'$. $\qquad\square$

Definition 13.6. The maps $P(A) \to A$ and $A \to N(A)$ are called the $\mathfrak{N}$-*projective approximation* and the $\mathfrak{P}$-*injective approximation* of A.

We may also localise a *functor* $F \colon \mathfrak{T} \to \mathfrak{C}$ at a thick subcategory $\mathfrak{N}$. This yields a functor $\mathfrak{T}/\mathfrak{N} \to \mathfrak{C}$, which is exact or homological if F is. Category theorists call it *Kan extension* of F (along the natural projection $\mathfrak{T} \to \mathfrak{T}/\mathfrak{N}$) in honour of Daniel M. Kan. It exists up to the same set theoretic difficulty as with the category $\mathfrak{T}/\mathfrak{N}$ (see [93]). In the situation of a complementary pair $(\mathfrak{P}, \mathfrak{N})$, the localisations at $\mathfrak{N}$ and $\mathfrak{P}$ are naturally isomorphic to $F \circ P$ and $F \circ N$, respectively; that is, we apply our functor to a projective or injective approximation. This is the same recipe as in the computation of derived functors in homological algebra.

13.1 Examples of localisations

First we reconsider the passage from the stable homotopy category $\Sigma\mathrm{Ho}$ to $\mathrm{kk}^?$. We let $\mathfrak{N}^? \subseteq \Sigma\mathrm{Ho}$ for $? = \mathscr{S}, \mathcal{CK}, \mathscr{L}$ be the kernel of the appropriate exact functor

$$\mathcal{K}_{\mathscr{S}}, \mathcal{CK}^r, \mathcal{K}_{\mathscr{S}}\mathcal{K}_{\mathscr{L}^1} \colon \Sigma\mathrm{Ho} \to \Sigma\mathrm{Ho}.$$

Thus $\mathfrak{N}^{\mathscr{S}}$ is the class of all objects A of $\Sigma\mathrm{Ho}$ for which $\mathcal{K}_{\mathscr{S}}(A) \cong 0$ in $\Sigma\mathrm{Ho}$. We denote the relevant stabilisation functor by $\mathcal{K}_?$ in the following; notice that this functor is $\mathcal{K}_{\mathscr{S}}\mathcal{K}_{\mathscr{L}^1}$ for $? = \mathscr{L}$. We let $\mathfrak{P}^?$ be the essential range of $\mathcal{K}^?$, that is, the class of all objects isomorphic to $\mathcal{K}_?(A)$ for some object A of $\Sigma\mathrm{Ho}$.

Theorem 13.7. *The pair of subcategories $(\mathfrak{P}^?, \mathfrak{N}^?)$ is complementary. The localisation functor to $\Sigma\mathrm{Ho}/\mathfrak{N}^?$ is naturally isomorphic to the functor $\mathrm{kk}^? \colon \Sigma\mathrm{Ho} \to \mathrm{kk}^?$. The $\mathfrak{N}^?$-projective approximation functor is naturally isomorphic to $\mathcal{K}_?$, and the localisation of a functor F is $F \circ \mathcal{K}_?$.*

Proof. Notice that $\mathfrak{N}^?$ is a thick subcategory by construction. It is clear that $\mathfrak{P}^?$ is closed under suspensions and desuspensions and under isomorphism. First we prove the orthogonality relation that $\Sigma\mathrm{Ho}(A, \mathcal{K}_?B) = 0$ if $A \in \mathfrak{N}^?$, $B \in \Sigma\mathrm{Ho}$. For this, we observe that the map

$$\Sigma\mathrm{Ho}(A, \mathcal{K}_?B) \xrightarrow{\mathcal{K}_?} \Sigma\mathrm{Ho}(\mathcal{K}_?A, \mathcal{K}_?\mathcal{K}_?B) \xrightarrow{\cong} \Sigma\mathrm{Ho}(\mathcal{K}_?A, \mathcal{K}_?B) \xrightarrow{(\iota_A)_*} \Sigma\mathrm{Ho}(A, \mathcal{K}_?B)$$

is the identity map because it effectively composes f with an inner endomorphism of $\mathcal{K}_?(B)$, and such homomorphisms act identically on $\Sigma\mathrm{Ho}(A, \mathcal{K}_?B)$ because this functor is $\mathbb{M}_2$-stable (Proposition 3.16). Since $\mathcal{K}_?A \cong 0$ this map factors through the zero group, forcing $\Sigma\mathrm{Ho}(A, \mathcal{K}_?B) = 0$.

Next we claim that the $\mathfrak{N}^?$-projective approximation functor agrees with $\mathcal{K}_?$. By Axiom 6.50 (TR1), we may embed the stabilisation homomorphism $A \to \mathcal{K}_?(A)$ in an exact triangle of the form $\Sigma N \to A \to \mathcal{K}_?(A) \to N$. The map $\mathcal{K}_?(A) \to \mathcal{K}_?\mathcal{K}_?(A)$ induced by the stabilisation homomorphism is an isomorphism in $\Sigma\mathrm{Ho}$ by Lemma 7.20. By Lemma 6.61, this implies that $N \in \ker\mathcal{K}_? = \mathfrak{N}^?$. Hence the entries of our exact triangle belong to the required subcategories. Therefore, the pair of subcategories $(\mathfrak{P}^?, \mathfrak{N}^?)$ is complementary, and the $\mathfrak{N}^?$-projective approximation functor is naturally isomorphic to $\mathcal{K}_?$. Now the assertions follow from Proposition 13.5. It yields that the localisation of a functor F at $\mathfrak{N}^?$ is $F \circ \mathcal{K}_?$ and that the morphisms in the localisation at $\mathfrak{N}^?$ are given by

$$\Sigma\mathrm{Ho}/\mathfrak{N}^?(A, B) \cong \Sigma\mathrm{Ho}(\mathcal{K}_?A, \mathcal{K}_?B) \cong \Sigma\mathrm{Ho}(A, \mathcal{K}_?B). \qquad \square$$

An entirely similar discussion applies to the passage from $\Sigma\mathrm{Ho}^{C^*}$ to kk^{C^*}, see §8.5. The natural functor $\mathrm{KK} \to \mathrm{E}$ is a localisation functor as well. This follows in a routine fashion from the universal property of E-theory. But it is unclear whether this localisation comes from a complementary pair of subcategories.

13.1.1   The Universal Coefficient Theorem

The Universal Coefficient Theorem (UCT) approximates bivariant K-theory in terms of ordinary K-theory. It is discussed in detail in [10]. Let A and B be separable C^*-algebras. We need some preparation to formulate the UCT.

The composition in KK yields a natural map of $\mathbb{Z}/2$-graded Abelian groups

$$\gamma\colon \mathrm{KK}_*(A, B) \to \mathrm{Hom}\big(\mathrm{K}_*(A), \mathrm{K}_*(B)\big) := \prod_{m,n \in \mathbb{Z}/2} \mathrm{Hom}\big(\mathrm{K}_n(A), \mathrm{K}_m(B)\big).$$

We may represent any $f \in \mathrm{KK}_*(A, B) \cong \mathrm{Ext}\big(A, C_0(\mathbb{R}^{*+1}, B)\big)$ by a C^*-algebra extension $C_0(\mathbb{R}^{*+1}, B) \otimes \mathcal{K} \rightarrowtail E \twoheadrightarrow A$. This yields an exact sequence

$$
\begin{array}{ccc}
\mathrm{K}_{*+1}(B) & \longrightarrow \mathrm{K}_0(E) \longrightarrow & \mathrm{K}_0(A) \\
{\scriptstyle f_*}\big\uparrow & & \big\downarrow{\scriptstyle f_*} \\
\mathrm{K}_1(A) & \longleftarrow \mathrm{K}_1(E) \longleftarrow & \mathrm{K}_*(B).
\end{array}
\tag{13.8}
$$

The vertical maps in (13.8) are the two components of $\gamma(f)$. Hence (13.8) splits into two extensions of Abelian groups if $f \in \ker \gamma$. Thus we get a map of $\mathbb{Z}/2$-graded Abelian groups

$$\kappa \colon \ker \gamma \to \operatorname{Ext}\big(\mathrm{K}_*(A), \mathrm{K}_{*+1}(B)\big) := \prod_{m, n \in \mathbb{Z}/2} \operatorname{Ext}\big(\mathrm{K}_m(A), \mathrm{K}_{n+1}(B)\big).$$

This map is due to Lawrence Brown (see [110]).

Definition 13.9. The *UCT holds for* $\mathrm{KK}_*(A, B)$ if γ is surjective and κ is bijective. Thus we obtain a short exact sequence

$$\operatorname{Ext}\big(\mathrm{K}_*(A), \mathrm{K}_{*+1}(B)\big) \rightarrowtail \mathrm{KK}_*(A, B) \twoheadrightarrow \operatorname{Hom}\big(\mathrm{K}_*(A), \mathrm{K}_*(B)\big)$$

of $\mathbb{Z}/2$-graded Abelian groups. We say that A satisfies the UCT if the UCT holds for $\mathrm{KK}_*(A, B)$ for all separable C^*-algebras B.

Lemma 13.10. *If* $\mathrm{K}_*(A) \cong \mathrm{K}_*(B)$ *and* A *and* B *both satisfy the UCT, then* A *and* B *are KK-equivalent.*

Proof. Since the maps

$$\gamma \colon \mathrm{KK}_*(A, B) \to \operatorname{Hom}\big(\mathrm{K}_*(A), \mathrm{K}_*(B)\big), \qquad \mathrm{KK}_*(B, A) \to \operatorname{Hom}\big(\mathrm{K}_*(B), \mathrm{K}_*(A)\big)$$

are surjective, we may lift the isomorphism $\mathrm{K}_*(A) \cong \mathrm{K}_*(B)$ to elements $\alpha \in \mathrm{KK}_0(A, B)$ and $\beta \in \mathrm{KK}_0(B, A)$. The composites $\beta\alpha$ and $\alpha\beta$ differ from 1 by elements of $\operatorname{Ext}(\ldots)$. The UCT implies that this is a nilpotent ideal in KK. Hence $\beta\alpha$ and $\alpha\beta$ are invertible. This implies that α and β are invertible, so that A and B are KK-equivalent. $\qquad\square$

It is known that all commutative separable C^*-algebras and, more generally, all separable type I C^*-algebras satisfy the UCT [10]. Since any pair of countable Abelian groups arises as the K-theory for a locally compact space, Lemma 13.10 implies that a separable C^*-algebra satisfies the UCT if and only if it is KK-equivalent to a commutative separable C^*-algebra.

If we combine Lemma 13.10 with the universal property of Kasparov theory, we conclude that two separable C^*-algebras with the same K-theory that satisfy the UCT cannot be distinguished by any C^*-stable split-exact functor.

The results of the last two paragraphs may suggest that few C^*-algebras satisfy the UCT. But, to the contrary, this property is very common. At the moment, we know no nuclear C^*-algebra for which the UCT fails.

Now we reformulate the UCT in terms of localisation. Let $\mathfrak{P}$ be the class of all separable C^*-algebras that satisfy the UCT, and let

$$\mathfrak{N} := \{B \mid \mathrm{K}_*(B) = 0\}.$$

Theorem 13.11. *The pair of subcategories* $(\mathfrak{P}, \mathfrak{N})$ *is complementary.*

Proof. It is clear that $KK_*(A, B) = 0$ if A satisfies the UCT and $K_*(B) = 0$. To finish the proof, we must construct for each separable C^*-algebra A an exact triangle $\Sigma N \to P \to A \to N$ with $P \in \mathfrak{P}$, $N \in \mathfrak{N}$. By the K-theory long exact sequence, we have $N \in \mathfrak{N}$ if and only if the map $P \to A$ induces an isomorphism on K-theory. We construct such a map by lifting a free resolution of $K_*(A)$ to KK.

Since subgroups of free Abelian groups are again free, there is a free resolution of $K_*(A)$ of the form $F_1 \rightarrowtail F_0 \twoheadrightarrow K_*(A)$. Let I_k^+ and I_k^- for $k = 0, 1$ be generating sets for the even and odd parts of F_k. Let

$$\hat{F}_k := \bigoplus_{i \in I_k^+} \mathbb{C} \oplus \bigoplus_{i \in I_k^-} C_0(\mathbb{R}).$$

Then we have

$$KK(\hat{F}_k, B) \cong \prod_{i \in I_k^+} KK(\mathbb{C}, B) \times \prod_{i \in I_k^-} KK(C_0(\mathbb{R}), B)$$

$$\cong \prod_{i \in I_k^+} K_0(B) \times \prod_{i \in I_k^-} K_1(B) \cong \operatorname{Hom}(F_k^+, K_0(B)) \times \operatorname{Hom}(F_k^-, K_1(B)). \quad (13.12)$$

Hence the map $F_0 \to K_*(A)$ yields an element in $KK(\hat{F}_0, A)$. Since $K_*(\hat{F}_0) \cong F_0$, the map $F_1 \to F_0$ also lifts to an element in $KK(\hat{F}_1, \hat{F}_0)$. Thus we get morphisms $\hat{F}_1 \to \hat{F}_0 \to A$ in KK that lift the maps $F_1 \rightarrowtail F_0 \twoheadrightarrow K_*(A)$. Since these liftings are unique and $F_1 \to F_0 \to K_*(A)$ vanishes, the composite map $\hat{F}_1 \to \hat{F}_0 \to A$ vanishes as well.

Embed the morphism $\hat{F}_1 \to \hat{F}_0$ in an exact triangle $\Sigma P \to \hat{F}_1 \to \hat{F}_0 \to P$. The long exact homology sequence allows us to lift the map $\hat{F}_0 \to A$ to a map $P \to A$ because the composite map $\hat{F}_1 \to \hat{F}_0 \to A$ vanishes. The Five Lemma implies that the map $P \to A$ induces an isomorphism on K-theory.

Equation 13.12 shows that $\hat{F}_1$ and $\hat{F}_0$ satisfy the UCT. It is known that the class of separable C^*-algebras that satisfy the UCT is closed under extensions. Hence P satisfies the UCT as well. This finishes the proof. $\qquad \square$

The localisation $KK/\mathfrak{N}$ is equivalent to $\mathfrak{P}$ by Proposition 13.5. Since the UCT applies to $KK(P, B)$ whenever $P \in \mathfrak{P}$, we can compute $KK/\mathfrak{N}(A, B)$ by a UCT as in Definition 13.9 for all A, B. Roughly speaking, this localisation, unlike KK, satisfies the UCT in complete generality. It agrees with HKK as defined in §8.3.

Exercise 13.13. In the situation of the proof of Theorem 13.11, use the long exact sequence for the functor $KK(\llcorner, B)$ and the exact triangle $\Sigma P \to \hat{F}_1 \to \hat{F}_0 \to P$ to construct directly an exact sequence

$$\operatorname{Ext}(K_*(A), K_{*+1}(B)) \rightarrowtail KK_*(P, B) \twoheadrightarrow \operatorname{Hom}(K_*(A), K_*(B)).$$

Hence P satisfies the UCT.

If B is a C^*-algebra, construct an exact sequence

$$\mathrm{K}_*(A) \otimes \mathrm{K}_*(B) \rightarrowtail \mathrm{K}_*(P \otimes B) \twoheadrightarrow \mathrm{Tor}_1\left(\mathrm{K}_{*+1}(A), \mathrm{K}_*(B)\right);$$

this is the Künneth Formula for $\mathrm{K}_*(A \otimes B)$.

More generally, we get a certain exact sequence that computes $H(P)$ for any (co)homological functor $H\colon \mathrm{KK} \to \mathfrak{C}$. The only issue that remains is to identify the kernel and cokernel with suitable derived functors.

Exercise 13.14. The class $\mathfrak{P}$ is the smallest thick subcategory of KK that is closed under direct sums and contains $\mathbb{C}$.

Since all objects of $\mathfrak{P}$ can be constructed from $\mathbb{C}$ by some simple operations (cpc-split extensions, suspensions, direct sums), this class of C^*-algebras is also called *bootstrap category*.

Most of the above argument carries over literally to the categories $\Sigma\mathrm{Ho}$ and $\mathrm{kk}^?$. But we fail eventually because of the following technical problem: we need

$$\Sigma\mathrm{Ho}\Big(\bigoplus_{i\in\mathbb{N}} \mathbb{C}, B\Big) \cong \prod_{i\in\mathbb{N}} \Sigma\mathrm{Ho}(\mathbb{C}, B), \qquad \mathrm{kk}^?\Big(\bigoplus_{i\in\mathbb{N}} \mathbb{C}, B\Big) \cong \prod_{i\in\mathbb{N}} \mathrm{kk}^?(\mathbb{C}, B).$$

We have discussed in §6.3.1 why such assertions are problematic for $\Sigma\mathrm{Ho}$. The same problems are still present in $\mathrm{kk}^?$. This prevents us from stating a Universal Coefficient Theorem for $\mathrm{kk}^?$.

13.1.2   The Baum–Connes assembly map via localisation

Now we consider a pair of complementary subcategories that is related to the construction of the Baum–Connes assembly map in §5.3. We work with equivariant Kasparov theory for C^*-algebras. The problem with infinite direct sums discussed in §6.3.1 prevents us from treating the corresponding constructions for bornological algebras.

Recall that $\mathfrak{CI} \subseteq \mathrm{KK}^G$ is the class of all retracts of direct sums of compactly induced separable G-C^*-algebras. We let $\langle\mathfrak{CI}\rangle$ be the smallest triangulated subcategory of KK^G that contains $\mathfrak{CI}$ and is closed under (countable) direct sums. Equivalently, $\langle\mathfrak{CI}\rangle$ is the smallest thick subcategory of KK^G that is closed under countable direct sums and contains all compactly induced actions of G. We may think of $\langle\mathfrak{CI}\rangle$ as an analogue of the bootstrap category in KK.

In §5.3, we have used the class $\mathfrak{N} \subseteq \mathrm{KK}^G$ defined by the condition $\mathrm{K}_*(H \ltimes A) = 0$ for all compact subgroups $H \subseteq G$. But the pair of subcategories $(\langle\mathfrak{CI}\rangle, \mathfrak{N})$ is not complementary.

Example 13.15. Let G be the trivial group, so that $\mathrm{KK}^G = \mathrm{KK}$. Then $\langle\mathfrak{CI}\rangle \subseteq \mathrm{KK}$ contains all objects and $\mathfrak{N} \subseteq \mathrm{KK}$ is the class of separable C^*-algebras with vanishing K-theory, which is non-trivial because the UCT does not hold for $\mathrm{KK}(A, B)$ in complete generality (compare §13.1.1). Hence the pair of subcategories $(\langle\mathfrak{CI}\rangle, \mathfrak{N})$ is never complementary, for any group G.

To apply our machinery of localisation, we have to replace $\mathfrak{N}$ by a smaller subcategory $\mathfrak{CC}$ [87]; we let $\mathfrak{CC}$ be the class of all $A \in \mathrm{KK}^G$ with $\mathrm{Res}_G^H(A) \cong 0$ in KK^H for all compact subgroups $H \subseteq G$. Here $\mathrm{Res}_G^H \colon \mathrm{KK}^G \to \mathrm{KK}^H$ is the forgetful functor that restricts the G-action on A to an H-action.

Theorem 13.16. *The pair of subcategories* $(\langle \mathfrak{CI} \rangle, \mathfrak{CC})$ *in* KK^G *is complementary. The total left derived functor of the functor* $A \mapsto \mathrm{K}_*\big(C^*_{\mathrm{red}}(G, A)\big)$ *at* $\mathfrak{CC}$ *is the domain* $\mathrm{K}_*^{\mathrm{top}}(G, A)$ *of the Baum–Connes assembly map.*

Proof. We only sketch the proof of the first assertion for discrete G. The second assertion follows from formal properties of the Baum–Connes assembly map.

We have $\mathrm{KK}(A, B) = 0$ for $A \in \mathfrak{CI}$, $B \in \mathfrak{CC}$ and hence for $A \in \langle \mathfrak{CI} \rangle$, $B \in \mathfrak{CC}$ because there are natural isomorphisms

$$\mathrm{KK}^G(\mathrm{Ind}_H^G A, B) \cong \mathrm{KK}^H(A, \mathrm{Res}_G^H B)$$

for all $A \in \mathrm{KK}^H$, $B \in \mathrm{KK}^G$, and $H \subseteq G$ open. Given $A \in \mathrm{KK}^G$, we claim that the extension triangle of the extension $\mathfrak{F}_0 B \rightarrowtail B \twoheadrightarrow B/\mathfrak{F}_0 B$ in Theorem 5.18 has $\mathfrak{F}_0 B \cong A$ in KK^G, $B \in \mathfrak{CC}$, and $B/\mathfrak{F}_0 B \in \langle \mathfrak{CI} \rangle$. This finishes the proof.

Since $\mathfrak{F}_0 B \cong A \,\widehat{\otimes}_{C^*}\, \mathcal{K}(\mathcal{H})$, the first claim follows from the stability of KK^G. The argument in the proof of Theorem 5.18.(3) shows more: the maps $\mathfrak{F}_n B \to \mathfrak{F}_{n+1} B$ vanish in KK^H for all $n \in \mathbb{N}$ and all compact subgroups $H \subseteq G$. This implies $\mathrm{Res}_G^H(B) \cong 0$ in KK^H because $B = \varinjlim \mathfrak{F}_m B$ and all relevant exact sequences are cpc-split. We have $\mathfrak{F}_{n+m} B/\mathfrak{F}_m B \in \langle \mathfrak{CI} \rangle$ for all $m, n \in \mathbb{N}$; this follows by induction on m. Taking another inductive limit, we get $B/\mathfrak{F}_0 B \in \langle \mathfrak{CI} \rangle$ as well. $\qquad\square$

13.2   The Octahedral Axiom

The *Octahedral Axiom*—which is due to Jean-Louis Verdier—received its name because the various commuting triangles and squares and exact triangles that it involves can be drawn on the surface of an octahedron. The following is a planar representation of this axiom:

Axiom 13.17 (TR4). *Let* $\alpha \colon A \to A'$ *and* $f' \colon A' \to B$ *be composable morphisms and put* $f := f' \circ \alpha$. *Then there are commuting diagrams as in Figure* 13.1 *such that the rows and columns in the big diagram are exact triangles.*

There are various reformulations of the Octahedral Axiom. We discuss a particularly simple one from [97, Theorem 1.11] and [75, Appendix A]. It requires the following definition:

Definition 13.18. A commuting square

$$\begin{array}{ccc} X & \xrightarrow{\ \alpha\ } & Y \\ {\scriptstyle \alpha'}\big\downarrow & & \big\downarrow{\scriptstyle \beta} \\ X' & \xrightarrow{\ \beta'\ } & Y' \end{array}$$

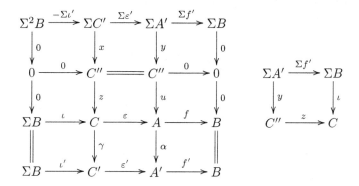

Figure 13.1: Axiom TR4

is called *homotopy Cartesian* if there is an exact triangle

$$\Sigma Y' \xrightarrow{\gamma} X \xrightarrow{\begin{pmatrix} \alpha \\ \alpha' \end{pmatrix}} Y \oplus X' \xrightarrow{(\beta, -\beta')} Y'.$$

The map γ is called a *differential* of the homotopy Cartesian square; it is not unique.

Axiom 6.50 (TR1) shows that any pair of maps $X \to X', Y$ is part of a homotopy Cartesian square. Moreover, this square is unique up to (non-canonical) isomorphism.

Axiom 13.19 (TR4′). *Any pair of maps $X \to Y$ and $X \to X'$ can be completed to a morphism of exact triangles*

$$
\begin{array}{ccccccc}
\Sigma Z & \longrightarrow & X & \longrightarrow & Y & \longrightarrow & Z \\
\| & & \downarrow & & \downarrow & & \| \\
\Sigma Z & \longrightarrow & X' & \longrightarrow & Y' & \longrightarrow & Z,
\end{array}
$$

such that the middle square is homotopy Cartesian and the composite map $\Sigma Y' \to \Sigma Z \to X$ is a differential.

The advantage of Axiom (TR4′) is its simple formulation, which makes it easy to check in examples. The following proposition follows from the Octahedral Axiom; it is not clear whether it is equivalent to it. It formulates an exactness property of the mapping cone construction in general triangulated categories.

Proposition 13.20. *Given a commuting diagram*

$$
\begin{array}{ccc}
A & \xrightarrow{f} & B \\
\downarrow{\scriptstyle\alpha} & & \downarrow{\scriptstyle\beta} \\
A' & \xrightarrow{f'} & B',
\end{array}
$$

(13.21)

there exists a diagram

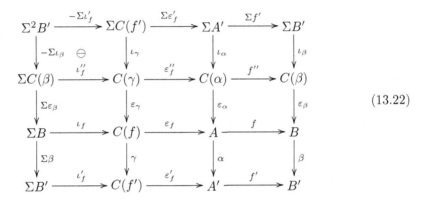

(13.22)

which commutes except for the triangle marked ⊖*, which anti-commutes, and whose rows and columns are exact triangles.*

Axiom 13.17 (TR4) makes a stronger assertion than Proposition 13.20 in a special case, namely, for $\beta = \mathrm{id}_B$. Proposition 13.20 is not itself an axiom because it follows from the Octahedral Axiom (see [9, Proposition 1.1.11]). Again, this proof is elementary but confusing.

We do not present the arguments that relate the axioms (TR4) and (TR4') and Proposition 13.20 because they are not very illuminating. The equivalence of Axioms (TR4) and (TR4') is proved in [75, Appendix A]. Inspection shows that (TR4) can be strengthened: it can be achieved in addition that the square

$$
\begin{array}{ccc}
C & \to & A \\
\downarrow & & \downarrow \\
C' & \to & A'
\end{array}
$$

in Figure 13.1 is homotopy Cartesian; its differential is the map $\Sigma A' \to C$ in the small diagram in Figure 13.1.

Now we verify Axiom (TR4') for the category $\Sigma\mathrm{Ho}$. The same argument also works for $\mathrm{kk}^?$, $\Sigma\mathrm{Ho}^{C^*}$, and kk^{C^*} because it only uses formal properties of mapping cones and cylinders. We shall use the opposite of Axiom (TR4'):

Axiom 13.23 (TR4$'_{op}$). *Every pair of maps $X \to Y$ and $Y' \to Y$ can be completed to a morphism*

$$
\begin{array}{ccccccc}
\Sigma Z & \longrightarrow & X' & \longrightarrow & Y' & \longrightarrow & Z \\
\| & & \downarrow & & \downarrow & & \| \\
\Sigma Z & \longrightarrow & X & \longrightarrow & Y & \longrightarrow & Z
\end{array}
$$

between exact triangles such that the middle square is homotopy Cartesian and the composite map $\Sigma Y \to \Sigma Z \to X'$ is a differential.

This is yet another equivalent formulation of the Octahedral Axiom. It is more convenient for us because the category ΣHo as a category of algebras behaves more like the opposite category to the stable homotopy category of spectra.

Let $f\colon X \to Y$ and $\beta\colon Y' \to Y$ be maps in ΣHo. First we improve this data: we claim that it suffices to treat the case where f and β are bounded algebra homomorphisms between bornological algebras. Here we use that we may replace the given diagram $X \to Y \leftarrow Y'$ by an isomorphic diagram or a (de)suspension without changing anything substantial.

Let $X = (A, m)$, $Y = (B, n)$, $Y' = (B', n')$ with bornological algebras A, B, and B' and $m, n, n' \in \mathbb{Z}$, and represent f and β by bounded algebra homomorphisms $J^{m+k} A \to S^{n+k} B$ and $J^{n'+k} B' \to S^{n+k} B$. Since $X \cong (J^{m+k} A, -k)$, $Y \cong (S^{n+k} B, -k)$, $Y' \cong (S^{n'+k} B', -k)$, our given diagram is isomorphic to the kth desuspension of a diagram $A \overset{f}{\to} B \overset{\beta}{\leftarrow} B'$ with bounded algebra homomorphisms f and β. Hence we may assume from now on that we are dealing with a diagram of this special form.

The crucial ingredient of the proof is the mapping cylinder $Z(f)$, which is defined in (6.64). Recall that there is a natural homotopy equivalence between A and $Z(f)$, which intertwines $f\colon A \to B$ and the natural map $\tilde{f}\colon Z(f) \to B$. Moreover, there is a natural semi-split extension $C(f) \rightarrowtail Z(f) \twoheadrightarrow B$, whose extension triangle is isomorphic to the mapping cone triangle for f. Pulling back this extension via β, we get a morphism of semi-split extensions

$$
\begin{array}{ccccc}
C(f) & \rightarrowtail & Z(f) & \overset{\tilde{f}}{\twoheadrightarrow} & B \\
\| & & \uparrow & & \uparrow{\scriptstyle\beta} \\
C(f) & \rightarrowtail & Z(f, \beta) & \longrightarrow & B',
\end{array}
$$

where

$$
Z(f, \beta) := \{(a, b, b') \in A \oplus B[0, 1] \oplus B' \mid f(a) = b(0),\ \beta(b') = b(1)\}.
$$

The projection $Z(f, \beta) \to A \oplus B'$, $(a, b, b') \mapsto (a, b')$, is a split surjection with kernel $B(0, 1)$. Hence we get a semi-split extension $SB \rightarrowtail Z(f, \beta) \twoheadrightarrow A \oplus B'$. The map $SA \oplus SB' \to SB$ in the resulting extension triangle

$$
SA \oplus SB' \to SB \to Z(f, \beta) \to A \oplus B' \tag{13.24}
$$

is equal to the suspension of $(-f, \beta)$. This follows from the naturality of extension triangles because there are obvious morphisms of extensions from the cone extensions $A(0,1) \rightarrowtail A[0,1) \twoheadrightarrow A$ and $B'(0,1) \rightarrowtail B'(0,1] \twoheadrightarrow B'$ to the extension $B(0,1) \rightarrowtail Z(f, \beta) \twoheadrightarrow A \oplus B'$. Using Axiom 6.53 (TR3) to rotate (13.24), we see that the maps $Z(f, \beta) \to A, B' \to B$ form a homotopy Cartesian square in ΣHo, whose differential is the inclusion map $SB \to Z(f, \beta)$. The composite map $SB \to C(f) \to Z(f, \beta)$ is equal to the standard embedding as well. Hence we get a diagram as in Axiom 13.19 (TR4').

This finishes the proof that ΣHo is a triangulated category. Recall that the proofs of Theorem 6.48 and Proposition 7.22 were incomplete so far because we postponed the treatment of the Octahedral Axiom.

Next we prove Proposition 13.20 and Axiom 13.17 (TR4) directly for the category ΣHo in order to see what they assert. This is not logically necessary because they follow from Axiom 13.19. Both results are proved by essentially the same argument, which also yields the corresponding statements for $kk^?$.

In a first step, we modify a given commuting diagram as in Proposition 13.20 such that f, f', α, β are bounded algebra homomorphisms and $\beta \circ f = f' \circ \alpha$ holds exactly, not just in ΣHo. The following modifications are allowed because we merely replace the given commuting square by one that is isomorphic in ΣHo.

We have already seen during the verification of Axiom 6.50 (TR1) that the morphisms f and f' may be replaced by equivalent bounded algebra homomorphisms. During the verification of Axiom 6.53 (TR3), we have seen that we can also achieve that α and β become bounded algebra homomorphisms, such that $\langle \beta \circ f \rangle = \langle f' \circ \alpha \rangle$; let $H \colon A \to B'[0,1]$ be the smooth homotopy with $H_0 = \beta \circ f$ and $H_1 = f' \circ \alpha$.

To achieve the relation $\beta \circ f = f' \circ \alpha$ exactly, we replace A' by the mapping cylinder $Z(f')$. We may combine $\alpha \colon A \to A'$ and $H \colon A \to B'[0,1]$ to a bounded homomorphism $\tilde{\alpha} \colon A \to Z(f')$. By construction $\tilde{f'} \circ \tilde{\alpha} = H_0 = \beta \circ f$. Therefore, any commuting square (13.21) in ΣHo is isomorphic to a commuting square in which f, f', α, β are bounded algebra homomorphisms and $f' \circ \alpha = \beta \circ f$ holds exactly. (Here we denote objects (A, m) of ΣHo simply by A to avoid clutter.)

Now we construct mapping cone triangles over f, f', α, β. Since $f' \circ \alpha = \beta \circ f$ holds exactly, we get bounded homomorphisms

$$\gamma \colon C(f) \to C(f'), \qquad \gamma(a, b) := \big(\alpha(a), C\beta(b)\big),$$
$$f'' \colon C(\alpha) \to C(\beta), \qquad f''(a, a') := \big(f(a), Cf'(a')\big).$$

We form the mapping cones $C(\gamma)$ and $C(f'')$ of these two maps. By construction,

$$C(\gamma) = \{(c, c') \in C(f) \oplus C\big(C(f')\big) \mid \gamma(c) = c'(1)\}$$
$$= \big\{(a, b, a', b') \in A \oplus C(B) \oplus C(A') \oplus C^2(B') \mid$$
$$b(1) = f(a), \; b'(s, 1) = f'\big(a'(s)\big), \; a'(1) = \alpha(a), \; b'(1, t) = \beta\big(b(t)\big)\big\},$$

and

$$C(f'') = \{(c,c') \in C(\alpha) \oplus C\big(C(\beta)\big) \mid f''(c) = c'(1)\}$$
$$= \big\{(a,a',b,b') \in A \oplus C(A') \oplus C(B) \oplus C^2(B') \mid$$
$$b(1) = f(a),\ b'(1,t) = f'\big(a'(t)\big),\ a'(1) = \alpha(a),\ b'(s,1) = \beta\big(b(s)\big)\big\}.$$

The crucial observation is that $C(\gamma) \cong C(f'')$ via $(a,b,a',b') \mapsto (a,a',b,\Phi b')$, where $\Phi b'(s,t) := b'(t,s)$. Hence we obtain a diagram as in (13.22) in which all the maps are bounded homomorphisms and all the rows and columns are mapping cone triangles. All squares commute exactly, except for the one marked $\ominus$ which commutes up to the flip automorphism Φ. Since the latter generates a sign, this square anti-commutes. Hence Proposition 13.20 holds for ΣHo.

The Octahedral Axiom deals with the special case $\beta = \mathrm{id}$. As above, we can replace f' and α by bounded homomorphisms; then we construct the various mapping cones and get the diagram (13.22). Now $C(\beta) \cong C(B)$ is smoothly contractible (compare Axiom (TR0)). The map $\varepsilon_f'' \colon C(\gamma) \to C(\alpha)$ is a smooth homotopy equivalence; the homotopy inverse, which we denote by $(\varepsilon_f'')^{-1}$, maps $(a,a') \in C(\alpha)$ to $\big(a,(Cf')(a'),a',\omega(a')\big)$, where $\omega(a')(s,t)$ is $f'(a')(s+t-1)$ for $s+t \geq 1$ and 0 otherwise (a similar map appears in the proof of Theorem 6.63). By definition, ω is a section for ε_f''. We leave it to the reader to smoothly deform the composition $(\varepsilon_f'')^{-1} \circ \varepsilon_f''$ to the identity map on $C(\gamma)$. Hence the second row in (13.22) is isomorphic to $0 \to C'' = C'' \to 0$. It remains to check that the small diagram in Figure 13.1 commutes. This follows because the two maps

$$S(A') \xrightarrow{S(f')} S(B') = S(B) \xrightarrow{\iota_f} C(f), \qquad S(A') \xrightarrow{\iota_\alpha} C(\alpha) \xrightarrow{(\varepsilon_f'')^{-1}} C(\gamma) \xrightarrow{\varepsilon_f} C(f)$$

coincide. This finishes the proof of Axiom 13.17 for ΣHo.

Exercise 13.25. In the above situation, check that the square

$$
\begin{array}{ccc}
C(f) & \xrightarrow{\ \varepsilon_f\ } & A \\
\downarrow{\scriptstyle \gamma} & & \downarrow{\scriptstyle \alpha} \\
C(f') & \xrightarrow{\ \varepsilon_f'\ } & A'
\end{array}
$$

in Figure 13.1 is homotopy Cartesian and that the map $\Sigma A' \to C(f)$ in the small diagram in Figure 13.1 is a differential for it.

Bibliography

[1]   J. F. Adams, *Stable homotopy and generalised homology*, University of Chicago Press, Chicago, Ill., 1974. Chicago Lectures in Mathematics.

[2]   Joel Anderson, *A C^*-algebra A for which* $\text{Ext}(A)$ *is not a group*, Ann. of Math. (2) **107** (1978), no. 3, 455–458.

[3]   William Arveson, *Notes on extensions of C^*-algebras*, Duke Math. J. **44** (1977), no. 2, 329–355.

[4]   M. F. Atiyah and F. Hirzebruch, *Vector bundles and homogeneous spaces*, Proc. Sympos. Pure Math., Vol. III, American Mathematical Society, Providence, R.I., 1961, pp. 7–38.

[5]   M. F. Atiyah and Graeme Segal, *Twisted K-theory*, Ukr. Mat. Visn. **1** (2004), no. 3, 287–330; English transl., Ukr. Math. Bull. **1** (2004), no. 3, 291–334.

[6]   M. F. Atiyah and I. M. Singer, *The index of elliptic operators. I*, Ann. of Math. (2) **87** (1968), 484–530.

[7]   Paul Baum and Ronald G. Douglas, *Index theory, bordism, and K-homology*, Operator Algebras and K-theory (San Francisco, Calif., 1981), Contemp. Math., 10, 1982, pp. 1–31.

[8]   Paul Baum, Ronald G. Douglas, and Michael E. Taylor, *Cycles and relative cycles in analytic K-homology*, J. Differential Geom. **30** (1989), no. 3, 761–804.

[9]   A. A. Beĭlinson, J. Bernstein, and P. Deligne, *Faisceaux pervers*, Analysis and topology on singular spaces, I (Luminy, 1981), Astérisque, vol. 100, Soc. Math. France, Paris, 1982, pp. 5–171.

[10] Bruce Blackadar, *K-theory for operator algebras*, 2nd ed., Mathematical Sciences Research Institute Publications, vol. 5, Cambridge University Press, Cambridge, 1998.

[11] A. K. Bousfield, *A classification of K-local spectra*, J. Pure Appl. Algebra **66** (1990), no. 2, 121–163.

[12] L. Boutet de Monvel, *On the index of Toeplitz operators of several complex variables*, Invent. Math. **50** (1978/79), no. 3, 249–272.

[13] L. Boutet de Monvel and V. Guillemin, *The spectral theory of Toeplitz operators*, Annals of Mathematics Studies, vol. 99, Princeton University Press, Princeton, NJ, 1981.

[14] Peter Bouwknegt, Alan L. Carey, Varghese Mathai, Michael K. Murray, and Danny Stevenson, *Twisted K-theory and K-theory of bundle gerbes*, Comm. Math. Phys. **228** (2002), no. 1, 17–45.

[15] Peter Bouwknegt, Jarah Evslin, and Varghese Mathai, *T-duality: topology change from H-flux*, Comm. Math. Phys. **249** (2004), no. 2, 383–415.

[16] Peter Bouwknegt and Varghese Mathai, *D-branes, B-fields and twisted K-theory*, J. High Energy Phys. (2000), no. 3, Paper 7, 11 pp. (electronic).

[17]   Lawrence G. Brown, *Operator algebras and algebraic K-theory*, Bull. Amer. Math. Soc. **81** (1975), no. 6, 1119–1121.

[18]   _____, *Extensions and the structure of C*-algebras*, Symposia mathematica, vol. XX (Convegno sulle Algebre C* e loro Applicazioni in Fisica Teorica, Convegno sulla Teoria degli Operatori Indice e Teoria K, INDAM, Rome, 1975), 1976, pp. 539–566.

[19]   _____, *Characterizing* Ext(X), *K-theory and operator algebras* (Proc. Conf., Univ. Georgia, Athens, Ga., 1975), 1977, pp. 10–18. Lecture Notes in Math., Vol. 575.

[20]   _____, *Stable isomorphism of hereditary subalgebras of C*-algebras*, Pacific J. Math. **71** (1977), no. 2, 335–348.

[21]   _____, *The universal coefficient theorem for Ext and quasidiagonality*, Operator algebras and group representations, vol. I (Neptun, 1980), 1984, pp. 60–64.

[22]   L. G. Brown, R. G. Douglas, and P. A. Fillmore, *Unitary equivalence modulo the compact operators and extensions of C*-algebras*, Proceedings of a conference on operator theory (Dalhousie Univ., Halifax, N.S., 1973), 1973, pp. 58–128. Lecture Notes in Math., Vol. 345.

[23]   _____, *Extensions of C*-algebras and K-homology*, Ann. of Math. (2) **105** (1977), no. 2, 265–324.

[24]   Ulrich Bunke, Michael Joachim, and Stephan Stolz, *Classifying spaces and spectra representing the K-theory of a graded C*-algebra*, High-dimensional manifold topology, World Sci. Publ., 2003, pp. 80–102.

[25]   Man Duen Choi and Edward G. Effros, *The completely positive lifting problem for C*-algebras*, Ann. of Math. (2) **104** (1976), no. 3, 585–609.

[26]   J. Daniel Christensen, *Ideals in triangulated categories: phantoms, ghosts and skeleta*, Adv. Math. **136** (1998), no. 2, 284–339.

[27]   Pierre Colmez and Jean-Pierre Serre (eds.), *Correspondance Grothendieck–Serre*, Documents Mathématiques (Paris), 2, Société Mathématique de France, Paris, 2001. translated by C. Maclean as *Grothendieck–Serre Correspondence*, Amer. Math. Soc., Providence, RI, 2004, ISBN 0-8218-3424-X.

[28]   Alain Connes, *An analogue of the Thom isomorphism for crossed products of a C*-algebra by an action of* **R**, Adv. in Math. **39** (1981), no. 1, 31–55.

[29]   _____, *Noncommutative geometry*, Academic Press Inc., San Diego, CA, 1994.

[30]   Alain Connes and Nigel Higson, *Déformations, morphismes asymptotiques et K-théorie bivariante*, C. R. Acad. Sci. Paris Sér. I Math. **311** (1990), no. 2, 101–106 (French, with English summary).

[31]   A. Connes and G. Skandalis, *The longitudinal index theorem for foliations*, Publ. Res. Inst. Math. Sci. **20** (1984), no. 6, 1139–1183.

[32]   Guillermo Cortiñas and Andreas Thom, *Bivariant algebraic K-theory*, J. Reine Angew. Math. **610** (2007), to appear, available at http://www.arxiv.org/math.KT/0603531.

[33]   Joachim Cuntz, *K-theory and C*-algebras*, Conference on K-theory (Bielefeld, 1982), Springer Lecture Notes in Mathematics, vol. 1046, pp. 55–79.

[34]   _____, *Generalized homomorphisms between C*-algebras and KK-theory*, Dynamics and processes (Bielefeld, 1981), 1983, pp. 31–45.

[35]   _____, *A new look at KK-theory*, K-Theory **1** (1987), no. 1, 31–51.

[36]   _____, *Bivariante K-Theorie für lokalkonvexe Algebren und der Chern–Connes–Charakter*, Doc. Math. **2** (1997), 139–182 (electronic) (German, with English summary).

[37]   _____, *Bivariant K-theory and the Weyl algebra*, K-Theory **35** (2005), no. 1-2, 93–137.

[38]   _____, *An algebraic description of boundary maps used in index theory*, Operator Algebras: The Abel Symposium 2004, Abel Symp., vol. 1, Springer, Berlin, 2006, pp. 61–86.

[39] Joachim Cuntz and Andreas Thom, *Algebraic K-theory and locally convex algebras*, Math. Ann. **334** (2006), no. 2, 339–371.

[40] Kenneth R. Davidson, *C*-algebras by example*, Fields Institute Monographs, vol. 6, American Mathematical Society, Providence, RI, 1996.

[41] Andreas Defant and Klaus Floret, *Tensor norms and operator ideals*, North-Holland Mathematics Studies, vol. 176, North-Holland Publishing Co., Amsterdam, 1993.

[42] L. E. Dickson, *Linear associative algebras and abelian equations*, Trans. Amer. Math. Soc. **15** (1914), no. 1, 31–46.

[43] Jacques Dixmier, *Sur les algèbres de Weyl*, Bull. Soc. Math. France **96** (1968), 209–242 (French).

[44] _____, *C*-algebras*, North-Holland Publishing Co., Amsterdam, 1977. Translated from the French by Francis Jellett, North-Holland Mathematical Library, Vol. 15.

[45] P. Donovan and M. Karoubi, *Graded Brauer groups and K-theory with local coefficients*, Inst. Hautes Études Sci. Publ. Math. (1970), no. 38, 5–25.

[46] R. G. Douglas, *The relation of* Ext *to K-theory*, Symposia Mathematica, vol. XX (Convegno sulle Algebre *C** e loro Applicazioni in Fisica Teorica, Convegno sulla Teoria degli Operatori Indice e Teoria *K*, INDAM, Rome, 1975), 1976, pp. 513–529.

[47] Jarah Evslin, *Twisted K-theory from monodromies*, J. High Energy Phys. (2003), no. 5, 030, 23 pp. (electronic).

[48] Daniel S. Freed and Michael Hopkins, *On Ramond–Ramond fields and K-theory*, J. High Energy Phys. (2000), no. 5, Paper 44, 14 pp. (electronic).

[49] Daniel S. Freed, *K-theory in quantum field theory*, Current developments in mathematics, 2001, 2002, pp. 41–87.

[50] James G. Glimm, *On a certain class of operator algebras*, Trans. Amer. Math. Soc. **95** (1960), 318–340.

[51] Philip Green, *The Brauer group of a commutative C*-algebra*, 1978. unpublished lecture notes.

[52] _____, *The local structure of twisted covariance algebras*, Acta Math. **140** (1978), no. 3-4, 191–250.

[53] _____, *Equivariant K-theory and crossed product C*-algebras*, 28th Summer Institute of the American Mathematical Society (Queen's University, Kingston, Ont., July 14–August 2, 1980), Operator algebras and applications. Part 1, Proceedings of Symposia in Pure Mathematics, vol. 38, American Mathematical Society, Providence, R.I., 1982, pp. 337–338.

[54] Alexander Grothendieck, *Produits tensoriels topologiques et espaces nucléaires*, Mem. Amer. Math. Soc., vol. 16, 1955 (French).

[55] _____, *Le groupe de Brauer. I. Algèbres d'Azumaya et interprétations diverses*, Séminaire Bourbaki, vol. 9, année 1964/65, exp. no. 290, 1995, pp. 199–219.

[56] Erik Guentner and Nigel Higson, *A note on Toeplitz operators*, Internat. J. Math. **7** (1996), no. 4, 501–513.

[57] Ulrich Haag, *Some algebraic features of Z_2-graded KK-theory*, K-Theory **13** (1998), no. 1, 81–108.

[58] _____, *On $\mathbf{Z}/2\mathbf{Z}$-graded KK-theory and its relation with the graded Ext-functor*, J. Operator Theory **42** (1999), no. 1, 3–36.

[59] Nigel Higson, *A characterization of KK-theory*, Pacific J. Math. **126** (1987), no. 2, 253–276.

[60] _____, *Algebraic K-theory of stable C*-algebras*, Adv. in Math. **67** (1988), no. 1, 140.

[61] _____, *A primer on KK-theory*, Operator theory: Operator Algebras and Applications, Part 1 (Durham, NH, 1988), Proc. Sympos. Pure Math., 51, 1990, pp. 239–283.

[62] Nigel Higson and John Roe, *Analytic K-homology*, Oxford Mathematical Monographs, Oxford University Press, Oxford, 2000. Oxford Science Publications.

[63] ———(ed.), *Surveys in noncommutative geometry*, Clay Mathematics Proceedings, vol. 6, Amer. Math. Soc., Providence, RI, 2006.

[64] Lars Hörmander, *Pseudo-differential operators*, Comm. Pure Appl. Math. **18** (1965), 501–517.

[65] Henri Hogbe-Nlend, *Les fondements de la théorie spectrale des algèbres bornologiques*, Bol. Soc. Brasil. Mat. **3** (1972), no. 1, 19–56 (French).

[66] ———, *Bornologies and functional analysis*, North-Holland Publishing Co., Amsterdam, 1977.

[67] Kjeld Knudsen Jensen and Klaus Thomsen, *Elements of KK-theory*, Mathematics: Theory & Applications, Birkhäuser Boston Inc., Boston, MA, 1991.

[68] Pierre Julg, *K-théorie équivariante et produits croisés*, C. R. Acad. Sci. Paris Sér. I Math. **292** (1981), no. 13, 629–632 (French, with English summary).

[69] Anton Kapustin, *D-branes in a topologically nontrivial B-field*, Adv. Theor. Math. Phys. **4** (2000), no. 1, 127–154.

[70] G. G. Kasparov, *Topological invariants of elliptic operators. I. K-homology*, Izv. Akad. Nauk SSSR Ser. Mat. **39** (1975), no. 4, 796–838 (Russian); English transl., Math. USSR-Izv. **9** (1975), no. 4, 751–792 (1976).

[71] ———, *The operator K-functor and extensions of C^*-algebras*, Izv. Akad. Nauk SSSR Ser. Mat. **44** (1980), no. 3, 571–636, 719 (Russian); English transl., Math. USSR-Izv. **16** (1981), 513–572.

[72] ———, *Operator K-theory and its applications: elliptic operators, group representations, higher signatures, C^*-extensions*, 2 (Warsaw, 1983), PWN, Warsaw, 1984, pp. 987–1000.

[73] ———, *Equivariant KK-theory and the Novikov conjecture*, Invent. Math. **91** (1988), no. 1, 147–201.

[74] Eberhard Kirchberg and Simon Wassermann, *Permanence properties of C^*-exact groups*, Doc. Math. **4** (1999), 513–558 (electronic).

[75] Henning Krause, *Derived categories, resolutions, and Brown representability*, Proceedings of the Summer School "Interactions between Homotopy Theory and Algebra", Chicago, available at http://www.arxiv.org/math.KT/0511047.

[76] Vincent Lafforgue, *K-théorie bivariante pour les algèbres de Banach et conjecture de Baum–Connes*, Invent. Math. **149** (2002), no. 1, 1–95 (French).

[77] Pierre-Yves Le Gall, *Théorie de Kasparov équivariante et groupoïdes. I*, K-Theory **16** (1999), no. 4, 361–390 (French).

[78] Saunders Mac Lane, *Homology*, Classics in Mathematics, Springer, Berlin, 1995. Reprint of the 1975 edition.

[79] Ib Madsen and Jonathan Rosenberg, *The universal coefficient theorem for equivariant K-theory of real and complex C^*-algebras*, Index theory of elliptic operators, foliations, and operator algebras (New Orleans, LA/Indianapolis, IN, 1986), Contemp. Math., 70, 1988, pp. 145–173.

[80] Juan Maldacena, Gregory Moore, and Nathan Seiberg, *D-brane instantons and K-theory charges*, J. High Energy Phys. (2001), no. 11, Paper 62, 42 pp. (electronic).

[81] J. P. May, *A concise course in algebraic topology*, Chicago Lectures in Mathematics, University of Chicago Press, Chicago, IL, 1999.

[82] Ralf Meyer, *Analytic cyclic cohomology*, Ph.D. Thesis, Westf. Wilhelms–Universität Münster, 1999, http://www.arxiv.org/math.KT/9906205.

[83] _____, *Equivariant Kasparov theory and generalized homomorphisms*, K-Theory **21** (2000), no. 3, 201–228.

[84] _____, *Bornological versus topological analysis in metrizable spaces*, Banach algebras and their applications (Anthony To-Ming Lau and Volker Runde, eds.), Contemporary Mathematics, vol. 363, American Mathematical Society, Providence, RI, 2004, pp. 249–278.

[85] _____, *Smooth group representations on bornological vector spaces*, Bull. Sci. Math. **128** (2004), no. 2, 127–166 (English, with English and French summaries).

[86] _____, *Local and Analytic Cyclic Homology*, EMS Tracts in Math., vol. 3, EMS Publ. House, Zürich, 2007.

[87] Ralf Meyer and Ryszard Nest, *The Baum–Connes conjecture via localisation of categories*, Topology **45** (2006), no. 2, 209–259.

[88] John Milnor, *Construction of universal bundles. II*, Ann. of Math. (2) **63** (1956), 430–436.

[89] Ruben Minasian and Gregory Moore, *K-theory and Ramond–Ramond charge*, J. High Energy Phys. (1997), no. 11, Paper 2, 7 pp. (electronic).

[90] Calvin C. Moore, *Group extensions and cohomology for locally compact groups. III*, Trans. Amer. Math. Soc. **221** (1976), no. 1, 1–33.

[91] Gregory Moore, *K-theory from a physical perspective*, Topology, geometry and quantum field theory, London Math. Soc. Lecture Note Ser., 308, 2004, pp. 194–234.

[92] Gerard J. Murphy, *C*-algebras and operator theory*, Academic Press Inc., Boston, MA, 1990.

[93] Amnon Neeman, *Triangulated categories*, Annals of Mathematics Studies, vol. 148, Princeton University Press, Princeton, NJ, 2001.

[94] Donal P. O'Donovan, *Quasidiagonality in the Brown–Douglas–Fillmore theory*, Duke Math. J. **44** (1977), no. 4, 767–776.

[95] Judith A. Packer and Iain Raeburn, *Twisted crossed products of C*-algebras*, Math. Proc. Cambridge Philos. Soc. **106** (1989), no. 2, 293–311.

[96] Ellen Maycock Parker, *The Brauer group of graded continuous trace C*-algebras*, Trans. Amer. Math. Soc. **308** (1988), no. 1, 115–132.

[97] B. J. Parshall and L. L. Scott, *Derived categories, quasi-hereditary algebras, and algebraic groups*, Carleton U. Math. Notes **3** (1988), 1–144.

[98] Gert K. Pedersen, *C*-algebras and their automorphism groups*, London Mathematical Society Monographs, vol. 14, Academic Press Inc. [Harcourt Brace Jovanovich Publishers], London, 1979.

[99] N. Christopher Phillips, *The Atiyah–Segal completion theorem for C*-algebras*, K-Theory **3** (1989), no. 5, 479–504.

[100] _____, *K-theory for Fréchet algebras*, Internat. J. Math. **2** (1991), no. 1, 77–129.

[101] M. Pimsner and D. Voiculescu, *Exact sequences for K-groups and Ext-groups of certain cross-product C*-algebras*, J. Operator Theory **4** (1980), no. 1, 93–118.

[102] Iain Raeburn and Jonathan Rosenberg, *Crossed products of continuous-trace C*-algebras by smooth actions*, Trans. Amer. Math. Soc. **305** (1988), no. 1, 1–45.

[103] Iain Raeburn and Dana P. Williams, *Pull-backs of C*-algebras and crossed products by certain diagonal actions*, Trans. Amer. Math. Soc. **287** (1985), no. 2, 755–777.

[104] _____, *Morita equivalence and continuous-trace C*-algebras*, Mathematical Surveys and Monographs, vol. 60, American Mathematical Society, Providence, RI, 1998.

[105] Marc A. Rieffel, *Connes' analogue for crossed products of the Thom isomorphism*, Operator algebras and K-theory (San Francisco, Calif., 1981), 1982, pp. 143–154.

[106] John Roe, _Elliptic operators, topology and asymptotic methods_, Pitman Research Notes in Mathematics Series, vol. 179, Longman Scientific & Technical, Harlow, 1988.

[107] Jonathan Rosenberg, _Homological invariants of extensions of C^*-algebras_, 28th Summer Institute of the American Mathematical Society (Queen's University, Kingston, Ont., July 14–August 2, 1980), Operator algebras and applications. Part 1, Proceedings of Symposia in Pure Mathematics, vol. 38, American Mathematical Society, Providence, R.I., 1982, pp. 35–75.

[108] ———, _Continuous-trace algebras from the bundle theoretic point of view_, J. Austral. Math. Soc. Ser. A **47** (1989), no. 3, 368–381.

[109] ———, _Algebraic K-theory and its applications_, Graduate Texts in Mathematics, vol. 147, Springer, New York, 1994.

[110] Jonathan Rosenberg and Claude Schochet, _The Künneth theorem and the universal coefficient theorem for Kasparov's generalized K-functor_, Duke Math. J. **55** (1987), no. 2, 431–474.

[111] Norberto Salinas, _Homotopy invariance of_ Ext($\mathcal{A}$), Duke Math. J. **44** (1977), no. 4, 777–794.

[112] Claude Schochet, _Topological methods for C^*-algebras. I. Spectral sequences_, Pacific J. Math. **96** (1981), no. 1, 193–211.

[113] ———, _Topological methods for C^*-algebras. IV. Mod p homology_, Pacific J. Math. **114** (1984), no. 2, 447–468.

[114] Graeme Segal, _Equivariant K-theory_, Inst. Hautes Études Sci. Publ. Math. (1968), no. 34, 129–151 (French).

[115] Barry Simon, _Trace ideals and their applications_, London Mathematical Society Lecture Note Series, vol. 35, Cambridge University Press, Cambridge, 1979.

[116] Edwin H. Spanier, _Algebraic topology_, Springer-Verlag, New York, 1981. Corrected reprint.

[117] W. Forrest Stinespring, _Positive functions on C^*-algebras_, Proc. Amer. Math. Soc. **6** (1955), 211–216.

[118] Richard J. Szabo, _D-branes, tachyons and K-homology_, Modern Phys. Lett. A **17** (2002), no. 35, 2297–2315.

[119] Klaus Thomsen, _Hilbert C^*-modules, KK-theory and C^*-extensions_, Various Publications Series (Aarhus), vol. 38, Aarhus Universitet Matematisk Institut, Aarhus, 1988.

[120] ———, _The universal property of equivariant KK-theory_, J. Reine Angew. Math. **504** (1998), 55–71.

[121] François Trèves, _Topological vector spaces, distributions and kernels_, Academic Press, New York, 1967.

[122] Alain Valette, _Introduction to the Baum–Connes conjecture_, Lectures in Mathematics ETH Zürich, Birkhäuser Verlag, Basel, 2002. From notes taken by Indira Chatterji; with an appendix by Guido Mislin.

[123] Jean-Louis Verdier, _Des catégories dérivées des catégories abéliennes_, Astérisque (1996), no. 239 (French).

[124] Dan Voiculescu, _A non-commutative Weyl–von Neumann theorem_, Rev. Roumaine Math. Pures Appl. **21** (1976), no. 1, 97–113.

[125] Toni Aliza Watson, _Twisted cohomology groups_, Master's Thesis, University of Maryland, College Park, 2006.

[126] Dana P. Williams, _Crossed products of C^*-algebras_, Mathematical Surveys and Monographs, vol. 134, American Mathematical Society, Providence, RI, 2007.

[127] Edward Witten, *D-branes and K-theory*, J. High Energy Phys. (1998), no. 12, Paper 19, 41 pp. (electronic).

[128] ———, *Overview of K-theory applied to strings*, Strings 2000. Proceedings of the International Superstrings Conference (Ann Arbor, MI), 2001, pp. 693–706.

[129] Richard Zekri, *A new description of Kasparov's theory of C^*-algebra extensions*, J. Funct. Anal. **84** (1989), no. 2, 441–471.

Notation and Symbols

$\oplus$ orthogonal direct sum of idempotents or of homomorphisms, page 4

$+$ for orthogonal idempotents: sum; also used for the sum of orthogonal homomorphisms and quasi-homomorphisms, page 4

$\#$ reverse order composition product in the categories $\Sigma\mathrm{Ho}$ and $\mathrm{kk}^?$, page 100

$\bullet$ concatenation of smooth homotopies, page 92

$\widehat{\otimes}$ complete projective bornological tensor product, page 23

$\widehat{\otimes}_\pi$ complete projective topological tensor product, page 23

$\widehat{\otimes}_{C^*}$ C^*-algebra tensor product (in the nuclear case), page 78

$\widehat{\otimes}_{\min}$ minimal C^*-algebra tensor product, page 152

$\otimes_X$ tensor product over X, page 179

$\langle A, B\rangle$ set of smooth homotopy classes of bounded homomorphisms $A \to B$, page 92

$\langle f\rangle$ class of $f\colon A \to B$ in $\langle A, B\rangle$, page 92

$\ltimes$ (full) C^*-algebra crossed product, page 70

$\sim$ for idempotents: similarity, page 2

$A * B$ free product of two algebras, page 164

$A \rightrightarrows D \rhd B$ notation for a quasi-homomorphism from A to B via D, page 46

$\widehat{\alpha}$ dual action to α, page 187

$\delta(A)$ Dixmier–Douady class of A, page 178

Δ_G modular function of a group G, page 185

$\Gamma(V)$ space of continuous sections of a vector bundle, page 6

$\Gamma_0(X, \mathcal{A})$ algebra of sections vanishing at ∞ of a bundle of algebras $\mathcal{A}$ over X, page 178

γ_l the restriction $JA \to B$ of the bounded algebra homomorphism $TA \to B$ associated to a bounded linear map $l: A \to B$, page 214

$\kappa^k_{A,B}$ natural map $J^k(A \widehat{\otimes} B) \to (J^k A) \widehat{\otimes} B$, page 97

Λ canonical map $\langle A, B \rangle \to \langle JA, SB \rangle$, page 97

λ^k_A shorthand for $\Lambda^k(\mathrm{id}_A) \in \langle J^k A, S^k A \rangle$, page 97

ΛX free loop space of a space X, page 192

π_A the canonical projection $TA \to A$, page 94

$\varrho_A(S)$ the spectral radius of S in A, page 27

Σ symbol map $\Psi(M) \to C_0(S^* M)$ for pseudo-differential operators on M, page 203

σ symbol in $\mathrm{K}^0(T^* M)$ of a pseudo-differential operator on M, page 203

$\Sigma_A(x)$ spectrum of x in A, page 26

σ_V the canonical map $V \to TV$, page 94

$\Psi(M)$ C^*-algebra of pseudo-differential operators on M, page 203

$\widehat{A}$ dual space of a C^*-algebra A, page 173

$A \rtimes_\alpha G$ crossed product of A by an action α of G, page 185

$\mathrm{ad}\, H$ commutator with H, page 190

Ad_u inner automorphism $x \mapsto uxu^{-1}$ associated to an invertible multiplier u, page 49

$\mathrm{Ad}_{v,w}$ inner endomorphism $x \mapsto vxw$ associated to multipliers v, w with $wv = 1$, page 49

A^{op} opposite algebra to A, page 179

$\mathrm{Aut}(A)$ automorphism group of an algebra or $*$-algebra, page 75

Fréchet space This is a complete, metrisable, locally convex topological vector space; that is, it is a complete topological vector space whose topology is defined by an increasing sequence of semi-norms., page 21

$\widehat{G}$ Pontrjagin dual of a locally compact Abelian group G, page 187

$\mathrm{Gl}_m(R)$ invertible elements in $\mathbb{M}_m(R)$, page 13

$\mathrm{Gr}(M)$ Grothendieck group of a semigroup, page 8

$HC^\alpha(X,V)$ space of Hölder continuous functions $X \to V$, page 53

HKK   homotopy-theoretic KK-theory, page 148

Homeo X homeomorphism group of a topological space X, page 174

Idem R set of idempotents in R, page 2

ind   the index map $\mathrm{K}_1^{\mathrm{alg}}(Q) \to \mathrm{K}_0(I)$ for a ring extension $I \rightarrowtail E \twoheadrightarrow Q$, page 14

Ind_H^G induction functor from H- to G-C^*-algebras, page 86

$J_{\mathrm{cpc}}A$ kernel of the natural $*$-homomorphism $T_{\mathrm{cpc}}A \to A$, page 153

JA the kernel of the natural projection $\pi_A : TA \to A$, page 94

j_T canonical map $A \to \mathcal{T}_{\ldots}(A, \alpha)$, page 77

j_U canonical map $A \to U_{\ldots}(A, \alpha)$, page 75

$\mathbb{K}$ either $\mathbb{R}$ or $\mathbb{C}$

$\mathrm{K}_\delta^{-*}(X)$ twisted K-theory of a space X with twist $\delta \in H^3(X)$, page 182

$\mathrm{K}_*(\lrcorner; \mathbb{Z}/m)$ K-theory with coefficients in $\mathbb{Z}/m$, page 144

$\mathrm{K}_*(\lrcorner; \mathbb{Q})$ K-theory with coefficients in $\mathbb{Q}$, page 143

$\mathrm{K}_*(\lrcorner; \mathbb{Q}/\mathbb{Z})$ K-theory with coefficients in $\mathbb{Q}/\mathbb{Z}$, page 145

$\mathbf{K}(A)$ the topological K-theory spectrum of an algebra A, page 148

$\mathbf{K}$ the topological (complex) K-theory spectrum, page 148

$K(\mathbb{Z}, n)$ Eilenberg–Mac Lane space with $\pi_n = \mathbb{Z}$, page 178

$\mathrm{K}_G^*(X)$ G-equivariant K-theory of a locally compact G-space X, page 70

$kk^?$	one of the bivariant K-theories $kk^{\mathscr{S}}$, $kk^{C\mathcal{K}}$, or $kk^{\mathscr{L}}$, page 130
$\mathcal{K}_{\mathscr{S}}(A)$	smooth stabilisation of A, page 42
$\mathcal{K}^*(A)$	A-valued version of $\mathcal{K}^*$ for a bornological algebra A, page 52
$\mathcal{K}^*$	a certain algebra of compact operators on $\ell^1(\mathbb{N})$, page 52
$\mathcal{K}$	used for algebras of compact operators, page 52
$\mathcal{K}_{C^*}(A)$	C^*-stabilisation of A, page 42
$\mathcal{K}_{\mathscr{L}^p}(A)$	stabilisation of A by the Schatten ideal $\mathscr{L}^p$, page 44
$\mathcal{K}_V(A)$	stabilisation of A by $V \,\widehat{\otimes}\, V$, page 125
$\mathrm{K}_0^{\mathrm{top}}(A)$	stabilised version of K_0 with better properties, page 136
$\ell^1(I, V)$	space of absolutely summable maps $I \to V$, page 24
$\mathcal{L}(\mathcal{H})$	algebra of bounded linear operators on $\mathcal{H}$, page 149
$\mathcal{L}$	used for algebras of bounded operators, page 52
$\mathscr{L}^2(\mathcal{H})$	Hilbert–Schmidt operators, page 188
$\mathscr{L}^p$	Schatten ideal with exponent $p \in [1, \infty)$, page 43
$\mathcal{M}(R)$	the ring of multipliers of a ring R, page 50
$\mathbb{M}_n(R)$	ring of $n \times n$-matrices over R, page 2
$\mathbb{M}_\infty(R)$	ring of finite matrices over R, page 2
$\mathrm{Mod}(R)$	category of left modules over a unital ring R, page 2
$\mathbb{N}$	natural numbers, including 0
$\mathfrak{N}$	G-C^*-algebras with vanishing H-equivariant K-theory for compact subgroups $H \subseteq G$, page 86
$\mathfrak{N}$	generic thick triangulated subcategory of a triangulated category, page 86
N	number operator on sequence spaces, page 52
$\mathcal{O}(X)$	algebra of germs of holomorphic functions near $X \subseteq \mathbb{C}$, page 26

Prim A primitive ideal space of A, page 173

$PU(\mathcal{H})$ projective unitary group of a Hilbert space $\mathcal{H}$, page 174

projective resolution A projective resolution of a module M is an exact chain complex of the form

$$\cdots \to P_2 \to P_1 \to P_0 \to M \to 0 \to \cdots$$

with projective modules P_j for $j \in \mathbb{N}$., page 10

$\mathcal{Q}(\mathcal{H})$ Calkin algebra, page 149

QA free product $A * A$, page 165

qA ideal in the free product $A * A$, page 165

quasi-isomorphism chain map that induces an isomorphism on homology, page 9

$\mathbb{R}$ real numbers

R^{∞} countably generated free R-module, page 2

R^{op} opposite ring, see Definition 1.13, page 4

R^{+} ring obtained by adjoining a unit to R, page 10

$R_{\mathbb{C}}^{+}$ $\mathbb{C}$-algebra obtained by adjoining a unit to a $\mathbb{C}$-algebra R, page 10

$\mathscr{S}(\mathbb{N}, A)$ A-valued Schwartz space on $\mathbb{N}$, page 42

S   the unilateral shift operator, which generates the standard representation of the Toeplitz algebra, page 63

spec a spectrum of a Hilbert space operator a, page 177

SB the smooth suspension $B(0, 1)$, page 92

S^{∞} $S^{\infty} := \bigcup_{n=1}^{\infty} S^n$, page 26

$\mathbb{S}^m$ the m-dimensional sphere, page 9

Sq^3 a Steenrod cohomology operation, page 183

$\mathcal{T}^0$ a certain 1-codimensional ideal in the Toeplitz algebra, page 65

$\mathcal{T}$	This denotes Toeplitz algebras and crossed Toeplitz algebras, see §4.1 and §5.1.1; here $\mathcal{T}_{C^*}$, $\mathcal{T}_{\mathscr{S}}$, and $\mathcal{T}_{\mathrm{alg}}$ denote the Toeplitz C^*-algebra, the smooth Toeplitz algebra, and the algebraic Toeplitz algebra, respectively, page 63
$\mathbb{T}$	the circle group $\{z \in \mathbb{C} \mid |z| = 1\}$, page 174
$T_{\mathrm{cpc}}A$	C^*-algebraic variant of the tensor algebra that is universal for completely positive contractive linear maps, page 153
$\widehat{\mathcal{T}}_{\mathscr{S}}$	an auxiliary algebra constructed out of smooth Toeplitz algebras, page 66
TV	tensor algebra of V, page 94
$U(\mathcal{H})$	unitary group of a Hilbert space $\mathcal{H}$, page 174
$U_{\dots}(A, \alpha)$	crossed product for the action of $\mathbb{Z}$ on A given by the automorphism α; we define the variants U_{C^*}, U_{ℓ^1}, $U_{\mathscr{S}}$, and U_{alg} in §5.1, page 75
$\mathbf{V}(R)$	monoid of isomorphism classes of finitely generated projective R-modules, page 2
V_B	semi-normed space generated by the disk B in the vector space V, page 20
$\mathbf{V}_{\mathbb{K}}(X)$	monoid of K-vector bundles over a compact space X, page 6
w_3	third Stiefel–Whitney class, page 196
$\mathbb{Z}$	integer numbers

Index

Advanced Courses in Mathematics CRM Barcelona

Edited by
Manuel Castellet

Since 1995 the Centre de Recerca Matemàtica (CRM) in Barcelona has conducted a number of annual Summer Schools at the post-doctoral or advanced graduate level. Sponsored mainly by the European Community, these Advanced Courses have usually been held at the CRM in Bellaterra.
The books in this series consist essentially of the expanded and embellished material presented by the authors in their lectures.

Argyros, S. / Todorcevic, S.
Ramsey Methods in Analysis (2005)
ISBN 978-3-7643-7264-4

This book introduces graduate students and resarchers to the study of the geometry of Banach spaces using combinatorial methods. We show, for example, how to introduce a conditional structure to a given Banach space under construction that allows us to essentially prescribe the corresponding space of non-strictly singular operators. We also apply the Nash-Williams theory of fronts and barriers in the study of Cezaro summability and unconditionality present in basic sequences inside a given Banach space. We further provide a detailed exposition of the block-Ramsey theory and its recent deep adjustments relevant to the Banach space theory due to Gowers.

Audin, M. / Cannas da Silva, A. / Lerman, E.
Symplectic Geometry of Integrable Hamiltonian Systems (2003)
ISBN 978-3-7643-2167-3

Brady, N. / Riley, T. / Short, H.
The Geometry of the Word Problem for Finitely Generated Groups (2006)
ISBN 978-3-7643-7949-0

The origins of the word problem are in group theory, decidability and complexity, but, through the vision of Gromov and the language of filling functions, the topic now impacts the world of large-scale geometry, including topics such as soap films, isoperimetry, coarse invariants and curvature.
The first part introduces van Kampen diagrams in Cayley graphs of finitely generated, infinite groups; it discusses the van Kampen lemma, the isoperimetric functions or Dehn functions, the theory of small cancellation groups and an introduction to hyperbolic groups. The second part is dedicated to Dehn functions, negatively curved groups, in particular, CAT(0) groups, cubings and cubical complexes. In the last part, filling functions are presented from geometric, algebraic and algorithmic points of view. Many examples and open problems are included.

Brown, K.A. / Goodearl, K.R.
Lectures on Algebraic Quantum Groups (2002)
ISBN 978-3-7643-6714-5

Catalano, D. / Cramer, R. / Damgård, I. / Di Creszenso, G. / Pointcheval, D. / Takagi, T.
Contemporary Cryptology (2005)
ISBN 978-3-7643-7294-1

Christopher, C. / Li, C.
Limit Cycles of Differential Equations (2007)
ISBN 978-3-7643-8409-8

Cohen, R.L. / Hess, K. / Voronov, A.A.
String Topology and Cyclic Homology (2006)
ISBN 978-3-7643-2182-6

The subject of this book is string topology, Hochschild and cyclic homology. The first part consists of an excellent exposition of various approaches to string topology and the Chas-Sullivan loop product. The second gives a complete and clear construction of an algebraic model for computing topological cyclic homology. The book provides many references for the reader wishing to learn more about the subject, to which it gives a perfect introduction. It is therefore suitable for both graduate students and established researchers. It is certainly the best source of much information that was until now available only to specialists and covers material from the elementary bases to the most recent developments.

Da Prato, G.
Kolmogorov Equations for Stochastic PDEs (2004)
ISBN 978-3-7643-7216-3

Drensky, V. / Formanek, E.
Polynomial Identity Rings (2004)
ISBN 978-3-7643-7126-5

Dwyer, W.G. / Henn, H.-W.
Homotopy Theoretic Methods in Group Cohomology (2001)
ISBN 978-3-7643-6605-6

Markvorsen, S. / Min-Oo, M.
Global Riemannian Geometry: Curvature and Topology (2003)
ISBN 978-3-7643-2170-3

Mislin, G. / Valette, A.
Proper Group Actions and the Baum-Connes Conjecture (2003)
ISBN 978-3-7643-0408-9

Oberwolfach Seminars (OWS)

The workshops organized by the *Mathematisches Forschungsinstitut Oberwolfach* are intended to introduce students and young mathematicians to current fields of research. By means of these well-organized seminars, also scientists from other fields will be introduced to new mathematical ideas. The publication of these workshops in the series *Oberwolfach Seminars* (formerly *DMV seminar*) makes the material available to an even larger audience.

OWS 36: Cuntz, J. / Meyer, R. / Rosenberg, J.M., Topological and Bivariant K-theory (2007). ISBN 978-3-7643-8398-5

OWS 35: Itenberg, I. / Mikhalkin, G. / Shustin, E., Tropical Algebraic Geometry (2007). ISBN 978-3-7643-8309-1

Tropical geometry is algebraic geometry over the semifield of tropical numbers, i.e., the real numbers and negative infinity enhanced with the (max,+)-arithmetics. Geometrically, tropical varieties are much simpler than their classical counterparts. Yet they carry information about complex and real varieties.
These notes present an introduction to tropical geometry and contain some applications of this rapidly developing and attractive subject. It consists of three chapters which complete each other and give a possibility for non-specialists to make the first steps in the subject which is not yet well represented in the literature. The intended audience is graduate, post-graduate, and Ph.D. students as well as established researchers in mathematics.

OWS 34: Lieb, E.H. / Seiringer, R. / Solovej, J.P. / Yngvason, J., The Mathematics of the Bose Gas and its Condensation (2005). ISBN 978-3-7643-7336-8

This book contains a unique survey of the mathematically rigorous results about the quantum-mechanical many-body problem that have been obtained by the authors in the past seven years. It is a topic that is not only rich mathematically, using a large variety of techniques in mathematical analysis, but it is also one with strong ties to current experiments on ultra-cold Bose gases and Bose-Einstein condensation. It is an active subject of ongoing research, and this book provides a pedagogical entry into the field for graduate students and researchers. It is an outgrowth of a course given by the authors for graduate students and post-doctoral researchers at the Oberwolfach

Research Institute in 2004. The book also provides a coherent summary of the field and a reference for mathematicians and physicists active in research on quantum mechanics.

OWS 33: Kreck, M. / Lück, W., The Novikov Conjecture: Geometry and Algebra (2004). ISBN 978-3-7643-7141-8

These lecture notes contain a guided tour to the Novikov Conjecture and related conjectures due to Baum–Connes, Borel and Farrell–Jones. They begin with basics about higher signatures, Whitehead torsion and the s-Cobordism Theorem. Then an introduction to surgery theory and a version of the assembly map is presented. Using the solution of the Novikov conjecture for special groups some applications to the classification of low dimensional manifolds are given. Finally, the most recent developments concerning these conjectures are surveyed, including a detailed status report. The prerequisites consist of a solid knowledge of the basics about manifolds, vector bundles, (co-) homology and characteristic classes.

DMV 32: Bolthausen, E. / Sznitman, A.-S., Ten Lectures on Random Media (2002). ISBN 978-3-7643-6703-9

DMV 31: Huckleberry, A. / Wurzbacher, T. (Eds.), Infinite Dimensional Kähler Manifolds (2001). ISBN 978-3-7643-6602-5

DMV 30: Scholz, E. (Ed.), Hermann Weyl's *Raum—Zeit—Materie* and a General Introduction to His Scientific Work (2001). ISBN 978-3-7643-6476-2

DMV 29: Kalai, G. / Ziegler, G.M. (Eds.), Polytopes — Combinatorics and Computation (2000). ISBN 978-3-7643-6351-2

DMV 28: Cercignani, C. / Sattinger, D., Scaling Limits and Models in Physical Processes (1998). ISBN 978-3-7643-5985-0